Population size, with its effects on market size and economic and political cooperation; population densities and distribution, with their effects on economic integration and the degree of pressure on the land; population dynamics, with their effects on the rate of population growth and distribution; the often excessive flow of population to the cities with its effects on the society and polity of each country: these are some of the many cause and effect relationships associated with the population factor which are present to some degree in societies everywhere and are, as such, of transcendent interest to an unusually wide spectrum of students, scholars, and researchers.

William A. Hance is Chairman of the Department of Geography and Professor of Geography at Columbia University. Past president of the African Studies Association and an Honorary Fellow of the American Geographical Society, he is the author of *The Geography of Modern Africa* (1964) and editor of and contributor to *Southern Africa and the United States* (1968), both published by Columbia University Press.

Population, Migration, and Urbanization
in Africa

POPULATION, MIGRATION, AND URBANIZATION IN AFRICA

WILLIAM A. HANCE

COLUMBIA UNIVERSITY PRESS

New York and London   1970

William A. Hance is Chairman of the Department
of Geography at Columbia University.

Copyright © 1970 Columbia University Press
ISBN: *0-231-03166-1*
Library of Congress Catalog Card Number: 75-116378
Printed in the United States of America

To Margie, Jean, Bronwen,

Karen, and Laura

PREFACE

The purpose of this book is to examine a series of topics related to population in Africa. Chapter 1 presents a brief summary of the major demographic features of the continent, including a discussion of the shortcomings of available statistical data. Chapter 2 analyzes the significance of population densities, noting the very limited utility of crude density figures, and examines the major factors of population distributions in many African regions. Chapter 3 attempts to synthesize the role of population movements in Africa, stressing their impacts on the areas and people involved. Chapter 4 contains a range of data related to urbanization in Africa and a discussion of some of the problems it poses, which are presenting dilemmas of increasing concern to African countries. Chapter 5 consists of thumbnail sketches of a number of urban communities, indicating the wide diversity among them but also the commonality of many of their problems. And Chapter 6 focuses upon population pressure in Africa, a condition which is considered to be far more pervasive than has generally been acknowledged. An overall objective is to encourage greater awareness of the population factor in African studies, a factor which, despite the work of numbers of dedicated students, has not yet received the attention that its importance demands.

I am indebted to many more people than can be personally named. Particular thanks go to Professor Etienne van de Walle for valuable bibliographic assistance with demographic documents, to numerous students, including research assistants, to Miss Joan Breslin for assistance with typing and her always cheerful support, to staff members of the Columbia University Press, especially Mrs. Sue Bishop and Mr. Robert Tilley, to Vaughan Gray, cartographer, and to correspondents on three continents who responded generously to my queries. Finally, my greatest debt is to my dear wife, who continues to put up with a harried writer with supreme patience, willing help, and unwavering support.

Columbia University, February, 1970   William A. Hance

CONTENTS

MAPS

TABLES

Population, Migration, and Urbanization

in Africa

CHAPTER I

AFRICAN DEMOGRAPHY

A BETTER UNDERSTANDING of the population factor as it affects Africa is essential. Its impact on economic, social, and political development is pervasive, direct, and frequently complex. Population size is one important determinant of political power and prestige; it provides one measure of market size and thus strongly affects the economic potential of many of Africa's individual countries, in turn creating one of the greatest incentives to regional and subregional economic and political cooperation and consolidation. Population distribution affects political, social, and economic integration and the degree of land pressure which may have developed within a country; it greatly influences the costs and difficulties of providing adequate social and economic infrastructures. Population dynamics—fertility, mortality, and migration—determine the rate of population growth, the age distribution of the population, and internal and external movements which influence every aspect of a nation's economy and polity. Population growth has direct impacts on economic development. It may increase the size of the market for certain items and the number of operatives available for productive output, but rapid growth also tends to increase the difficulty of development in that a substantial part of investment must go to provide the population increment with essentially the same services enjoyed by the existent population. The characteristic age distribution of most African countries shows a very youthful population which means that a relatively small working component must support the large proportion below working age; the requirements for education in particular are disproportionately high. The high mortality which still characterizes most of Africa shortens the average working span. It creates a desire for a large number of children to assure that some will survive to take care of their parents in old age; yet reductions in child and infant mortality rates have not yet led to notable adjustments in this attitude. High mortality rates reflect the continuingly high incidence of disease and suffering, revealing the need for enlarged expenditures for health services.

Migrations have complex relations to source and destination areas, impacts too involved to summarize with brevity. One of the major currents leads to the study of urbanization, which brings us back to population distributions. Urbanization is progressing with remarkable rapidity in most parts of the continent, resulting in problems and dilemmas of which some seem almost insoluble in the short run. Nor can urbanization be comprehended without relation to the problems of rural areas. The interrelations of the population factor with physical, economic, social, and political considerations, all of which are themselves intricately interrelated, make the study of population a many-faceted one; the ramifications and intimate intermingling of so many variables greatly complicate the analysis but at the same time explain the growing significance and imperative of such study.

Population, migration, and urbanization studies are of concern to a wide variety of disciplines, not confined by any means to the social sciences. No one person can possess more than a segment of the expertise needed to comprehend the full breadth and complexity of the associated problems. The focus of this book is on those aspects which have a strong geographic component, narrowed further by an emphasis on economic geography as distinct from cultural, historical, political, physical, or other sub-branches of the discipline. The coverage in each section is, therefore, selective and is intended to be. The section on population, for example, contains only a fraction of what demographers would insist on examining; fortunately, several recent comprehensive surveys are available which do present much more detailed analyses of African demography.[1] Similarly, many of the elements of urbanization which are of interest to sociologists are either omitted or touched upon only briefly. It is nonetheless hoped that the material that has been selected for presentation, summarization, or analysis will be of interest and value to many nongeographers.

Population Data Needs

Since the population factor is of such pervasiveness it follows that there is need for demographic data in a great variety of theoretical and practical studies and plans. Censuses and surveys are needed by medical and health services in formulating programs for disease control, dietal improvements, family planning, and allocation of clinic and hospital facilities and staffs. Departments of education need demographic data to determine the present and future needs for school buildings, supplies, and teachers. Such data are also required for land-use studies, labor and manpower analyses, and urban planning. Indeed they provide the basis for effective social and economic planning and development and for the intelligent allocation of expenses among various departments and regionally within a country. Population data are also utilized for tax purposes and may be employed in the determination of political representation; the implications of these last uses do, unfortunately, provide incentives for distortion, the one tending toward underrecording, the other toward overrecording.

The types of data relating to population that are desirable for a nation to obtain are: the *de jure* and/or *de facto* population of the area; its population dynamics, that is, mortality, fertility, and migration; information on age, sex, and family structure, including the number of children per family and, where appropriate, the breakdown by monogamous and polygamous marriages; and a variety of data not strictly demographic in character such as employment by sector, wages and incomes, literacy and educational experience, ethnic affiliation, languages, and religions. If census data are properly recorded, informa-

---

[1] Caldwell and Okonjo, eds., *The Population of Tropical Africa*; Brass, Coale *et al.*, *The Demography of Tropical Africa*.

tion of great importance can be derived giving population distribution, densities, migration, and the trends in these and other matters.

Ideally, there should be periodic classic censuses either by individual or household count and recording of vital statistics (births, marriages, deaths). In fact very few countries in Africa have had successive censuses of equal validity, practically no area has an effective system of vital registration, and many countries, particularly in francophone Africa, have had sample censuses, which are usually based on a 5–10 percent enumeration. There is difference of opinion regarding the value of sample censuses, some experts maintaining that they may be more accurate than an attempt at a full census. Lorimer, for example, believes that many of the sample inquiries, notably the large-scale national or regional inquiries in former French and Belgian areas, were superior to many of the complete enumerations in Africa, though he notes that it would be better to have basic data on the entire population supplemented by fuller information obtained by sample. One undeniable advantage of the sample census is that fewer enumerators are required and hence it is likely that they can be better qualified.

Objections to the sample census are that they are subject to substantial error in estimating the total population, do not provide the required data on communities and populations, and are not necessarily notably cheaper than a full census, contrary to the generally accepted notion. Caldwell also makes the additional criticisms that they do not arouse the same interest and concern accorded full censuses, that they operate on the false assumption that subsistence farming areas are homogeneous, that they present greater difficulty in the enumeration of nomads, and that the original frame of sample censuses is likely to be quickly outdated and invalidated.[2] Despite these objections, one must be thankful that at least the results of sample censuses are available for a good many countries. Furthermore, as will be seen, many of the classic censuses taken in Africa are also subject to more or less serious reservations.

Other methods of obtaining population data are by administrative enumeration or survey, by estimate on the basis of hut tax or on male poll-tax-paying population, and by population count in a census assembly in which, for example, the population of an area is asked

[2] Caldwell, "Introduction," in Caldwell and Okonjo, eds., *The Population of Tropical Africa*, p. 5.

to assemble at a given point, each female depositing, say, a cowrie shell, each male perhaps a kernel of corn in the appropriate receptacle. Such a system was used as late as 1921 in remote areas of Ghana and in 1924 in Zanzibar.

Explanations for the inadequate number of censuses that have been held in Africa include their high cost and, at least in the past, the failure to realize the strategic importance of demographic statistics. Here it is well to remember that until recent years administrative allocations were made on a largely *ad-hoc* basis, planning on accepted modern approaches being a quite recent introduction in most countries. The reluctance to allot money to census-taking resulted both from the shortage of funds and from unwillingness to spend funds for this purpose. But it may be accepted that few if any African countries could afford to collect information on the scale taken for granted in Western Europe or the United States.

Illustrative of the costs of census-taking are the following: Ghana 1948—$120,000, 1960—$840,000; Morocco 1960—$356,000; Nigeria 1963—$7,000,000; and, for nine sample censuses in francophone countries from 1954 to 1966, $50,000 to $374,000. Ogunlesi has noted that the cost of the 1963 Nigerian census was "about as much as it costs to provide medical and health services for the whole of Western Nigeria for one year,"[3] suggesting that the decision to hold a census cannot be taken lightly.

Problems of Data Collection

A brief examination of the problems involved in the collection of census data in Africa will be helpful in assessing the validity of the available data. First are a number of problems associated with the organization of censuses.

ORGANIZATIONAL PROBLEMS

The recruitment of an adequate number of qualified enumerators often proves very difficult and for more than one census the literacy of some enumerators has not been as high as would be desired.

[3] T. O. Ogunlesi, "Before and After a Population Census Operation in Nigeria—a Physician's Experience," in Caldwell and Okonjo, eds., *The Population of Tropical Africa*, p. 118.

Securing adequately qualified census takers is a common problem in Africa. A census taker in the 1956 Sudan census.

Students, teachers, nurses, doctors, and others who could be expected to understand instructions and to take a real interest in the conduct of a census have been used effectively as enumerators. In the 1956 Sudan census a special problem was presented because men were not permitted to interview women in Muslim households, so women enumerators, preferably nurses, were employed. The results indicated that the suspected bias in results obtained by male enumerators in the previous census had indeed been present. The shortage of enumerators and also of supervisors has been the main reason for extending some censuses over a period of time, whereas a census should be taken as nearly simultaneously as possible. The 1952 census in Nigeria, for example, was conducted at different periods in the three regions and required three months in the North.[4] In the 1963 Sierra Leone census it was hoped that enumerators would be mature Africans with at least a sixth grade education, but these standards had to be relaxed in certain areas. This census was spread over a three-week period to permit time for the 3,260 enumerators to fill out each return. The Botswana (then

[4] The timing of the census-taking was also occasioned by climatic differences among and within regions, and, in the North, the necessity to avoid Ramadan and the major period of seasonal migration from October to May.

African Demography

Difficulties of census-taking in cities include the absence of street names and numbers and the problem of locating all households. A part of Mogadishu, Somalia.

Bechuanaland) census of 1964 was spread over a five-month period, but still involved the use of some unskilled enumerators. Elsewhere a specific day is set as census day, it is highly publicized, and then the people are asked to remember who was present in the household on that occasion even though they may be enumerated some days or weeks later. In the Addis Ababa census of 1961, for example, the night of September 10–11 was chosen because it was New Year's Eve in the Ethiopian calendar and hence easy to remember. It was considered that the advantage of selecting a notable date outweighed the disadvantages of its abnormal festival conditions and that the possibility of error resulting from abnormal movements would also be minimized by recording information on both a *de jure* and a *de facto* basis.

Another difficulty in organizing censuses has been the inadequacy of base maps and the necessity to delineate enumeration areas which could be readily recognized on the ground, reduce travel to the minimum, and assure coverage of all households. Few countries have had the advantage of full coverage by aerial photographic survey which can be very useful in delineating enumeration areas. In the 1960 Ghana census it was found that the existing topographic maps were so defective as to be useless until corrected, especially for the location of settlements.

That country had an additional problem related to the uncertain boundaries of local councils since it was common for two councils to claim adjacent areas and some land was not claimed at all, these boundary problems having defied the courts for years. To avoid exacerbating conflicts over land rights, mappers working for the census drew squiggly lines where the precise line was subject to debate. In the 1963 Sierra Leone census, mappers were instructed, where delineation was difficult, to have enumeration areas coincide with areas where villagers used a common dip tank for cattle, which provided a check which enumerators could later use. The 1962 and 1963 Nigerian censuses faced comparable problems since many villages and fairly large communities were not shown on any map. In this case it was possible to commission aerial photographs for use in census maps. In the 1964 Botswana census no maps existed showing the lesser villages and enumerators had to draw their own. To add to the difficulties, the 108 villages with over 1,000 persons had no system of street names or house numbers. Similarly it was not always easy in Nigeria to avoid confusion because some of the houses were so complex in shape, size, and location that even the definition of what constitutes a house, or how it is differentiated from a compound, proved difficult. The 1948 census in East Africa also faced problems of locating all the huts in rural areas, while the organizers of the 1967 census in Tanzania found that many of the maps used in earlier censuses could not be located.

The scattered nature of many rural population nuclei contributes to the cost and difficulty of covering some large regions. An extreme case is illustrated by the Botswana census which enumerated 543,105 people on an area almost three times the size of the United Kingdom with only five miles of tarred road, about 5,000 miles of gravel and earth roads, and 394 miles of railway. Enumerators in that country used donkey, camel, crocodile hunters' boats, and dugout canoes in covering some of the more difficult areas.

After the censuses are taken it has frequently been true that central staffs were inadequate to process the data quickly and thoroughly. Doubtless the use of sophisticated mechanical computers will expedite this task, but much valuable data remain to be extracted from most censuses for years after they have been taken and even the most basic compilations usually take many months to produce.

Another organizational problem is that of achieving a sufficient degree of international cooperation and standardization of data required to permit comparative studies. The Commission for Technical Cooperation in Africa (CCTA), the Economic Commission for Africa (ECA), and other United Nations agencies have worked to achieve better coordination and to suggest standards and procedures useful in all parts of the world, but the degree of uniformity in methods, definitions, and tabulations is not yet very impressive.

INTERVIEWING PROBLEMS

Another set of problems of census-taking in Africa is associated with the lack of understanding by persons being enumerated, stemming largely from illiteracy and an average low level of educational achievement, but also reflecting suspicions that the census is being taken or will be used for purposes detrimental to the respondent. Many early censuses ran into difficulties because of the belief of some that counts were being taken to provide a basis for increased taxation, recruitment for corvées or military service, relinquishing of wives of polygamous marriages, or suppression of a political nature. Pons, in a report on a social survey in Stanleyville (now Kisangani), wrote that

in a number of cases interviewees expressed fears concerning the motives of the investigation. Often these were in the realm of the supernatural. There were people who thought that the sponsors . . . were too friendly to be 'normal' Europeans and that they were ancestral spirits who, having lost touch with their descendants, were seeking to re-establish contact; others expressed the fear that by answering questions they were being recruited to work in Europe after death.[5]

In some censuses numerous people are thought to have left the enumeration areas in order to avoid being counted; in the 1947 survey of Dolisie, Congo (Brazzaville), for example, about 16 percent were believed to have escaped being counted. In the same country's sample census of 1960–1961, however, some heads of families could not understand why they were not being counted and, to avoid interminable palavers, these villages were fully counted. In the 1963–1964 Chad sample survey some village chiefs in the north refused to guide enumerators to the small hamlets or *dankouch* which were often very dispersed and which often contained as much as 80 percent of the population of a "village"

[5] V. G. Pons in UNESCO, *Social Implications of Industrialization and Urbanization in Africa South of the Sahara* (Paris, 1956), p. 249.

unit. For most recent censuses governments have taken pains to thoroughly prepare the populace for the forthcoming census. In the 1960 Ghana census, for example, a Census Education Committee mounted a half-year campaign before the census, using special films, leaflets, posters, radio talks, and mass meetings supported by party officials and local notables, whose cooperation was considered indispensable to success.[6] Modern census takers also publicize the need for censuses for hospital, school, and other planning, while in the 1962 and 1963 Nigerian censuses it was understood that budgetary allocations and political representation would be based on the findings and hence the incentives were clearly acceptable to those being interviewed. The trend, in other words, appears to be opposite that of earlier years, and future censuses may have to be more concerned with excessive willingness of interviewees than with desires to escape the enumeration. A somewhat original indication of the increasing acceptance of census-taking was illustrated during the 1964 Botswana count when it was found that a large number of children born during the census were given the name Census.

POLITICALLY ORIENTED PROBLEMS

As suggested above, an additional set of problems in taking censuses stems from political influences which may work either toward an upward or a downward bias in the results. The 1948 Ghana census, for example, was somewhat suspect because it took place at a time of considerable civil unrest, when the cocoa farmers were bitterly opposing the compulsory cutting of trees affected by Swollen Shoot disease, and with riots occurring in the cities. Similarly the 1960 census might have been sabotaged since some opponents of the government party chose to call it a "C.P.P. party project," there was bitterness in some districts over deportations and preventive detention, and rumors were spread that the census would be used to conscript for the military, to shift people from their present areas, increase taxes, and confiscate property.[7]

In Liberia the government chose temporarily to suppress the census of 1962, presumably because it indicated a lower total population than had been hoped and it was feared that this might prove to be a disincentive to investors concerned about an adequate supply of labor.

[6] St. Clair Drake, "Traditional Authority and Social Action in Former British West Africa," *Human Organization*, XIX, No. 3 (Fall, 1960), 150-58.
[7] *Ibid.*

African Demography

The 1965 census of the Central African Republic was discredited, apparently also for political reasons, necessitating a second count in July, 1968.

The 1962 and 1963 Nigerian censuses became so embroiled in politics that "the very survival of the Federation was threatened,"[8] the 1963 census itself having been scheduled because of the alleged politically motivated manipulation of the previous year's enumeration. Aluko, writing about Nigerian censuses, suggests that reluctance to allow census-taking gradually faded after 1952 as

people began to see that parliamentary and local council representation, government amenities, and the relative importance attached to towns, districts, provinces, or regions were largely dependent on the recorded population of each. The more literate people became overzealous about the value of a census and they were prepared to do anything, not only to enumerate all their people, but also, if possible, to engage in double or triple counts. The political leaders also became even more enthusiastic than others about the census returns, because they regarded them as an instrument of political power.[9]

Estimates vary regarding the degree of overcounting in the 1963 Nigerian census, but there is universal agreement that it was fraudulent. The results would have meant a 5½ percent increase per annum since the 1952/1953 census and increases of about 7 percent for the Western and Midwestern regions. Even allowing for possible undercounting in the 1952/1953 census, the 1963 census is thought to contain an overestimation of at least 10 percent. And it is now all too clear that it will be many years before it will prove possible to conduct a valid census in Nigeria.

DIFFICULTIES IN SECURING SPECIFIC TYPES OF INFORMATION

Turning from the general to the specific, we may next look at the difficulties associated with obtaining particular types of data. Estimates of the total population of an area are likely to be reduced by the traditional fears noted above, by distortions reflecting the belief of some groups that it is unlucky or sacrilegious to count anything, and by

[8] R. K. Udo, "Population and politics in Nigeria (Problems of census-taking in the Nigerian Federation)," in Caldwell and Okonjo, eds., The Population of Tropical Africa, p. 97.

[9] S. A. Aluko, "How Many Nigerians? An Analysis of Nigeria's Census Problems, 1901-63," Journal of Modern African Studies, III, No. 3 (October, 1965), 376-77.

political opposition to the census. A minor difficulty is that some peoples change their names at successive stages of life. The failure to locate all the households, huts, and encampments in remote areas or to delineate the complex compound boundaries in agglomerations may also lead to underrecording.

The downward bias of many earlier counts is clearly suggested by the variance between the projected estimate and the actual count of more recent censuses. In the following censuses, for example, the total populations turned out to be higher than the previous estimates by the percentages indicated:

Census year	Country	Percent above estimate	Census year	Country	Percent above estimate
1948	East Africa	22–31	1961	Dahomey—rural	9.3
1950	Portuguese Terr.	1.2–12.7		—5 towns	10–47
1952-1953	Nigerian Regions	20–34	1962	Kenya	13.5
1955-1956	Sudan	18	1962-1964	S. and E. Cameroon	7.1
1958	Uganda	10.5	1963	Malawi	21.2
1959-1969	C.A.R.	7	1963	Southern Rhodesia	40.4
1960	Ghana	36.2	1963	Zambia	35
1960	Niger	10	1967	Réunion	18.6
1960-1961	Upper Volta	22	1967	Tanzania	28.4

On the other hand, the 1963 Sierra Leone census recorded a population 12.8 percent lower than had been predicted. Perhaps the most serious situation resulted from the strong upward bias of the 1963 Nigeria census which is thought to have overrecorded the population by at least 10 percent or possibly by 25–30 percent, or by as much as 10–12 million people. As has already been suggested it is no longer clear whether the bias will be upward or downward in future censuses.

Despite the numerous problems there is no question that most of the more modern censuses have been reasonably and increasingly accurate and that much improved coverage of African populations is now available. Some data are available for almost all countries. The most serious deficiencies are in the estimates for Ethiopia, Somalia, and Nigeria, the last because of the distortions in the 1963 census. The U.N. *Demographic Yearbook 1967*[10] indicates that population estimates for

[10] United Nations, Department of Economic and Social Affairs, *Demographic Yearbook 1967*, pp. 98-103.

Africa were based on complete censuses in 44 cases, on sample surveys in 13 cases, on partial censuses for 4 political units, and on conjecture for 3 countries. Figures for 45 units were based on censuses taken within the previous ten years, but in almost all cases the extrapolation to the mid-1967 estimate was made by an assumed rate of increase rather than by continuous registration or by a calculated balance of births, deaths, and migration. Only Mauritius and Spanish North Africa are indicated as having 1967 population estimates not of questionable reliability.

With respect to population dynamics, fertility may be estimated by securing retrospective reports by women within broad classes on children ever born to them, by reports of births during the previous year, by calculation from age ratios, by continuous observation and recording in sample areas, or by hospital reports. Retrospective reports are subject to considerable error because of forgetfulness, unwillingness to report, or failure to count children who have died in infancy. Hospital reports are also of limited value in Africa, even in urban areas where women usually go to hospitals for births, because of the difficulty of subtracting those women who come from outside the study area. The U.N. provides estimates of crude birth rates for almost all African countries and on fertility rates (the number of live births reported in the calendar or survey year per 1000 female population aged 10–49 years in that year) for most, but only in a few cases are these estimates based on reliable registration or calculation. Brass, Coale, van de Walle, and others have developed sophisticated models and techniques for estimating fertility rates on the basis of inadequate data, but these experts are the first to caution regarding the degree of reliability of their estimates.[11]

Mortality can also be estimated by a variety of techniques, none of which are entirely satisfactory in the African context. Retrospective reports suffer the same disadvantages as when used for estimating fertility; reports in censuses or other surveys of deaths in the previous year may also be distorted because of the failure to count infant deaths or the unwillingness to speak of death. Extrapolation from two successive censuses where data are known for births and migration is seldom possible. Adequate registration is only rarely available. No country on the continent is considered to have reliable information on mortality.

[11] See Brass *et al.*, *The Demography of Tropical Africa*.

Data on migration are extremely spotty even though most major movements have been delineated. Recent censuses have been more useful by recording residents present, residents absent, and strangers or visitors present, and by securing information on place of birth. Special surveys and investigations of migrant streams have been more useful in providing information on migrations than have censuses, but they do not satisfy the need for maintaining records of immigration and emigration. On the whole, then, data regarding population dynamics in Africa are poor to very poor.

With respect to population trends any estimation must either be based on successive censuses or on data covering population dynamics. As there are very few countries with successive equally valid censuses, and as calculating growth rates from inaccurate censuses may compound the distortion, the first method of estimation is rarely adequate for African populations. The comments made regarding dynamics suggest that the second method is also subject to serious error. It follows that estimating growth rates of African populations is likely to be a hazardous exercise, as was suggested by the listing of the differences between extrapolated estimates from previous censuses and totals recorded in later censuses.

Despite all the shortcomings of African demographic statistics, however, experts have been able to provide increasingly valuable data and estimates on a variety of items and the trend, in most countries, is toward greater availability and increasing reliability. The current round of censuses, including ones for areas not previously covered, may be expected greatly to improve our knowledge of African populations.

Serious problems also exist in securing information on population characteristics such as age, ethnic classification, and religious affiliation. Information on age, for example, is likely to be distorted because many people do not know their exact age and the idea of time in years is often very hazy. In an attempt to achieve greater accuracy in censuses, interviewees are sometimes given a number of notable events to which they may relate, such as the year of a locust invasion, the construction of a new road or school, the investiture of a new chief, or the visit of an important individual. Another method is to seek to relate ages of those interviewed to others in the community whose ages have been well documented. Additional causes of possible error are: the practice of understating age to avoid taxation or service; when age groups are used, inflation of the category of women and deflation of the category of children due to the classification of every girl beyond puberty as a woman; a "quasi-universal tendency to assume a higher 'typical' age of marriage than actually prevails,"[12] and a common tendency to report ages at pivotal digits of the decimal system.

As a second example of the problems associated with obtaining information on specific population characteristics, it may be noted that ethnic affiliation is sometimes difficult to obtain because of confusion concerning just what a tribe is. The arbitrariness of tribal classifications is indicated by changes in the recognized groups from one census to another in various countries. Tribes sometimes get confused with sub-tribes, clans, or languages, while radically different spellings add further confusion, particularly in international comparisons.

Very considerable caution must be used in accepting and using almost all demographic data from Africa, including those given below and in succeeding chapters.

Demographic Data for Africa

In this section what are considered as the best available demographic data on Africa are presented. Estimates of the total population are followed by estimates on population dynamics.

POPULATION TOTALS

Africa and its appurtenances had an estimated 328 million inhabitants in mid-1967, or about 9.6 percent of the world total. Table 1 gives estimates of the total population of Africa from 1000 to 1967. The greatest likelihood of overestimation in the latest years relates to the use of the 1963 Nigerian census figures; underestimation may be expected in a number of countries on the basis of past experience. One might conclude that the balancing of over- and underestimates may support the recent U.N. estimates as about as close "guesstimates" as can reasonably be expected.

[12] Etienne van de Walle, "Characteristics of African Demographic Data," in Brass *et al.*, *The Demography of Tropical Africa*, p. 49.

Table 1

ESTIMATED POPULATION OF AFRICA AND THE WORLD
FOR SELECTED YEARS, 1000 TO 1967

| Year | Population (millions) | | African percent of world total |
	Africa	World	
1000	50	275	18.2
1200	61	384	17.5
1400	74	373	19.8
1600	90	486	18.6
1650	100	470–545	21.2–18.3
1750	95–100	694–728	13.6–13.7
1800	90–100	906–919	9.9–10.8
1850	95–100	1091–1171	8.7–8.5
1900	120–150	1571–1608	7.6–9.5
1920	140	1810	7.7
1930	164	2070	7.9
1940	191	2295	8.3
1950	222	2517	8.8
1960	278	3005	9.3
1963	297	3175	9.4
1966	320	3355	9.5
1967	328	3420	9.6

SOURCES:

Robert C. Cook, "How Many People Have Ever Lived on Earth?" *Population Bulletin* (February, 1962), pp. 10, 13.

J. D. Durand, "World Population Estimates, 1750–2000," *Proceedings of the World Population Conference, Belgrade, 1965* (New York, United Nations, 1967), II, 20.

U.N., *Demographic Yearbook 1967* (New York, 1968), p. 97.

—— *Population and Vital Statistics Report,* Statistical Papers, Series A, Vol. XX, No. 3 (1963).

POPULATION DYNAMICS

Fertility and Birth Rates. While there are very considerable ranges in the fertility and birth rates in Africa, both are on the average among the highest in the world. The range of total fertility rates (the number of children that would be born during the lifetime of each woman experiencing the given fertility rate) has been estimated by the Princeton population group for a total of 126,000 cases in 99 sub· divisions of 28 tropical African countries with full or partial data for periods ranging from 1950 to 1963, the only estimates which were con-

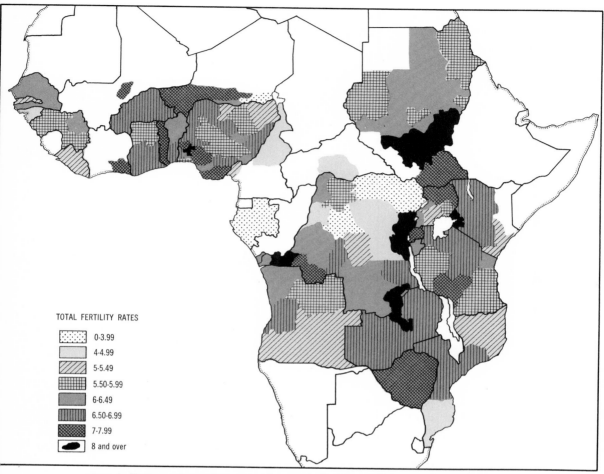

TOTAL FERTILITY RATES

- 0-3.99
- 4-4.99
- 5-5.49
- 5.50-5.99
- 6-6.49
- 6.50-6.99
- 7-7.99
- 8 and over

MAP 1. Estimated fertility rates in tropical Africa. (After William Brass *et al.*, *The Demography of Tropical Africa*, copyright © Princeton University Press, 1968, used by permission.)

sidered to be justified for that area.[13] These rates varied from 3.5 to 9.1, with the median being about 6.5, the highest among the continents.

Describing the distribution of fertility rates in tropical Africa (see Map 1), Coale and Lorimer write:

A ridge of high fertility in East Africa extends from the southeastern Sudan

[13] Brass *et al.*, *The Demography of Tropical Africa*, pp. 157-61.

African Demography   17

through parts of Uganda and Kenya, through Rwanda and Burundi and parts of Tanzania, and through the southern and eastern provinces of the Congo into Zambia, Southern Rhodesia, and the southern province of Mozambique. The provinces in this strip show a total fertility of 6.5 or higher. Somewhat lower fertility is found along the East African coast from the coastal province of Kenya to the northern part of Mozambique. The lowest fertility in Africa is in a region extending, apparently, from the west coast in Gabon through north central and northwestern provinces of the Congo into the southwestern region of the Sudan. . . . The high fertility ridge mentioned earlier appears to have a spur extending to the coast of West Africa through the southern provinces of the Congo and the northern provinces of Angola. Another strip of exceptionally high fertility is found along the coast in West Africa from coastal Nigeria to the Ivory Coast, with a branch extending up through the western part of Nigeria into parts of Niger and Upper Volta.[14]

Fertility rates, defined as the number of live births reported in the calendar or survey year per 1,000 female population aged 10–49 years in that year, are given in the U.N. *Demographic Yearbook 1967* for most African countries. At least twelve countries have estimated rates above the exceptionally high figure of 200. Even the median figure of about 175 is exceeded by only a few countries in the rest of the world.

African birth rates are estimated by the U.N. to have averaged 46 per M in the period 1960–67, with regional estimates ranging from a low of 40 in southern Africa to a high of 51 in western Africa. No other continent has comparably high birth rates, Latin America having an estimated average of 40 and Asia one of 38 per M. By comparison, western Europe's birth rate was put at 18 per M and northern America's at 21 per M in the same period. Rates for individual African countries show a considerable range, but most fall within the 35 to 55 per M levels.

Explanations for the generally very high birth and fertility rates include: (1) the typically low age of marriage; (2) the high percentage of women who are married, increased by the easy remarriage of widowed and divorced women; (3) the tolerance of illegitimacy; (4) the characteristic desire of women for as many children as possible. This desire is explained by the felt need to offset the high death rate by having enough children to assure that some will survive to adulthood, coupled with the traditional role of grown children in providing care for their parents

[14] Ansley J. Coale and Frank Lorimer, "Summary of Estimates of Fertility and Mortality," in Brass *et al.*, *The Demography of Tropical Africa*, pp. 166–67.

in old age; (5) the fear of wives that husbands may leave them if they do not bear children; (6) religious precepts, particularly among animist, Muslim, and Catholic families, and (7) the exceptionality of any interest in family limitation.

Alice Ogot presents very clearly the general attitudes which make for high fertility:

In the book, *East African Childhood*, edited by L. K. Fox, Mr. J. Lijembe from Western Kenya writes, 'The recognition my father holds in the society is directly dependent on the number of children he has been able to raise from his marriage.' In the same book Anna Agoko from Uganda writes, 'All married couples in rural Acholi . . . want to have as many children as they possibly can. Even with the coming of school education, which involves school fees, parents would rather go without clothes, sugar or soap . . . than to cut down the number of children they produce.'

Furthermore, in most African societies, up to this very moment, if a woman does not become pregnant soon after her marriage, or if three or four years pass before she becomes pregnant again, she invariably becomes the target of the village gossip and speculation, she is suspected of having contracted a bad disease, or of having committed a sin for which she is being punished by God. But if the woman cannot have a baby, she is doomed: she is despised and humiliated; her marriage can never be stable. If she is allowed to stay with her husband when he marries a second wife, she faces all the unjust accusations of causing sickness to the co-wife and her children.

Thus to the African woman, whether educated or uneducated, child-bearing is still the sole basis of a happy and permanent marriage.[15]

Explanations for the existence of some relatively low fertility rates are not entirely understood, but the available evidence suggests that the main reason is the high incidence of venereal disease in the regions where such rates prevail.[16] Sexual promiscuity is at least indirectly related to sterility. The practice of polygamy has been seen as reducing fertility because of less frequent mating by wife, more marked age differentials of partners, increased risk of venereal disease, and more frequent dissolutions of polygamous marriages, but the contrast between monogamous and polygamous marriages is probably not a

[15] Alice Ogot, "Family Planning for African Women," *East Africa Journal,* IV, No. 4 (July, 1967), 19-20.

[16] A. Romaniuk, "Infertility in Tropical Africa," in Caldwell and Okonjo, eds., *The Population of Tropical Africa*; see also his *La Fécondité des Populations Congolaises* (Paris, Mouton, 1967).

major explanation for varying fertility rates. Other factors tending to reduce fertility include certain sexual taboos, especially those connected with the prolonged breast-feeding of babies, sometimes extending as long as two to three years, the occasional use of contraceptives and abortifacients, and the separation of families due to the preponderance of males in many migratory movements.[17]

Future Trends in Birth and Fertility Rates. Estimation of the possible changes in birth and fertility rates is hazardous at best. Factors which would tend to increase the rates include: (1) the expected reduction in venereal diseases as health campaigns are expanded and relatively easy control measures become more widely available; (2) lowered death rates which would prolong the average procreative life of African women; and (3) greater stability in sexual relations, particularly through the decline of polygamous marriages.

Factors tending to reduce the birth and fertility rates are: increasing urbanization; improved educational levels; higher average incomes; later marriages; acceptance that the reduction in mortality assures greater survival of children and hence reduces the need of large families; and the extension or introduction of family planning whether encouraged or not by formal programs.

Evidence that urbanization is correlated with lower birth rates is available for Ghana, parts of Nigeria, Guinea, Mali, and elsewhere, but too little is known to tell just how significant and how rapid the changes may be as the rapid influx to urban areas continues. Caldwell suggests that changes in basic family relationships and practices are occurring in urban areas, including several which would be likely to reduce fertility, such as rising female marriage ages, a greater interest in family planning among the educated, and a lessened desire for large families due to the difficulty of supporting them, though he notes that the number of children considered ideal remains high.[18]

A demographic survey made in the Congo in 1955–1957, however, revealed urban rates to be higher than rural birth rates. Explanations given for this somewhat surprising result include the existence of

[17] See, for example, Aidan Southall, "The Demographic and Social Effects of Migration on the Populations of East Africa," paper presented in U.N., *Proceedings of the World Population Conference, Belgrade, 1965,* p. 4.

[18] Caldwell, "Introduction," part II, in Caldwell and Okonjo. eds., *The Population of Tropical Africa,* pp. 336-37.

better medical facilities, particularly for maternity cases, material support by the government through family allocations and housing priority to big families, a relaxation of sexual taboos, no indication that there was any desire to have fewer children, and the relative importance of young adults among the migrants to urban communities.

There is little evidence to date that declining death rates have had any significant impact on attitudes toward the number of children desired. Public awareness that mortality is low is considered by many experts to be a prerequisite to reducing fertility.

Family planning programs exist in only a few African countries and nowhere have they yet had a significant impact on reducing birth rates. In Egypt government interest dates from 1953 with the opening of four clinics by the National Population Commission. It was reported in 1966, however, that fewer than 20,000 of an estimated 14,880,000 females were using the loop, which was considered the main hope for success of an intensified campaign, to be promoted by some 400 clinics.

Tunisia began a family-planning program in 1956; the evolution of this program is described by Caldwell as having moved

through such stages as the emancipation of women, abolition of polygamy, the limitation of welfare support to the first four children, the removal of restrictions on the import, sale and use of contraceptives, and the legalization of abortions for women with five or more living children, to the implementation of a family planning action programme in June 1964.[19]

The same source notes that Algeria and Morocco have recently shown considerable interest in the Tunisian program, though there was little evidence of this in Algeria in 1969. Most anglophone countries have family planning clinics, but they are usually confined to only a few localities. Ghana, Nigeria, and Kenya, among anglophone countries, have shown particular interest in recent years in furthering family planning or national population policies. In Ghana the Nkrumah government had a pro-natalist policy, in line with the myth that size of population is equated with political prestige, but in March, 1969 the present Ghana government announced adoption of a policy designed to "bring down the rate of population growth to manageable limits."[20] Its program in-

[19] *Ibid.*, p. 336.
[20] Republic of Ghana, *Population Planning for National Progress and Prosperity: Ghana Population Policy* (Accra, March, 1969), p. 1.

cludes an intensive public information and education effort by both public and private agencies, demographic surveys, and modification of employment policies to reduce existing incentives for large families.

In Nigeria, the federal government has only recently shown an interest in fostering a family-planning program and in considering support of the private Family Planning Council of Nigeria (FPCN), but the Western and Midwestern states have already adopted several measures designed to promote dissemination of information on birth control, while three of the six predominantly Muslim northern states have shown interest in exploring a possible program.

The accomplishments of agencies in Ghana and Nigeria should not be exaggerated, however. As of mid-1968 only about 1200–1500 of about 1,500,000 Ghanaian women and only 7–8,000 of 10,000,000 Nigerian women of child-bearing age had received contraceptive services through organized family planning programs. Most of these women were in the two capital cities, Accra and Lagos, and only in 1968 was the FPCN able to undertake a concerted effort to extend facilities to other cities in Nigeria.

In Kenya a family-planning program for Africans began in the late 1950s. Kenya accepted the recommendations of an expert commission in 1966, becoming "the first country in mainland tropical Africa to adopt an interventionist population policy and to use central government resources in an attempt to reduce the birth rate."[21] The Family Planning Association of Kenya has been very active in trying to promote the idea of family planning but has encountered many objections both to the idea and to specific methods of contraception.[22]

On Mauritius the Mauritius Family Planning Program, founded in 1957, and Action Familiale, founded in 1963, took the first initiative in promoting birth control on this very densely populated island. The Titmuss report of 1960 [23] recommended the establishment of a nationwide family-planning service as a prerequisite to a comprehensive program for improving social conditions on that island. It specifically suggested that family allowances not provide bounties for large families,

[21] Caldwell, "Introduction," part II, in Caldwell and Okonjo, eds., *The Population of Tropical Africa*, p. 337.

[22] Ogot, "Family Planning for African Women," pp. 19-23.

[23] Richard M. Titmuss and Brian Abel-Smith, *Social Policies and Population Growth in Mauritius* (London, Methuen and Co., 1960); see also J. E. Meade *et al.*, *The Economic and Social Structure of Mauritius* (London, Methuen and Co., 1961).

that officials be required to refer couples with four or more children to some agency for family-planning advice, that higher old-age pensions be provided for families with three or fewer children, that the minimum marriage age be raised, that maternity benefits be paid so as to lengthen the period between successive children, and to provide free public and subsidized private advice and assistance on birth control. In fact, a national family-planning program was not organized until late 1966. It is too early to tell just how successful this program may be, although it is clear that much more effective measures will have to be applied in the not-too-distant future. Mauritius is one of the very few political entities in Africa which has a reliable population registration system and data compiled indicate that there is interest in restricting family size, since there has been a significant reduction in the birth rate in recent years (Table 2). The crude rate of increase remains high, however, having been four times the 1936–40 rate in 1967. Mauritius is not, of course, a characteristic African country since about three-quarters of its population is of Asian origin. Because it is an island with an area of only 720 square miles it was possible to conduct an antimalaria campaign rapidly and effectively. It will presumably be more difficult to re-

Table 2

SELECTED DEMOGRAPHIC DATA FOR MAURITIUS, 1936-1940 TO 1967

Year	Population (thousands)	Density per sq. mi.	Birth rate per thousand	Death rate per thousand	Infant mortality rate per thousand	Crude rate of increase (percent)
1936-1940[a]			33.1	27.7	155.7	0.54
1944	419	582	43.4	27.1	141.0	1.63
1948	439	610	43.4	23.8	186.2	1.96
1949	449	624	45.6	16.6	91.0	2.90
1950	466	647	49.7	13.9	76.3	3.58
1958	610	847	40.8	11.8	67.4	2.90
1962	682	947	38.0	9.3	60.1	2.87
1966	759	1054	34.9	8.8	64.2	2.61
1967	774	1075	30.4	8.5		2.19

SOURCES:
Richard M. Titmuss and Brian Abel-Smith, *Social Policies and Population Growth in Mauritius* (London, Methuen and Co., 1960).
U.N., *Demographic Yearbook 1967* (New York, 1968).
[a] Average annual rates.

Liberian mothers and children at a child clinic. In a number of countries such clinics are now contributing to family-planning programs.

duce both the death rate and the birth rate in the larger continental countries with more dispersed populations.

Family planning goes against deeply held convictions of African women. Objections reflect the fears, superstitions, and prejudices of populations with still high rates of illiteracy and generally low levels of educational achievement. Illustrative is the fear expressed by some Kenyan women that the pill was simply a new drug manufactured by Europeans to sterilize African women and the reaction in more than one country that foreign experts promoting family planning were either playing a kind of neo-colonialist trick to keep the African population down[24] or were supporting white imperialist efforts to curb the increase of black populations. In Egypt the birth control campaign has been partially sabotaged by midwives who were fearful of reduced incomes; to enlist the help of these women the authorities have elected to pay a fee to any midwife who persuaded a woman to visit a clinic.

More optimistic attitudes are reported from surveys in Lagos

[24] Ogot, "Family Planning for African Women," *East Africa Journal*, p. 23.

and in Katsina Province, Nigeria, which "seem to indicate that, for the majority of persons, deeply felt emotional factors in opposition to the idea of family planning are not at work, and, given sufficient education, the opening of sufficient clinics and the ready availability of contraceptive devices, one might anticipate a significant expansion in use."[25]

A 1965–1966 exploratory demographic survey dealing with fertility attitudes among various socioeconomic groups of married women in Ibadan, Nigeria, found that the higher the educational, income, and occupational levels, the more favorable was the attitude toward the use of contraceptives, suggesting that a certain standardization of fertility behavior was unfolding in the population studied.[26]

The fact that fertility-size ideals can change with considerable rapidity is suggested by the results of surveys in Lagos which revealed that, in 1964, 93 percent of the women questioned wished to have five or more children, whereas the proportion had fallen by 1968 to 70 percent. So-called KAP (Knowledge-Attitude-Practice) surveys have revealed that 40 percent of urban women and 23 percent of rural women interviewed in Ghana were interested in learning more about family limitation, while only 11 percent of the urban and 4 percent of the rural women surveyed said that they already knew some method to control births. A similar survey in Nigeria found that over three-fourths of those questioned expressed approval of contraception, 80 percent said they would be interested in more information about it, 37 percent accepted an appointment slip to a family planning clinic, and 85 percent believed that the government should provide family planning information to those who requested it.

One should not, of course, overestimate the effect of even a highly successful family planning program on reducing birth rates and achieving desirable growth rates in underdeveloped countries. Indeed, Kingsley Davis feels that the emphasis on family planning has been a major obstacle to population control because the use of contraceptives is compatible with high fertility in that the aim is usually for parents to have the number of children they want. It is more correctly "parent

[25] T. Daramola, R. D. Wright et al., "Surveys of attitudes in Nigeria towards family planning," in Caldwell and Okonjo, eds., The Population of Tropical Africa, p. 409.

[26] Francis Olu Okediji, "Some Social Psychological Aspects of Fertility among Married Women in an African City," The Nigerian Journal of Economic and Social Studies, IX, No. 1 (March, 1967), 67-79.

planning," not national control or planning. He calls for the use of more effective control methods and for reducing the desire for children, suggesting that zero growth would seem to be the appropriate goal, and noting that the goal of achieving a growth rate comparable to that of the industrial nations would still result in doubling populations in 50 years.[27]

While it is hazardous to predict what the future trend of African birth rates may be, the present evidence suggests that they are not likely to decline significantly in the next decade or so and that they may very likely continue to be the highest in the world at the turn of the century. Unless the factors that tend to reduce birth rates assume considerably greater weight than they now have, the average birth rate is not likely to be much below 40 at that time.

Death Rates. African mortality rates are also among the highest in the world, ranging by region from an estimated 16 per M for southern Africa to 27 per M for western Africa in the period 1960–1967, and averaging 22 per M for the continent as a whole. The last rate compares with 18 for Asia, 12 for Latin America, 9 for northern America, and 15 for the world. Only East Asia and Melanesia among major world regions come close to the African average. For individual African countries, mortality rates range from about 8 to 40 per M.

Estimates of African death rates are considerably less certain than those of birth rates and extrapolation is far more precarious since significant changes can occur in a very short period of years. The death rate on Mauritius, for example, fell 30.3 percent in one year, largely as a response to an intensive anti-malaria campaign (Table 2).[28]

The prevailing birth and death rates of Africa result in a very high percentage of young people and generally low life expectancies. In most countries over 40 percent of the population is under 15 years of age, making for a very great dependency load only partially offset by improvements in the health of the working force. Life expectancies at birth range from under 30 years for a few countries to over 60 for such atypical countries as Mauritius, Réunion and the Seychelles. Three-quarters of the political units for which estimates are available have life expectancies within the range of 32 to 47 years.

[27] Kingsley Davis, "Population Policy: Will Current Programs Succeed?" *Science*, CLVIII (10 November, 1967), 730-39.

[28] Titmuss and Abel-Smith, *Social Policies and Population Growth in Mauritius*, p. 47.

The characteristically high death rates of most African countries have commonly been explained by a variety of factors: low standards of living, low educational levels, the scarcity of medical and health facilities, poor sanitation practices, inadequate clothing and housing, malnutrition, a high incidence of diseases and, more in the past than at present, intertribal warfare, famine, and epidemics. But increasing evidence indicates that the incidence of disease has been far more important than the other factors.

Since the Second World War, a number of developing countries have succeeded in reducing their previously high mortality primarily by the application of new technology, such as vector control and mass immunization, without commensurate progress in classical medical services, sanitation, education, level of living, etc.[29]

Some extremely rapid declines in mortality have been recorded since the last war, with some areas adding a year each year to life expectancies. And several projections suggest an annual rise in life expectancies of a half year over the next two decades. The U.N. medium estimate for the period 1995–2000 is for an average death rate of 13.1 compared with 22.5 for the period 1960–1965.

As has already been suggested, the prime mover in this trend has been and is likely to be improved disease control. Particularly important will be reductions in infant and child mortality rates. Death rates in the first year now range in most of Africa from about 100 to over 200 per M. Rates under 100 are reported only for Algeria, the U.A.R. (Egypt), and many of the island appurtenances. Regional rates within some tropical African countries are thought to exceed 300 per M. Mortality from age 1 to 4 is also characteristically very high.

While disease-control campaigns may result in dramatic decreases in death rates these rates are likely to remain relatively high for many years, because the numerous factors noted above which affect death rates are not likely to be corrected with great rapidity. Even the incidence of disease, though subject to marked reductions, will continue as a potent force in keeping the death rates high. Only a small percentage of Africans are not handicapped by some endemic disease —more likely than not they have several diseases to which they have

[29] U.N. Department of Economic and Social Affairs. *1965 Report on the World Social Situation* (New York, 1966), p. 3.

acquired some tolerance but which nonetheless sap their vitality and make them subject to the acquisition of other diseases which may prove fatal.

The pestilential diseases—cholera, plague, relapsing fever, small-pox, typhus and yellow fever—have been attacked with considerable success. Malaria, despite some massive gains, remains the single most important disease problem in many African countries. While subject to complete eradication, the cost of the necessary programs remains beyond the reach of most peoples and over three-fourths of the population of tropical Africa have not yet benefited from an eradication program.

Communicable diseases such as leprosy, venereal diseases, and yaws are now also subject to effective control, although most victims still go untreated and the incidence of venereal disease appears to be rising in some areas. Other communicable diseases remain as serious problems: trachoma, bilharziasis, filariasis, trypanosomiasis, tuberculosis, and measles.

Nutritional problems contribute to ill health, particularly among children. Outright starvation may be confined to such situations as the

A government worker about to spray a hut near Fort Dauphin, Madagascar, with D.D.T. Antimalaria campaigns have reduced death rates dramatically in some African areas.

African Demography

Nigeria-Biafra war, but improper dietal intake is a major factor in the prevailingly high infant and child mortality rates. The main food and nutrition problems of Africa are: (1) seasonal food shortages, (2) nutritional deficiencies, (3) undernutrition of children, (4) food shortage and preservation (losses to locusts and birds before the harvest, to rodents, insects, and fungi thereafter), and (5) the lack of nutrition education, which would help to prevent some of the other problems.

It is estimated that in tropical Africa calorie intake varies from 70–75 percent to 120–140 percent of requirements, but that for children it is usually only 60–70 percent of actual requirements. Protein intake is characteristically low, sometimes only half of requirements. There is a heavy dependence on carbohydrates, the main sources being starchy roots such as manioc and yams and plantains in the rainy tropics and on cereals, particularly sorghum and millet, in the drier areas. Fishing tribes tend to have the highest protein intake but their diets are usually lacking in vegetables and fruit. The urban situations that have been studied reveal that dietal standards often do not improve in consonance with increased incomes and that the lower paid or under-employed workers are usually worse off than the rural residents.

The two general nutritional problems that affect all parts of Africa are protein-calorie malnutrition (P.C.M.) of early childhood and nutritional anaemia, which is a serious and common cause of illness and death.[30] Children from one to four need about double the protein intake in relation to body weight of adult men; if the calorie intake is inadequate the protein needs greatly increase, hence the linkage of proteins and calories in P.C.M.

Protein-calorie malnutrition takes three main forms: (1) kwashiorkor, (2) marasmus (extreme emaciation), and (3) nutritional growth failure. The incidence of kwashiorkor is estimated at 2–4 percent for all children in sub-Saharan Africa, but as high as 9 percent in Nigeria and 6–11 percent in Uganda. Marasmus is more common than kwashiorkor in urban areas, while nutritional growth failure affects an estimated 15 to 30 percent of children. There is also a strong interaction between P.C.M. and infection, each tending to increase the other.

There are many explanations for the prevalence of P.C.M. in

[30] See R. Cook, "The General Nutritional Problems of Africa," *African Affairs*, LXV, No. 261 (October, 1966), 329-40.

tropical Africa.[31] First are certain practices with respect to weaning. While breast-feeding traditionally goes on as long as possible, directly thereafter the child is put on a diet composed completely of starchy foods, animal products being excluded either because they are considered inappropriate, or the available supply goes mainly to men, or because they are forbidden by taboos, which is particularly true for eggs. Second, malnutrition results from the social disabilities of females, whose lower status gives them lower priority in the apportioning of food even during pregnancies, as it is seldom recognized that there are increased needs during these periods. Third, the incidence of malnutrition is affected by poverty, as has been shown by a number of surveys which indicate correlations between income level and malnutrition. A fourth factor tending to increase malnutrition is the trend to a smaller number of crops and hence some reduction in dietal variety.

Protein deficiency is explained in part by a general deficiency of animal products. Beef is eaten by many ethnic groups only at festivals and celebrations or when cattle die. Chickens, which are usually pathetically scrawny, and eggs, which are small, are frequently reserved for gifts. The availability of game animals, which could be a more important source under proper management, has been greatly reduced or virtually eliminated over large areas. Milk is drunk but very little is traded, and it tends to be high cost and in short supply in urban areas, a situation which is only partially ameliorated by the importation of powdered milk. Fish are greatly appreciated by all but a few tribes which have taboos against them, but fish are not always available and tend to be too expensive as soon as one gets some distance from the place of catch. Proteins are derived by some groups by eating such things as termites, grasshoppers, rodents, snakes, and lizards or by drinking blood taken from the veins of domestic cattle, but these sources scarcely provide the answer to the problem of malnutrition.

A greater intake of fruits and vegetables would be highly desirable in many areas, particularly to provide vitamins. And a wide variety of fruits can be grown, particularly in the wetter areas, but the interest in doing so is often lacking, especially in the planting of fruit-bearing trees whose yields are perforce delayed. This kind of attitude, plus vari-

[31] See R. G. Hendrickse, "Some Observations on the Social Background to Malnutrition in Tropical Africa," *African Affairs*, LXV, No. 261 (October, 1966), 341-49.

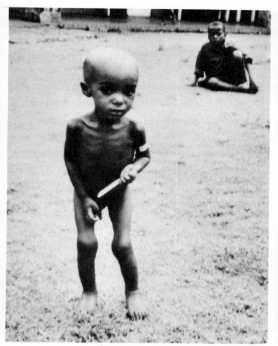

Child suffering from kwashiorkor, Kivu, Congo (Kinshasa).

Medical aide with child in famine-stricken area, Kasai, Congo (Kinshasa).

ous food taboos, are tending to recede, but nonetheless remain as formidable barriers to achieving a proper dietal balance.

A relatively new but highly significant consideration regarding P.C.M. is the increasing evidence "that malnutrition during the first few years of life does have an adverse effect on subsequent learning and behavior."[32] Dr. Scrimshaw also states that

the reduced physical growth and development and costly morbidity and mortality of preschool children . . . is already reason enough for giving high priority to programs for improving the nutritional status of the preschool child.

[32] Nevin S. Scrimshaw, "Malnutrition, Learning and Behavior," *The American Journal of Clinical Nutrition*, XX, No. 5 (May, 1967), 493. See also Walsh McDermott, "Modern Medicine and the Demographic-Disease Pattern of Overly Traditional Societies: A Technologic Misfit," in *Manpower for the World's Health*, Pt 2 of *The Journal of Medical Education*, XLI, No. 9 (September, 1966), 145.

. . . The probability that early malnutrition can cause significant retardation of mental development is an important added reason for emphasizing the universal prevention of malnutrition in the preschool child.[33]

Inadequacies in health services must also remain for some decades as a serious deterrent to reducing the death rate. Particularly serious is the shortage of doctors, as is evidenced by the fact that some new hospital facilities cannot be fully utilized because of staff shortages, and by the low ratios of doctors to population in various African countries, summarized below for years ranging from 1960 to 1966. These ratios compare with that of 1 doctor per 670 inhabitants in the United States in 1965.

1 doctor per 1900-4999	1 doctor per 5000-9999	1 doctor per 10,000-14,999	1 doctor per 15,000-19,999
South Africa	Gabon	Madagascar	Port Guinea
Egypt	Equat. Guinea	Liberia	Sierra Leone
Réunion	Swaziland	Uganda	Comoro Is.
São Tome, Príncipe	Rhodesia	Congo (Brazzaville)	Mozambique
Libya	Algeria	Morocco	Tanganyika
Seychelles	Tunisia	Kenya	Senegal
Mauritius		Angola	Ivory Coast

1 doctor per 20,000-29,999	1 doctor per 30,000-49,999	1 doctor per 50,000-76,200
Guinea	Somalia	Burundi
Botswana	Mauritania	Upper Volta
Zambia	Sudan	Niger
Gambia	Congo (Kinshasa)	Ethiopia
Lesotho	Central African Rep.	Chad
Togo	Cameroon	Rwanda
	Nigeria	
	Malawi	

SOURCE: U.N., *Statistical Yearbook 1967* (New York, 1968), p. 698.

The trends with respect to mortality, in summary, may be expected to reveal a continuing decline, sometimes notably rapid, particularly in response to massive antimalarial campaigns, but a variety of negative factors may make further gains much more difficult to achieve.

Migration. The third population dynamic, migration, is not significant as far as continental totals are concerned. The same gen-

[33] Scrimshaw, "Malnutrition, Learning and Behavior," p. 500.

eralization applies to many individual countries, where external movements have been small in relation to total populations of sending and receiving areas. But immigration has contributed to population growth in some countries such as Ghana, Ivory Coast, Uganda, Rhodesia, and South Africa, while emigration has probably reduced growth rates in such countries as Guinea, Mali, Upper Volta, Mozambique, Lesotho, and, at least temporarily, Algeria. Estimates of migration are extremely sparse and can rarely be entered into the growth equation with any accuracy.

Population Growth. Since birth-rate, death-rate, migration, and census figures are all subject to serious inadequacies and distortions, it follows that estimates of the crude or actual rates of increase must be accepted with great caution. Table 3 gives estimates for various periods for the continent as a whole. The estimated annual rate for 1963–1967 of 2.5 percent is exceeded only by Latin America with an estimated rate of 2.9 percent; and this rate also compares with rates of 1.3 for northern America, 0.8 for Europe excluding the U.S.S.R., and 1.9 for the world as a whole.

Map 2 gives the estimated annual rates of increase for the period 1963–1967 for those countries for which data are available. Ten countries have rates in excess of 3.0 but they do not form any recognizable pattern. The high rates in a few countries, such as Rhodesia, may be explained in part by immigration; other high rates reflect declining death rates coupled with persistently high birth rates.

Trends in growth rates are, as is seen in Table 3, distinctly upward, and only a very few countries show either a slowdown in the trend or a declining rate of increase. The major components are, of course, continuing high birth rates but decreasing death rates, and the expectation that numerous countries will continue to experience declining death rates suggests that the average growth rate will rise for at least the next two decades.

The theory of demographic transition hypothesizes that the high birth-rate–high death-rate stage characteristic of agrarian low-income economies is succeeded by a high birth-rate–declining death-rate phase as the economy changes and improved health conditions are introduced. Somewhat later, the birth rate begins to fall but not as rapidly as the death rate. Finally, the birth rate nears equality with the death rate and

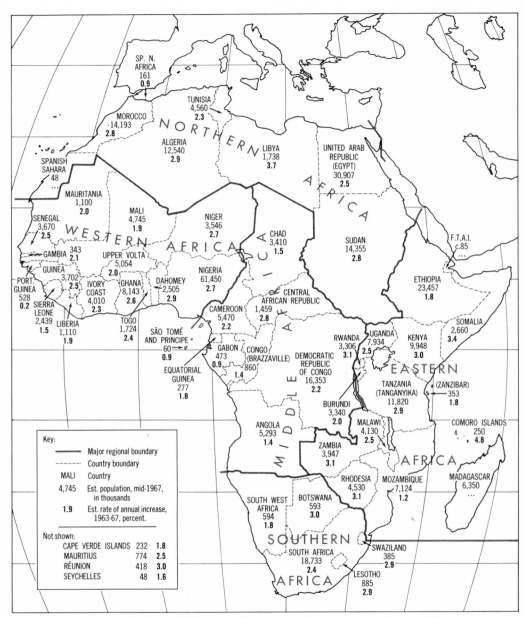

MAP 2. Total 1967 population and estimated rate of annual increase of African countries, 1963-1967; major regions of Africa as delineated by the United Nations.

African Demography

Table 3

ESTIMATED AND PROJECTED POPULATION GROWTH RATES FOR
AFRICA FOR SELECTED PERIODS, 1750–1800 TO 1960–2000

Period	Source	Annual growth rate (%)	Period	Source	Annual growth rate (%)
1750–1800	1	0.0	1930–1950	5	1.3
1800–1850	1	0.1	1900–1960	4	1.0
1850–1900	1	0.4	1958–1965	6	2.3
	2	0.5	1960–1967	7	2.4
	3	0.7	1963–1967	7	2.5
	4	0.8	1950–2000	1	2.5
1900–1930	3	0.6	1960–2000	4	2.6
1900–1950	1	1.0			

SOURCES:
1. John D. Durand, "The Modern Expansion of World Population," *Proceedings of the American Philosophical Society*, CXI, No. 3 (June 22, 1967), 137.
2. A. M. Carr-Saunders, *World Population: Past Growth and Present Trends* (Oxford, 1936), pp. 34–35, 42.
3. Walter F. Willcox, "Increase in the Population of the Earth and of the Continents since 1650," in Willcox, ed., *International Migrations*, Vol. II, *Interpretations* (New York, 1931), 78.
4. J. D. Durand, "World Population Estimates, 1750–2000," in *Proceedings of the World Population Conference, Belgrade, 1965*, II (New York, United Nations, 1967), 20.
5. U.N., *The Determinants and Consequences of Population Trends* (New York, 1953), p. 11.
6. U.N., *Statistical Yearbook 1966* (New York, 1967), p. 26.
7. U.N., *Demographic Yearbook 1967* (New York, 1968), p. 97.

a more gradual rate of growth is established with low mortality rates and small families being typical.[34]

Most African countries would fall either in the first or second of these stages and hence it may be concluded that the continent is in an early period of rapidly accelerating population growth. The present rates of growth, it should be noted, exceed those experienced in Europe during its demographic transition, being about double that of industrial Europe in the nineteenth century. The expectation is that Africa may have the highest growth rate in the world in the year 2000, though the

[34] See, for example, Ansley J. Coale and Edgar M. Hoover, *Population Growth and Economic Development in Low-income Countries* (Princeton, Princeton University Press, 1958), p. 13.

rate may no longer be accelerating at that time. Thirty years is a long time under modern conditions of change, however, and today's demographic predictions may prove to be as innacurate as those of the 1930s which saw declining populations in the developed world within two or three decades.

Estimates of the total African population at the turn of the present century vary considerably. U.N. projections made in 1963 for the year 2000 A.D. were as follows: Low 684 million, Medium 768 million, and High 865 million.[35] Relating these figures to world figures may help to give them significance. It took all of recorded time to the middle of the nineteenth century to achieve a population of 1 billion, less than 100 years to add the second billion and only 30 years to add the third. At today's rate of increase there will be nearly 7 billion by the year 2000. For Africa it took all of recorded time to about 1650 to achieve a population of 100 million, the next 300 years to add the second 100 million, and only about 15 years to add the third 100 million. By the year 2000 and using the U.N. medium projection for that year, the population may be increasing 100 million almost every 5 years. By about 2010 Africa's population may well exceed 1 billion, the level of the world total not more than about 120–140 years ago. The implications of these extremely rapid increases on planning, development, pressure on the land, and urbanization are immense. Some of these implications will be examined in the following chapters.

Bibliography

PART A: CENSUSES AND OTHER SOURCES OF DEMOGRAPHIC DATA

Algeria   Algeria, Commissariat National au Recensement de la Population. *Résultats Préliminaires du Recensement Général de la Population Effectué en 1966*. Oran, 1966.

Angola   Angola. *População segundo os grupos étnicos, estado civil e idades*, I. Luanda, Imprensa Nacional, 1960.
——. *3° Recenseamento geral da população 1960*, I. Luanda, Imprensa Nacional, 1964.

Botswana   Bechuanaland, Population Census Office. *Report on the Census of the Bechuanaland Protectorate, 1964*. Mafeking, Government Printer, 1965.

[35] U.N., *World Population Prospects as Assessed in 1963*. Population Studies No. 41 (New York, 1966).

Burundi   United Nations, Department of Social Affairs, Population Division. *The Population of Ruanda-Urundi.* New York, 1953.

Cameroon   Durupt, Marie-Josèphe, et François Turbot. *La Population du Cameroun Occidental.* Paris, S.E.D.E.S./Yaoundé, Ministère du Plan, 1965.

République du Cameroun, Service de la Statistique. *Enquête Démographique au Cameroun—Résultats Définitifs pour la Région Sud-Est 1962-1964; Résultats Définitifs pour la Région Nord 1962-1964.* Paris, République Française, Secrétariat d'État aux Affaires Étrangères/I.N.S.E.E., 1968.

Cape Verde Islands   Cabo Verde. *Boletim Trimestral de Estatística, 4° Trimestre de 1961,* No. 4. Praia, Imprensa Nacional, 1962.

Central African Republic   République Centrafricaine, Service de la Statistique Générale. *Enquête Démographique en République Centrafricaine 1959–1960.* Paris, République Française, Ministère de la Coopération/I.N.S.E.E., April, 1964.

——. *Recensement Général de la Population de la République Centrafricaine,* 1er fasc. *Résultats pour la Région de Lobaye Haute-Sangha 1961–1963.* Paris, République Française, Ministère de la Coopération/I.N.S.E.E., July, 1964.

——. *Recensement Général de la Population de la République Centrafricaine,* 2e fasc. *Résultats pour la Région de Ouham, Ouham-Pende, Nana-Membere (Baboua).* Paris, République Française, Secrétariat d'État aux Affaires Étrangères Chargé de la Coopération/I.N.S.E.E., March, 1968.

Chad   République du Tchad, Service de Statistique. *Enquête Démographique au Tchad 1964.* 2 vols. Paris, Secrétariat d'État aux Affaires Étrangères Chargé de la Coopération/S.E.D.E.S./I.N.S.E.E., June, 1966.

Congo (Brazzaville)   République du Congo, Service de Statistique. *Enquête Démographique 1960-1961.* Paris, République Française, Ministère de la Coopération/I.N.S.E.E., February, 1965.

Congo (Kinshasa)   République du Congo. *Tableau Général de la Démographie Congolaise, Enquête Démographique par Sondage 1955-1957.* Léopoldville, July, 1961.

Dahomey   République du Dahomey. *Enquête Démographique au Dahomey, 1961: Résultats Définitifs.* Paris, République Française, Ministère de la Coopération/I.N.S.E.E., 1964.

Ethiopia   Ethiopia, Central Statistical Office. *Statistical Abstract 1966.* Addis Ababa, 1967.

Gabon   République Gabonaise, Service de Statistique. *Recensement et Enquête Démographique 1960-1961.* Paris, République Française, Ministère de la Coopération/I.N.S.E.E., 1965.

Gambia   The Gambia. *Population Census of the Gambia, 17th/18th April 1963.* Bathurst, 1963.

Ghana   Ghana, Census Office. *1960 Population Census of Ghana.* 5 vols. Accra, 1962-1964.

Ghana, Survey of Ghana and Census Office. *1960 Population Census of Ghana: Atlas of Population Characteristics*. Accra, 1964.

Guinea   France, Administration Générale des Services de la France d'Outre-Mer/Haut Commissariat Générale de l'A.O.F. *Étude Démographique par Sondage en Guinée 1954-1955*. 3 vols. Paris, 1959-1961.

Ivory Coast   France, Ministère de la France d'Outre-Mer. *Étude Démographique du 1er Secteur Agricole de la Côte d'Ivoire 1957-1958*. Paris, December, 1958.

Kenya   Kenya, Ministry of Finance and Economic Planning, Directorate of Economic Planning. *Kenya Population Census 1962*. 2 vols. Nairobi, 1964-1965.

Lesotho   Lesotho, Bureau of Statistics. *Basutoland Census*. Maseru, November, 1966.

Liberia   Republic of Liberia, Office of National Planning, Bureau of Statistics. *1962 Population Census*. Monrovia, 1965.

Libya   Kingdom of Libya, Ministry of Economy and Trade, Census and Statistical Department. *General Population Census 1964*. Tripoli, 1965.

Madagascar   Centre Scientifique et Médical de l'Université Libre de Bruxelles en Afrique Centrale/Office de la Recherche Scientifique et Technique Outre-Mer. *Madagascar: Cartes de Densité et de Localisation de la Population*. Brussels, 1967.

République Malgache, Ministère des Finances et du Commerce, Institut National de la Statistique et de la Recherche Économique. *Population de Madagascar au 1er Janvier, 1964 et au 1er Janvier, 1965*. Tananarive, 1966.

——. *Recensements Urbains*. 3 vols. Tananarive, 1965-1966.

Malawi   Malawi, National Statistical Office, Population Division. *Malawi Population Census 1966, Provisional Report*. Zomba, Government Printer, 1966.

Mali   République du Mali, Mission Socio-Économique du Soudan 1956-1958. *Enquête Démographique dans le Delta Central Nigerien*, 2e fasc. *Résultats Détaillés*. Paris, République Française, Ministère de la Coopération/I.N.S.E.E., n.d.

Mauritania   Mauritania, Service de la Statistique. *Recensement Démographique des Agglomérations (Enquête 1961-1962)*. *Bulletin Statistique et Économique*, No. 3. Nouakchott, 1964.

——. *Statistique Générales Années 1963-1964*. *Bulletin Statistique et Économique*, Nos. 4 and 5-6. Nouakchott, 1965.

Mauritius   Mauritius, Central Statistical Office. *1962 Population Census of Mauritius and Its Dependencies*. Port Louis, Government Printer, 1963.

Morocco   Royaume du Maroc, Service Central des Statistiques. *Recensement Démographique: Population Légale du Maroc*, Rabat, 1961.

——. *Recensement Démographique (Juin 1960): Population Rurale du Maroc*. Casablanca, 1962.

Mozambique   Portugal, Província de Moçambique, Direção dos Serviços de Economia e Estatística Geral, Repartiçao de Estatística Geral. "Alguns Aspector do IV Recenseamiento Geral da População do

	Moçambique (1960)," in *Boletim Mensal de Estatística*, ano 4, No. 3 (March, 1963). Lourenço Marques.
Niger	République du Niger, Mission Économique du Niger. *Étude Démographique du Niger*. Paris, République Française, Ministère de la Coopération/I.N.S.E.E., 1963.
	———. Mission Économique et Pastorale 1963. *Étude Démographique et Économique en Milieu Nomade*. Paris, République Française, Ministère de la Coopération/S.E.D.E.S./I.N.S.E.E., 1966.
Nigeria	Federation of Nigeria. *Population Census of Nigeria: Summary of Tables*. Lagos, 1964.
Portuguese Guinea	Portugal, Província da Guiné. *Anuário Estatístico, 1956-1958*. Lisbon, 1961.
Réunion	République Française, Institut National de la Statistique et des Études Économiques. *Résultats Statistiques du Recensement Général de la Population des Départments d'Outre-Mer Effectué le 9 Octobre, 1961, Réunion*. Paris, Imprimerie Nationale, 1962.
Rhodesia	Southern Rhodesia, Central Statistical Office. *Final Report of the April/May, 1962 Census of Africans in Southern Rhodesia*. Salisbury, June, 1964.
	———. *Census of the European, Asian and Coloured Population, 1961*. Salisbury, 1963.
Rwanda	United Nations, Department of Social Affairs, Population Division. *The Population of Ruanda-Urundi*. New York, 1953.
Senegal	Cantrelle, Pierre. *Étude Démographique dans la Région du Sine Saloum (Sénégal)*. Dakar-Hann, ORSTOM, 1965.
	Verrière, Louis. *La Population du Sénégal*. Dakar, Université de Dakar, 1963.
Seychelles	Colony of Seychelles, Office of the Census Commissioner. *Population Census of the Seychelles Colony: Report and Tables for 1960*, by A. W. T. Webb. Kenya, Government Printer, 1960.
Sierra Leone	Sierra Leone Government, Central Statistics Office. *1963 Population Census of Sierra Leone*. 3 vols. Freetown, 1965.
South Africa	Republic of South Africa, Bureau of Statistics. *Population Census, 6th September, 1960*. 9 vols. Pretoria, Government Printer, 1963-1968.
South West Africa	Union of South Africa, Bureau of Census and Statistics. *South West Africa: Population Census, 8, May, 1951*. Pretoria, Government Printer, 1951.
Sudan	Sudan, Department of State, Population Census Office. *First Population Census of Sudan 1955-1956: Final Report*. 2 vols. Khartoum, 1961-1962.
Swaziland	Swaziland Government. *Report on the 1966 Swaziland Population Census*. Mbabane, 1968.
Tanzania	University College, Dar es Salaam, Bureau of Resource Assessment and Land Use Planning. *Population Density in Tanzania, 1967*. Dar es Salaam, 1968.
	United Republic of Tanzania, Central Statistical Bureau. *Provisional Estimates of Fertility, Mortality and Population Growth of*

	Tanzania. Dar es Salaam, December, 1968.
	———. *Preliminary Results of the Population Census August 1967.* Dar es Salaam, December, 1967.
Togo	République du Togo, Ministère des Finances et des Affaires Économiques, Service de la Statistique Générale du Togo. *Recensement Général de la Population du Togo 1958-1960.* 7 vols. Lomé, 1961-1963.
	———. *Le Recensement de la Population Urbaine au Togo (1958).* Lomé, 1960.
	———. "Estimation de la population au 1ᵉʳ janvier 1966," *Bulletin de Statistique de la République Togolaise,* No. 2, 1966, pp. 1-111.
Tunisia	République Tunisienne, Secrétariat d'État au Plan et à l'Économie Nationale, Service des Statistiques Démographiques. *Recensement Général de la Population, 3 Mai, 1966.* Tunis, 1966.
Uganda	Uganda, Ministry of Economic Affairs, Statistical Branch. *Uganda Census, 1959, African Population.* Entebbe, 1961.
	———. *Uganda Census, 1959, Non-African Population.* Entebbe, 1960.
	———. *Uganda: General African Census.* 2 vols. Entebbe, 1960.
U.A.R. (Egypt)	United Arab Republic (Egypt), Presidency, Statistical Department. *Population Census, 1960, Preliminary Results.* Cairo, 1961 (In Arabic).
Upper Volta	République du Haute-Volta, Service de Statistique. *La Situation Démographique en Haute-Volta: Résultats Partiels de l'Enquête Démographique 1960-1961.* Paris, République Française, Ministère de la Coopération/I.N.S.E.E., 1962.
Zambia	Government of the Republic of Zambia, Central Statistical Office. *Final Report of the September 1961 Censuses of Non-Africans and Employees.* Lusaka, August, 1965.

PART B: SOURCES ON AFRICAN DEMOGRAPHY

Aluko, S. A. "How Many Nigerians? An Analysis of Nigeria's Census Problems, 1901-63," *The Journal of Modern African Studies,* III, No. 3 (October, 1965), 371-92.

Barbour, Kenneth M. *Population in Africa: A Geographer's Approach.* Ibadan, Ibadan University Press, 1963.

———, and R. Mansell Prothero, eds. *Essays on African Population.* London, Routledge and Kegan Paul, 1961.

Beaujeu-Garnier, J. *The Geography of Population.* New York, St. Martin's Press, 1967.

Brass, William, Ansley J. Coale *et al.* *The Demography of Tropical Africa.* Princeton, Princeton University Press, 1968.

Brenez, J. "Une expérience de recensement en zone nomade: enquête démographique en Mauritanie," *Coopération et Developpement,* No. 9 (March, 1966), pp. 33-41.

Caldwell, John C., and Chukuka Okonjo, eds., *The Population of Tropical Africa.* New York, Columbia University Press, 1968.

Cook, R. "The General Nutritional Problems of Africa," *African Affairs,* LXV, No. 261 (October, 1966), 329-40.

Davis, Kingsley. "Population Policy:

Will Current Programs Succeed?" *Science*, CLVIII (November 10, 1967), 730-39.

France. Institut National de la Statistique et des Études Économiques— Service de Coopération, Institut National d'Études Démographiques. *Afrique Noire, Madagascar, Comores: Démographie Comparée*. 10 vols. Paris, Délégation Générale à la Recherche Scientifique et Technique, 1967.

——. *Bibliographie Démographique, 1945-1964*. Paris, 1965.

——. *Situation des Enquêtes Statistiques et Socio-Économiques dans les États Africains et Malgache au 1er Janvier 1964*. Paris, 1964.

——. S.E.D.E.S. *Étude Démographique et Économique en Milieu Nomade*. 2e fasc. Paris, 1966.

Hendrickse, R. G. "Some Observations on the Social Background to Malnutrition in Tropical Africa," *African Affairs*, LXV, No. 261 (October, 1966), 341-49.

Kayser, B. "La Démographie de l'Afrique Occidentale et Centrale," *Les Cahiers d'Outre-Mer*, XVIII, No. 69 (January-March, 1965), 73-86.

Kuczynski, Robert R. *Demographic Survey of the British Colonial Empire*. 2 vols. London, Oxford University Press, 1948, 1949.

Lorimer, Frank. *Demographic Information on Tropical Africa*. Boston, Boston University Press, 1961.

——, William Brass, and Etienne van de Walle. "Demography," in Robert Lystad, ed. *The African World: A Survey of Social Research*. New York, Praeger, 1965, pp. 271-303.

——, and Mark Karp, eds. *Population in Africa: Report of a Seminar Held at Boston University*. Boston, Boston University Press, 1960.

Lury, D. A. "Population Data in East Africa," Discussion Paper No. 18, Institute for Development Studies, University College, Nairobi, 1966.

McDermott, Walsh. "Modern Medicine and the Demographic-Disease Pattern of Overly Traditional Societies: A Technologic Misfit," *Manpower for the World's Health*, Pt. 2 of *The Journal of Medical Education*, XLI, No. 9 (September, 1966), 137-62.

Molnos, Angela. *Attitudes towards Family Planning in East Africa*. New York, Humanities Press, 1968.

Ogot, Alice. "Family Planning for African Women," *East Africa Journal*, IV, No. 4 (July, 1967), 19-23.

Okediji, Francis Olu. "Some Social Psychological Aspects of Fertility among Married Women in an African City," *The Nigerian Journal of Economic and Social Studies*, IX, No. 1 (March, 1967, 67-79.

Romaniuk, A. *La Fécondité des Populations Congolaises*. Paris, Mouton, 1967.

——. "Évolution et Perspectives Démographiques de la Population du Congo," *Zaïre*, XIII, No. 6 (1959), 563-626.

Scrimshaw, Nevin S. "Malnutrition, Learning and Behavior," *The American Journal of Clinical Nutrition*, XX, No. 5 (May, 1967), 493-502.

Smith, T. E., and J. G. C. Blacker. *Population Characteristics of the Commonwealth Countries of Tropical Africa*. London, Athlone Press for the Institute of Commonwealth Studies, 1963.

Texas University, Department of Sociology, Population Research Center. *International Population Census Bibliography: Africa*. Austin, 1965.

Union Internationale pour l'Étude Scientifique de la Population. *Problems in African Demography*. Paris, 1960.

United Nations. *Demographic Yearbook 1967*. New York, 1968.

——. *The Determinants and Conse-*

quences of *Population Trends*. 2d ed., rev. New York, 1963.

——. *Future Growth of World Population.* New York, 1958.

——. *Proceedings of the World Population Conference, Belgrade, 1965.* New York, 1967.

——. *Statistical Yearbook 1967.* New York, 1968.

——. *World Population Prospects as Assessed in 1963.* Population Studies No. 41 New York, 1966.

——. Economic Commission for Africa. *Report of the Seminar on Population Problems in Africa.* Addis Ababa, 1963.

——. UNESCO. *Social Implications of Industrialization and Urbanization in Africa South of the Sahara.* Paris, 1956.

Zelinsky, Wilbur. *A Bibliographic Guide to Population Geography.* University of Chicago, Department of Geography Research Paper, No. 80, Chicago, 1962.

CHAPTER 2

POPULATION DENSITIES AND DISTRIBUTION

THE PURPOSES of this chapter are to present data regarding population densities in Africa and to examine some of the major elements of its population distribution, as knowledge of both is essential to an understanding of the population factor and its complex influences. The question as to whether Africa suffers from population pressure is postponed to Chapter 6, while the relations between population factors and economic development are discussed in several of the chapters succeeding this one.

Population Densities

The average population density of Africa in mid-1969 may be estimated at about 29.5 per square mile. This figure has very little significance other than to suggest a relatively low overall density which in turn tends

Table 4

POPULATION DENSITY RANGE BY AFRICAN POLITICAL UNITS, MID-1967

Density per square mile	Number of political units	Percent of total population	Percent of total area	Cumulative percent	
				Population	Area
−10	10	4.2	27.3		
10−	8	18.3	36.8	22.5	64.1
20−	4	4.5	5.0	27.0	69.1
30−	7	14.9	11.8	41.9	80.9
40−	4	12.8	7.5	54.7	88.4
50−	2	0.9	0.4	55.6	88.8
60−	—	—	—	—	—
70−	3	1.7	0.7	57.3	89.5
80−	6	19.9	6.8	77.2	96.3
90−	2	1.3	0.4	78.5	96.7
100−	3	18.8	3.1	97.3	99.8
200−	1	0.1	. . .	97.4	
300−	4	2.2	0.2	99.6	
400–500	1	0.1	. . .	99.7	
1000+	2	0.3	. . .	100.0	100.0
Total	57	100.0	100.0		

Crude density—28.4

SOURCE: U.N., *Demographic Yearbook 1967* (New York, 1968), pp. 98-103.
. . . = Data not available.
— = Nil or negligible.

to contribute to the conception that Africa does not have a population problem. The claim that the "low density of population is the most important demographic fact about Africa"[1] must be seriously queried, for it is difficult to see how a misleading and rather meaningless statistic can be so dignified.

The crude densities of the five major continental subdivisions delineated by the U.N. (see Map 2) are also relatively useless, revealing a range of densities for mid-1967 from about 13.8 for Middle Africa to about 43.8 per square mile for western Africa.

The objection to the crude density figure is that the emphasis is on area, not people. The need is to use some measure that is more revealing with respect to the densities which people actually experience, possibly the density above and below which halves of the population live.

[1] Hodder and Harris, eds., *Africa in Transition: Geographical Essays*, p. 11.

Population Densities and Distribution

But if there is no obligation to provide just one figure, it is surely desirable to show the population of a country by various density ranges, even though such tables may also conceal essential details.

DENSITY RANGE BY POLITICAL UNIT

A look at the density range of Africa by political unit reveals why overall density figures must be questioned. Table 4 summarizes these figures for 57 political units, indicating that only about a quarter of Africa's population lived in countries or territories with densities at or below the average in mid-1967, that about 43 percent of the population lived at national densities double the average, and that about a quarter lived in countries with densities three and more times the continental average. The table also indicates that about 43 percent of the population lived in political units covering only 10.5 percent of the total area, and that the thirty-five most densely populated units occupied only about 31 percent of the total area but accounted for 73 percent of the total of 328 million people.

DENSITY RANGES WITHIN POLITICAL UNITS

The range of densities for individual countries brings out more clearly the desirabilitiy of having data for smaller subdivisions, either natural regions or administrative units. In practice, it is seldom possible to use other than political subdivisions, and entirely satisfactory data are available for only a limited number of countries in Africa.

These data are not entirely comparable from country to country because the political subdivisions vary greatly by size and relative importance; because densities vary by and within subdivisions depending on how boundaries are drawn; and because differing practices are followed with respect to urban populations. Some of the anomalies will be revealed in discussion of Tables 5 to 9.

The utility of density-range figures varies inversely with the size and directly with the number of the subdivisions for which density data are available. The coverage in Table 6 is, thus, more valuable for such countries as Congo (Kinshasa), the three East African countries, Madagascar, South Africa, and Swaziland, The data for some of the other countries are not entirely adequate. Those for Sierra Leone, for example, are for districts, which conceal considerable variations by chiefdoms within these districts; none of the districts has a density below 28 per square mile, though it is known that eight chiefdoms have densities well below this figure and that these account for at least 7 percent of the area of the country.

The summaries presented in Table 5 for the mainland portion of Tanzania at the 1967 census illustrate the disparities which develop when subdivisions of different orders are used. Using regions, of which there are eighteen, it appears that only 3.3 percent of the area had densities under ten per square mile, but calculating by districts (97) 25.2 percent of the area has densities below ten, and using divisions and subdivisions (c. 1298) we find that no less than 46.3 percent has under ten

Table 5

TANGANYIKA: POPULATION DENSITY RANGES, 1967, BY DIFFERING ORDER OF POLITICAL UNIT

Density per square mile	By region (18)				By districts and towns of district status (83 + 14)				By division and subdivision (c.1298)			
	Pop. %	Area %	Cumulative Pop. %	Cumulative Area %	Pop. %	Area %	Cumulative Pop. %	Cumulative Area %	Pop. %	Area %	Cumulative Pop. %	Cumulative Area %
–10	0.7	3.3			3.2	25.2			4.5	46.3		
10–	13.5	30.7	14.2	34.0	6.4	16.7	9.6	41.9	6.1	14.6	10.6	60.9
20–	5.9	8.5	20.1	42.5	9.5	15.1	19.1	57.0	5.6	7.9	16.2	68.8
30–	31.8	34.0	51.9	76.5	13.0	12.8	32.1	69.8	8.6	7.8	24.8	76.6
40–	13.9	10.7	65.8	87.2	13.2	10.6	45.3	80.4	5.7	4.4	30.5	81.0
50–	5.7	3.3	71.5	90.5	9.7	6.7	55.0	87.1	5.2	3.4	35.7	84.4
60–	4.7	2.5	76.2	93.0	5.3	3.2	60.3	90.3	4.3	2.3	40.0	86.7
70–	6.7	3.1	82.9	96.1	1.0	0.5	61.3	90.8	3.0	1.5	43.0	88.2
80–					3.7	1.6	65.0	92.4	4.2	1.8	47.2	90.0
90–					1.2	0.5	66.2	92.9	4.2	1.6	51.4	91.6
100–	14.8	3.8	97.9	99.9	23.6	6.4	89.8	99.3	24.5	6.2	75.9	97.8
200–					4.0	0.6	93.8	99.9	8.2	1.2	84.1	99.0
300–									4.0	0.4	88.1	99.4
400–					0.9	0.1	94.7	100.0	3.6	0.3	91.7	99.7
500–									1.2	0.1	92.9	99.8
600–									0.7	—	93.6	
700–									0.3	—	93.9	
800–									0.1	—	94.0	
900–									0.1	—	94.1	
1000+	2.4	—	100.1	100.0	5.2	0.1	99.9	100.1	5.8	0.2	99.9	100.0

Crude density—34.6

SOURCE: University College Dar es Salaam, Bureau of Resource Assessment and Land Use Planning, *Population Density in Tanzania, 1967* (Dar es Salaam, 1968).
— = Less than 0.05.

Population Densities and Distribution

persons per square mile. As a second example the breakdown by regions suggests that only 17.1 percent of the population lived at densities above 100 per square mile, while the breakdown by lowest order unit indicates that 48.6 percent of the population lived at such densities.

Even use of the lowest-order political subdivision can conceal marked differences as, for example, if the district includes uninhabited or sparsely inhabited sections such as lakes, swamps, forest reserves, rugged terrain, or high lands. Gulliver cites the example of the Arusha Chiefdom in Tanganyika which had a crude density of 244 per square mile in the 1957 census; in fact, the slopes of Mt. Meru within the chiefdom had 65 percent of the population living at an average density of 1,069 per square mile while the peripheral area had the remaining 35 percent living at an average density of 127 per square mile.[2]

The treatment of urban figures can greatly influence the pattern of density ranges for individual countries. Some censuses give urban and rural totals by political subdivision, others extract only one or a limited number of urban centers, and others include urban populations in such sized political units as to distort greatly the actual density patterns.

Other difficulties encountered in attempting to secure density-range tables for individual countries, particularly for some of those not included in Table 6 but calculated for the summaries shown in Tables 8 and 9, resulted from:

1. The failure of some censuses to give data on either area or density for the enumeration areas. In some cases areas can be secured from other sources or be calculated from maps; in others the boundaries of enumeration districts are simply not given and cannot be calculated with any acceptable degree of accuracy.

2. The lack of reliability of some censuses or population estimates. Ethiopian figures, for example, were based on extrapolated administrative estimates by province, estimates which are of very questionable validity, while the use of provincial totals undoubtedly conceals marked contrasts which are readily observable on the land. Use of the 1963 Nigerian census also incorporates whatever errors are included in that highly suspect count.

3. The inclusion of outright errors in census tables. Errors were,

[2] P. H. Gulliver, "The Population of the Arusha Chiefdom: A High Density Area in East Africa," *Rhodes-Livingstone Journal*, No. 28 (1961), p. 1-21.

Table 6

POPULATION DENSITY RANGES FOR SELECTED AFRICAN COUNTRIES

NORTHERN AFRICA

Density per square mile	MOROCCO, 1960				ALGERIA, 1966				SUDAN, 1956			
			Cumulative				Cumulative				Cumulative	
	Pop. %	Area %	Pop. %	Area %	Pop. %	Area %	Pop. %	Area %	Pop. %	Area %	Pop. %	Area %
−10	3.6	30.1			6.9	89.2			21.0	62.3		
10−	2.3	10.7	5.9	40.8	2.3	1.9	9.2	91.1	35.1	28.2	56.2	90.!
20−	6.0	15.8	11.9	56.6	1.6	0.8	10.8	91.9	7.6	3.6	63.8	94.
30−	0.7	1.3	12.6	57.9	6.1	2.2	16.9	94.1	5.8	1.8	69.6	95.*
40−	2.1	2.9	14.7	60.8	2.2	0.7	19.1	94.8	6.6	1.5	76.2	97.
50−	1.0	1.2	15.7	62.0	1.8	0.4	20.9	95.2	4.0	0.7	80.2	98.
60−	4.7	4.7	20.4	66.7	1.0	0.2	21.9	95.4	5.7	0.9	85.9	99.
70−	6.7	5.8	27.1	72.5	2.9	0.5	24.8	95.9	2.7	0.4	88.5	99.
80−	5.6	4.2	32.7	76.7	0.9	0.1	25.7	96.0	1.0	0.1	89.5	99.
90−	2.8	1.8	35.5	78.5	1.5	0.2	27.2	96.2				
100−	38.5	17.9	74.0	96.4	25.1	2.2	52.3	98.4	4.0	0.3	93.5	99.
200−	4.7	1.3	78.7	97.7	21.3	1.1	73.6	99.5	1.8	0.1	95.3	
300−	2.5	0.5	81.2	98.2	2.1	0.1	75.7	99.6				
400−	5.5	0.8	86.7	99.0	6.6	0.2	82.3	99.8				
500−	3.3	0.4	90.0	99.4	6.5	0.1	88.8	99.9				
600−					2.4	—	91.2					
700−												
800−												
900−					0.9	—	92.1					
1000+	10.0	0.6	100.0	100.0	7.8	—	99.9	100.0	4.8	0.1	100.1	100.

Crude density—63.4	Crude density—13.5	Crude density—10.6
By cercle (73)	By commune (79)	By census area (96)

Columns and internal additions may lack consistency because of rounding.
Sources: See Bibliography, Chapter 1, Part A.
— = Less than 0.05.

in fact, detected in almost all censuses examined—misplaced commas and decimal points, errors in addition, and failure of total area or total population figures to jibe with the sum of the parts as given in detailed tables. In a number of cases, flagrant mistakes, which were probably typographical, were detected and could be corrected. One must assume, however, that other errors were not detected and also that errors were committed by the author in extraction of data and in the following calculations despite efforts to double check each step.

GAMBIA, 1963				SIERRA LEONE, 1963				GHANA, 1960			
Pop. %	Area %	Cumulative Pop. %	Area. %	Pop. %	Area %	Cumulative Pop. %	Area. %	Pop. %	Area %	Cumulative Pop. %	Area %
—	0.3							0.9	10.2		
				5.9	16.5			4.3	20.7	5.2	30.9
1.7	3.5		3.8					3.1	8.5	8.3	39.4
10.5	19.7	12.2	23.5					7.1	14.7	15.4	54.1
13.9	19.7	26.1	43.2	7.6	10.6	13.5	27.1	3.3	5.4	18.7	59.5
8.5	10.5	34.6	53.7	25.2	30.4	38.7	57.5	6.5	9.1	25.2	68.6
15.8	16.3	50.4	70.0	7.7	7.7	46.4	65.2	3.2	3.6	28.4	72.2
12.8	11.6	63.2	81.6					1.1	1.1	29.5	73.3
9.0	7.6	72.2	89.2	17.3	14.0	63.7	79.2	3.4	3.0	32.9	76.3
15.1	10.3	87.3	99.5	27.3	19.9	91.0	99.1	4.7	3.8	37.6	80.1
				3.1	0.9	94.1	99.9	20.6	11.6	58.2	91.7
								17.3	5.7	78.5	97.4
								7.6	1.7	83.1	99.1
1.8	0.2	89.1	99.7					0.8	0.1	83.9	99.2
								0.8	0.1	84.7	99.3
2.1	0.2	91.2	99.9					1.3	0.1	86.0	99.4
								0.7	0.1	86.7	99.5
8.8	—	100.0	100.0	5.9	—	100.0	100.0	13.2	0.4	99.9	99.9

Crude density—78.7   Crude density—78.5   Crude density—73.0
By district (39)   By district (13)   By local authority area (69)

Despite the hazards involved it is highly desirable that efforts be made to secure and to utilize more fully the kinds of density data shown in the accompanying tables. Just as socioeconomic surveys which give only national totals present a kind of faceless country, one which the reader may all-too-readily conclude is a homogeneous unit, so the use of crude national population densities conceals regional differences of immense importance in understanding a country and in planning for its growth and development. It may be hoped that future censuses will

Population Densities and Distribution

TABLE 6 (*Continued*)

| | WESTERN AFRICA | | | | MIDDLE AFRICA | | | | EASTERN AFRICA | | | |
| | DAHOMEY, 1961 | | | | CONGO (KINSHASA) 1958 | | | | KENYA, 1962 | | | |
Density per square mile	Pop. %	Area %	Cumulative Pop. %	Area %	Pop. %	Area %	Cumulative Pop. %	Area %	Pop. %	Area %	Cumulative Pop. %	Area %
−10					30.4	57.9			8.0	75.9		
10−	13.5	45.3			22.6	24.9	53.0	82.8	1.4	3.8	9.4	79.
20−	17.3	36.6	30.8	81.9	9.6	6.2	62.6	89.0	2.3	3.4	11.7	83.
30−					10.6	4.7	73.2	93.7	1.7	2.1	13.4	85.
40−	1.9	2.0	32.7	83.9	5.7	2.0	78.9	95.7	1.8	1.8	15.2	87.
50−					6.3	1.8	85.2	97.5	1.3	1.0	16.5	88
60−					2.0	0.4	87.2	97.9	1.6	1.0	18.1	89
70−	2.9	2.0	35.6	85.9	2.7	1.0	89.9	98.9	1.8	1.0	19.9	90.
80−					1.2	0.2	91.1	99.1	1.2	0.6	21.1	90.
90−	1.8	1.0	37.4	86.9					0.9	0.4	22.0	91
100−	27.6	9.2	65.0	96.1	4.5	0.6	95.6	99.7	12.8	3.5	34.8	94
200−	8.5	1.8	73.5	97.9	2.4	0.2	98.0	99.9	10.8	1.8	45.6	96
300−	11.9	1.6	85.4	99.5	0.8	—	98.8		11.5	1.4	57.1	97
400−									5.4	0.5	62.5	98
500−									9.3	0.7	71.8	98
600−									5.9	0.4	77.7	99
700−					0.5	—	99.3		4.0	0.2	81.7	99
800−									2.8	0.1	84.5	99
900−	10.9	0.5	96.3	99.9					3.6	0.1	88.1	99
1000+	3.7	—	100.0	100.0	0.7	—	100.0	100.0	11.9	0.3	100.0	100
	Crude density—48.3				*Crude density*—15.4				*Crude density*—41.1			
	By department (29)				By territory (137)				By location (497)			

combine the expertise of geographers and demographers more effectively and thus provide the data required to examine densities and distributions and to compile various regional analyses.

Censuses for francophone areas frequently do give population and area by density ranges. The Bureau of Resource Assessment and Land Use Planning of the University College of Dar es Salaam has made good use of the 1967 Tanzania census from this standpoint, and valuable information is still being culled from the 1960 Ghana census. The

UGANDA, 1962[a]				TANGANYIKA, 1967				ZANZIBAR, 1967			
		Cumulative				Cumulative				Cumulative	
Pop. %	Area %	Pop. %	Area. %	Pop. %	Area %	Pop. %	Area %	Pop. %	Area %	Pop. %	Area %
1.0	14.3			4.5	46.3				0.8		
2.5	15.7	3.5	30.0	6.1	14.6	10.6	60.9	0.1	1.7	0.1	2.5
3.2	11.8	6.7	41.8	5.6	7.9	16.2	68.8	0.4	5.3	0.5	7.8
4.1	11.0	10.8	52.8	8.6	7.8	24.8	76.6	0.4	3.8	0.9	11.6
2.4	5.0	13.2	57.8	5.7	4.4	30.5	81.0	0.3	2.5	1.2	14.1
2.8	4.7	16.0	62.5	5.2	3.4	35.7	84.4	0.1	0.7	1.3	14.8
2.5	3.2	18.5	65.7	4.3	2.3	40.0	86.7	0.3	1.6	1.6	16.4
2.8	3.4	21.3	69.1	3.0	1.5	43.0	88.2	0.5	2.5	2.1	18.9
1.8	2.0	23.1	71.1	4.2	1.8	47.2	90.0	0.5	2.2	2.6	21.1
2.8	2.6	25.9	73.7	4.2	1.6	51.4	91.6	0.6	2.3	3.2	23.4
18.7	11.8	44.6	85.5	24.5	6.2	75.9	97.8	12.7	26.0	15.9	49.4
21.9	8.1	66.5	93.6	8.2	1.2	84.1	99.0	10.1	13.4	26.0	62.8
12.4	3.3	78.9	96.9	4.0	0.4	88.1	99.4	16.6	16.8	42.6	79.6
9.2	1.9	88.1	98.8	3.6	0.3	91.7	99.7	7.5	5.8	50.1	85.4
4.6	0.8	92.7	99.6	1.2	0.1	92.9	99.8	9.0	5.6	59.1	91.0
1.2	0.2	93.9	99.8	0.7	—	93.6		4.4	2.5	63.5	93.5
0.9	0.1	94.8		0.3	—	93.9		3.8	1.8	67.3	95.3
1.2	0.1	96.0	99.9	0.1	—	94.0					
0.2	—	96.2		0.1	—	94.1		4.5	1.6	71.8	96.9
3.7	—	99.9	100.0	5.8	0.2	99.9	100.0	28.1	3.2	99.9	100.1

Crude density—90.0
By gombolola (612)

Crude density—34.6
By division and subdivision (c. 1298)

Crude density—346.5
By division (140)

[a] African population, Non-African population in 1962 was c. 1.34 percent of total.

1966 Swaziland census also deserves commendation for its discussion of the significance of density and distribution as well as for the large number of enumeration districts for which such data are provided.

Density Ranges in Northern Africa. The North African countries illustrate density-range patterns existing in countries which have substantial portions of their territories in desert and steppe regions (Table 6). The crude density of Algeria at the 1966 census, for example, was about 13.5 per square mile. But if one extracts the two Saharan

Population Densities and Distribution

TABLE 6 (*Concluded*)

<table>
<tr><td></td><td colspan="12">EASTERN AFRICA</td></tr>
<tr><td></td><td colspan="4">MALAWI, 1966</td><td colspan="4">MADAGASCAR
1955-1958</td><td colspan="4">RÉUNION, 1961</td></tr>
<tr><td rowspan="2">Density
per
square
mile</td><td colspan="2"></td><td colspan="2">Cumulative</td><td colspan="2"></td><td colspan="2">Cumulative</td><td colspan="2"></td><td colspan="2">Cumulative</td></tr>
<tr><td>Pop.
%</td><td>Area
%</td><td>Pop.
%</td><td>Area
%</td><td>Pop.
%</td><td>Area
%</td><td>Pop.
%</td><td>Area
%</td><td>Pop.
%</td><td>Area
%</td><td>Pop.
%</td><td>Area
%</td></tr>
<tr><td>-10</td><td></td><td></td><td></td><td></td><td>15.5</td><td>54.6</td><td></td><td></td><td></td><td></td><td></td><td></td></tr>
<tr><td>10-</td><td></td><td></td><td></td><td></td><td>15.9</td><td>22.6</td><td>31.4</td><td>77.2</td><td></td><td></td><td></td><td></td></tr>
<tr><td>20-</td><td>1.2</td><td>5.1</td><td></td><td></td><td>9.8</td><td>8.0</td><td>41.2</td><td>85.2</td><td></td><td></td><td></td><td></td></tr>
<tr><td>30-</td><td>5.4</td><td>17.4</td><td>6.6</td><td>22.5</td><td>8.1</td><td>4.7</td><td>49.3</td><td>89.9</td><td></td><td></td><td></td><td></td></tr>
<tr><td>40-</td><td></td><td></td><td></td><td></td><td>6.4</td><td>2.7</td><td>55.7</td><td>92.6</td><td>0.8</td><td>6.3</td><td></td><td></td></tr>
<tr><td>50-</td><td>7.8</td><td>15.4</td><td>14.4</td><td>37.9</td><td>4.5</td><td>1.7</td><td>60.2</td><td>94.3</td><td>1.9</td><td>10.3</td><td>2.7</td><td>16.(</td></tr>
<tr><td>60-</td><td>2.0</td><td>3.6</td><td>16.4</td><td>41.5</td><td>4.4</td><td>1.3</td><td>64.6</td><td>95.6</td><td></td><td></td><td></td><td></td></tr>
<tr><td>70-</td><td>2.3</td><td>3.5</td><td>18.7</td><td>45.0</td><td>3.6</td><td>1.0</td><td>68.2</td><td>96.6</td><td></td><td></td><td></td><td></td></tr>
<tr><td>80-</td><td>3.9</td><td>5.2</td><td>22.6</td><td>50.2</td><td>2.9</td><td>0.7</td><td>71.1</td><td>97.3</td><td></td><td></td><td></td><td></td></tr>
<tr><td>90-</td><td>11.3</td><td>13.0</td><td>33.9</td><td>63.2</td><td>1.9</td><td>0.4</td><td>73.0</td><td>97.7</td><td></td><td></td><td></td><td></td></tr>
<tr><td>100-</td><td>27.4</td><td>21.3</td><td>61.3</td><td>84.5</td><td>13.4</td><td>2.0</td><td>86.4</td><td>99.7</td><td>6.2</td><td>15.0</td><td>8.9</td><td>31.(</td></tr>
<tr><td>200-</td><td>19.3</td><td>9.3</td><td>80.6</td><td>93.8</td><td>2.2</td><td>0.2</td><td>88.6</td><td>99.9</td><td>26.7</td><td>25.9</td><td>47.4</td><td>74.(</td></tr>
<tr><td>300-</td><td>15.9</td><td>5.4</td><td>96.5</td><td>99.2</td><td>11.8</td><td>17.2</td><td>20.7</td><td>48.8</td><td>13.1</td><td>11.4</td><td>60.5</td><td>86.</td></tr>
<tr><td>400-</td><td>3.5</td><td>0.8</td><td>100.0</td><td>100.0</td><td>2.1</td><td>0.1</td><td>90.7</td><td></td><td>1.2</td><td>0.8</td><td>61.7</td><td>86.</td></tr>
<tr><td>500-</td><td></td><td></td><td></td><td></td><td>0.5</td><td>—</td><td>91.2</td><td></td><td></td><td></td><td></td><td></td></tr>
<tr><td>600-</td><td></td><td></td><td></td><td></td><td>0.1</td><td>—</td><td>91.3</td><td></td><td></td><td></td><td></td><td></td></tr>
<tr><td>700-</td><td></td><td></td><td></td><td></td><td>0.2</td><td>—</td><td>91.5</td><td></td><td>15.2</td><td>6.5</td><td>76.9</td><td>93.</td></tr>
<tr><td>800-</td><td></td><td></td><td></td><td></td><td>0.3</td><td>—</td><td>91.8</td><td></td><td></td><td></td><td></td><td></td></tr>
<tr><td>900-</td><td></td><td></td><td></td><td></td><td></td><td></td><td></td><td></td><td>23.1</td><td>6.6</td><td>100.0</td><td>100.(</td></tr>
<tr><td>1000+</td><td></td><td></td><td></td><td></td><td>0.9</td><td>—</td><td>92.7</td><td></td><td></td><td></td><td></td><td></td></tr>
<tr><td></td><td>7.3</td><td>—</td><td>100.0</td><td>100.0</td><td></td><td></td><td></td><td></td><td></td><td></td><td></td><td></td></tr>
</table>

Crude density—111.3
By region (24)

Crude density—19.9
By canton and commune rurale (737)

Crude density—359
By commune (23)

departments it is found that the thirteen northern departments had 94 percent of the population on 11.7 percent of the area and an average density of 105.6 per square mile. Using communes as the unit area, however, it appears that about three quarters of the population lived at densities above that figure and that nearly half of the population resided at densities above 200 per square mile.

Even in the Saharan departments it is likely that the bulk of the population lives at high densities in oasis communities, while the nomadic population also probably lives at higher densities than the de-

SEYCHELLES, 1960

Pop. %	Area %	Cumulative Pop. %	Cumulative Area. %
	31.9		
0.2	3.4		35.3
0.1	1.2	0.3	36.5
0.1	1.0	0.4	37.5
0.5	2.2	0.9	39.7
0.6	2.1	1.5	41.8
0.2	0.9	1.7	42.7
0.5	1.6	2.2	44.3
3.1	7.6	5.3	51.9
9.4	9.7	14.7	61.6
85.3	38.4	100.0	100.0

Crude density—267
By island (24)

SOUTH AFRICA, 1960

Pop. %	Area %	Cumulative Pop. %	Cumulative Area. %
5.1	44.5		
5.2	12.4	10.3	56.9
12.0	16.7	22.3	73.6
7.2	6.9	29.5	80.5
9.5	7.3	39.0	87.8
3.6	2.1	42.6	89.9
4.6	2.4	47.2	92.3
2.8	1.3	50.0	93.6
2.8	1.1	52.8	94.7
2.1	0.7	54.9	95.4
10.9	2.9	65.8	98.3
4.8	0.7	70.6	99.0
4.0	0.4	74.6	99.4
4.7	0.3	79.3	99.7
0.8	—	80.1	
19.8	0.3	99.9	100.0

Crude density—33.8
By magisterial district (277)

SWAZILAND, 1966

Pop. %	Area %	Cumulative Pop. %	Cumulative Area %
1.6	15.0		
4.7	18.4	6.3	33.5
6.4	13.7	12.6	47.2
6.6	10.8	19.3	58.0
5.0	6.1	24.3	64.2
7.1	7.2	31.4	71.4
5.3	4.6	36.7	76.0
5.1	3.9	41.8	79.8
4.4	2.9	46.2	82.7
6.4	3.8	52.6	86.5
25.1	10.5	77.7	97.0
8.2	1.9	85.9	98.9
1.2	0.2	87.1	99.1
1.0	0.1	88.0	99.2
1.5	0.2	89.5	99.4
2.1	0.2	91.6	99.6
0.3	—	91.9	99.6
3.7	0.2	95.6	99.8
4.4	0.1	100.0	99.9

Crude density—55.7
By enumeration area (c. 747)

partment averages reveal since large parts of the desert are uninhabited. In the north of Algeria many departmental averages are similarly unrealistically low because they conceal the generally lower densities on difficult mountainous areas and steppe zones and the exceptionally heavy concentrations on the discontinuous but favored Mediterranean lowlands. In any case, the example of Algeria, with over 90 percent of the population living at densities above the average and about 72.8 percent living on 3.8 percent of the area, illustrates well the meager value of a crude national density figure.

Population Densities and Distribution

The oasis of Bou Saada, Algeria. Desert areas have very low crude population densities, but very high densities frequently occur in oasis communities.

Egypt presents one of the most striking situations. Only about 3.2 percent of its area is occupied; the average density in that portion of the country was 1,887 per square mile in 1960, which was no less than 31 times the crude density for the nation.

The columns in Table 8 summarizing the estimated situation for northern Africa (see Map 2 for the definition of major regions) in mid-1967 reveal that about three-fourths of the area has only 4.1 percent of the population, that about seven-eighths of the population resides in political subdivisions at densities above the average crude density, that the mean experienced density is roughly seventeen times the crude figure, and that about 42.5 per cent of the population resides at densities in excess of 1,000 per square mile or at over 42 times the crude density.

Density Ranges in sub-Saharan Africa. Density-range patterns vary very considerably among tropical and southern African countries. As would be expected, the countries which share the Sahara with northern Africa have high proportions of their areas with low densities, but

Population Densities and Distribution

The Nile oasis and adjacent desert at Giza, U.A.R. (Egypt). The inhabited portions of Egypt have densities more than thirty-one times the average density for the country.

none of them have regions of very high density comparable to those in the north.

The wetter tropical countries have considerably more even distributions but still have substantial portions of their total populations residing on small parts of the total area (Table 6). Dahomey, for example, had 35.0 percent of its population in 1961 residing at densities above 100 per square mile on 3.9 percent of the area. Togo had 69.5 percent of its population in 1959 living at densities above 100 per square mile on 29.7 percent of its area. If the 1963 Nigerian census is used, that country had 44.2 percent of its population living on 13 percent of its area with densities above 300 per square mile. Similarly, Congo (Kinshasa) had 47.0 percent of its population in 1958 residing on only 17.2 percent of its area.

The East African countries, particularly Kenya and Tanzania, again show the minimal value of crude density figures. Using density figures for hundreds of "locations", the lowest-order political subdivision, the 1962 Kenya data indicate that at one extreme 8.0 percent of the pop-

Table 7

PERCENT OF POPULATION OF SELECTED AFRICAN COUNTRIES ON THE

Percent of Area

	Morocco 1960	Algeria 1966	Tunisia 1966	U.A.R. (Egypt) 1960	Sudan 1956	Gambia 1963	Sierra Leone 1963	Ghana 1960	Dahomey 1961	Congo (Kinshasa) 1958	Angola 1960
0		47.6	17.1	24.1	19.8	12.7	9.0	16.8	14.6	10.1	12.3
	26.0	72.7		89.7 / 100.0	30.4				35.0	21.1	22.4
		89.1	34.2					41.7			
10		94.0			43.8	27.8				37.4	36.3
									67.3	47.0	
						36.8		62.3			
20	64.5						36.3				
			71.4					70.4			62.3
30			78.3			49.6					
	79.6		83.0				53.6				
					78.9						
40	87.4						61.3	81.2		69.6	
			90.1			65.4		84.5			
50											86.1
						73.9			86.5		
60	94.1							91.6			
70	96.4							94.7			
							86.5				
						87.8					
80							94.1				
90								99.0			
100						98.3					

Population Densities and Distribution

Kenya 1962	Uganda 1962	Tanganyika 1967	Malawi 1966	Zambia 1956	Rhodesia 1961-62	Madagascar 1955-58	Réunion 1961	Seychelles 1960	South Africa 1960	Swaziland 1966	AFRICA 1967
• 28.2	• 11.8	• 8.2	• 3.5	• 18.6	• 17.5				• 29.3	• 13.0	• 30.9
		• 24.0		• 30.5		• 27.0				• 22.4	
• 65.2	• 33.4		• 19.4				• 23.1		• 49.9		• 57.6
		• 48.5				50.7					• 66.1
80.1											
	• 55.3	• 59.9	• 38.7			• 58.8	• 38.3		• 60.9	• 47.5	
		• 69.4		• 53.5	• 47.9				• 70.4		• 75.6
90.6										58.3	• 82.2
• 92.0		• 75.1				• 68.6					
	• 74.0						• 52.6		• 77.6		• 87.4
										• 68.7	
		• 83.7			• 63.9						
	• 83.9		• 66.1					• 85.3			
		• 89.3									
									• 89.6	• 80.8	• 94.2
	• 89.1					• 84.5					
			77.4		• 81.9			• 94.7			
		• 95.4					• 79.3			• 87.4	
								• 97.8	• 94.8		
	• 93.2		• 83.6								
								• 99.1			
			• 85.6								
					• 91.5					• 93.8	
							• 91.1	• 100.0			
	96.4										
99.5											
			• 93.4								
							• 97.3				
	• 98.9									• 98.5	
									99.5		
			• 98.8			• 99.2					

ulation occupied 75.9 percent of the area at densities below 10 per square mile, while at the other extreme 78.0 percent of the population lived on only 9.0 percent of the area at densities above 100 per square mile. The crude national density in 1962 was 41.1 per square mile.

It is interesting to look briefly at two small countries, Swaziland and Réunion, which appear on most maps to have uniformly distributed populations. Swaziland, according to the 1966 census, had a crude density of 55.7 per square mile. The unusually large number of subdivisions (885) for which population and area data were available permits a more accurate recording of density ranges than for any other African country. These data reveal that 47.4 percent of the population was residing at densities above 100 per square mile on only 13.5 percent of the area and that 22.3 percent resided at densities above 200 per square mile on 3.0 percent of the total area.

Réunion, which is too small (969 square miles) to permit delineating internal patterns except on large scale maps, actually had 38.3 percent of its population on 13.1 percent of its area in 1961 based on average densities in 23 communes and, at the other end of the density range, 8.9 percent of the population on 31.6 percent of the area. Calculating from the density map prepared by Defos du Rau (see Map 20, a consolidated version of that map),[3] which shows the large portions of Réunion which are unoccupied, however, it may be estimated that about 48 percent of the island is unoccupied, which would mean that the average density of the settled areas would have been about 900 per square mile in 1968.

Summary on Population Densities. Table 7 summarizes selected data from Table 6, indicating under each country the percentage of the population occupying the percentage of area in that country as given in the common left-hand column. The table cumulates the population percentages in inverse order from the cumulative columns shown in Table 6. As may be seen, most countries have one-seventh or less of their population on the least populated half of their areas, while two-thirds to three-quarters of their populations reside on the most densely populated quarter, and 40 to 50 percent commonly inhabit the most densely populated tenth of the country.

Table 8 presents estimates of the mid-1967 population and area

[3] Jean Defos du Rau, *L'Ile de la Réunion,* pocket in rear.

Population Densities and Distribution

by density ranges for the major regions of Africa, while Table 9 gives comparable data for the continent and its island appurtenances. Fifty-seven political units are included. Estimates were made, where possible, by extrapolation from most recent census data by smallest political sub-division for which statistics were given. Estimates for four countries were made by calculation from the best available population maps. The summaries are perforce crude, but they are undoubtedly conservative as far as the percentages of population shown in the upper ranges are concerned. Mid-1967 population figures were taken from the U.N. *Demographic Yearbook 1967*.

The summary tables suggest that the vast bulk of the area of Africa is indeed sparsely populated, but, of far greater significance, that a very substantial percentage of the total population is actually experiencing high densities. In fact, at least 31 percent of the total population resides at densities exceeding 300 per square mile on little more than 1 percent of the total area. Over four-fifths of the population live on about a fifth of the total area at densities exceeding the average for the continent, while 60 percent of the total area contains less than 6 percent of the total population.

If the point has been made that crude density figures are of very questionable utility a major purpose of this section will have been achieved. As put in the 1966 Swaziland census, crude density of a state "is the most popular and at the same time the most dangerous spatial expression without stringent qualifications."[4] Such figures disguise patterns of concentration and dispersion and often bear little if any relation to the densities which the bulk of the population actually experiences.

While they are useful for comparative purposes, crude density figures for small units must also be used with great caution. Taken by themselves they, too, mean very little. They cannot be used to adduce pressure of population or lack thereof, even when relating one local density figure to another, although unusually high rural density figures may at least suggest that the relation of population to resources is worthy of investigation. Comparative density figures are, of course, useful in describing population distribution, but they take on meaning only when related to the opportunities and limitations of the ecological and economic environments.

[4] Swaziland Government, *Report on the 1966 Swaziland Population Census*, Mbabane, 1968, Chapter 13, p. 1.

Table 8

ESTIMATED POPULATION DENSITY RANGES FOR
MAJOR AFRICAN REGIONS, MID-1967

Density per square mile	NORTHERN AFRICA				WESTERN AFRICA			
	Pop. %	Area %	Cumulative Pop. %	Cumulative Area %	Pop. %	Area %	Cumulative Pop. %	Cumulative Area %
−10	4.1	74.9			3.3	52.5		
10−	7.2	11.8	11.3	86.7	3.1	9.8	6.4	62.3
20−	3.5	4.2	14.7	91.0	2.9	5.5	9.3	67.8
30−	2.3	1.6	17.0	92.6	4.4	5.8	13.7	73.6
40−	2.3	1.2	19.3	93.8	4.8	4.8	18.5	78.4
50−	1.0	0.4	20.3	94.2	3.3	2.6	21.8	81.1
60−	2.0	0.7	22.3	94.9	5.7	3.8	27.5	84.8
70−	1.3	0.4	23.5	95.3	1.7	1.0	29.2	85.9
80−	1.4	0.4	25.0	95.6	2.7	1.5	31.9	87.3
90−	2.3	0.6	27.3	96.2	4.1	1.9	36.1	89.2
100−	13.6	2.2	40.9	98.4	19.8	6.4	55.9	95.7
200−	7.1	0.7	48.0	99.2	11.0	1.9	66.9	97.6
300−	1.6	0.1	49.7	99.3	4.2	0.6	71.1	98.2
400−	2.0	0.1	51.6	99.4	6.7	0.7	77.7	98.8
500−	2.4	0.1	54.0	99.5	2.4	0.2	80.1	99.0
600−	1.0	—	55.0	99.5	2.8	0.2	82.8	99.2
700−					2.5	0.1	85.3	99.3
800−	1.5	—	56.5	99.6	0.2	—	85.5	
900−	1.0	—	57.5		3.2	0.1	88.7	99.5
1000+	42.5	0.5	100.0	100.1	11.3	0.5	100.0	100.0
Crude density	23.7				43.8			

— = Less than 0.05.

Population Distribution in Africa

There is a clear distinction between the concepts of density, which has an areal significance, and distribution, which has a point significance. Density figures are, however, very useful in describing distribution, and in the first portion of this chapter certain elements of distribution were at least alluded to, sometimes almost inadvertently. In this section, the patterns of population distribution of the continent and its subdivisions are briefly presented and an effort is made to assess the significance of the various factors that help to explain these patterns.

Several caveats must immediately be entered. First, both con-

MIDDLE AFRICA				EASTERN AFRICA				SOUTHERN AFRICA			
		Cumulative				Cumulative				Cumulative	
Pop. %	Area %	Pop. %	Area %	Pop. %	Area %	Pop. %	Area %	Pop. %	Area %	Pop. %	Area %
19.1	59.5			4.8	35.8			8.7	71.3		
20.7	21.4	39.8	80.9	5.8	15.7	10.5	51.5	4.7	6.4	13.4	77.7
15.9	9.4	55.7	90.3	4.6	7.9	15.1	59.4	8.7	6.7	22.1	84.3
5.8	2.4	61.4	92.8	13.3	15.2	28.4	74.6	7.4	4.2	29.5	88.5
10.3	3.2	71.8	95.9	6.1	5.3	34.5	79.9	6.5	3.0	36.0	91.5
5.2	1.2	76.9	97.1	7.6	5.7	42.1	85.6	6.6	2.4	42.6	93.8
3.8	0.8	80.7	98.0	3.8	2.4	45.9	87.9	4.1	1.2	46.7	95.0
3.3	0.7	84.1	98.7	1.0	0.6	46.9	88.5	4.3	1.1	51.0	96.2
1.3	0.2	85.3	98.9	4.4	2.0	51.3	90.5	2.9	0.7	53.9	96.8
0.9	0.1	86.2	99.0	9.0	3.5	60.3	94.0	2.7	0.6	56.6	97.4
9.4	0.9	95.6	99.9	12.0	3.5	72.3	97.5	11.3	1.7	67.9	99.1
2.0	0.1	97.5		6.7	1.0	79.0	98.5	4.6	0.4	72.5	99.5
				4.6	0.6	83.6	99.1	3.6	0.2	76.0	99.6
0.4	—	97.9		2.3	0.2	85.9	99.3	3.6	0.1	79.6	99.8
0.3	—	98.3		3.8	0.3	89.7	99.6	1.7	0.1	81.3	
				1.3	0.1	91.0	99.7				
				1.7	0.1	92.7	99.8	0.7	—	82.0	99.9
				0.5	—	93.2					
				0.6	—	93.8					
1.8	—	100.1	100.0	6.2	0.1	100.0	100.0	18.0	0.1	100.0	100.0
	13.8				40.8				20.1		

cerns of this section are worthy of book-length treatment; their coverage here must, therefore, be very general and lacking in the detail which would be appropriate to studies of the population geography of individual countries or subregions. Second, the available maps, despite some very commendable efforts, are not entirely satisfactory for the type of analysis proposed. Third, information on many of the correlative factors noted is much too sparse and uncertain to permit accurate assessments of the linkages between them and population distributions. This is especially so with respect to such factors as soil quality, hydrologic resources, the incidence of disease, and prehistoric and historic population movements which have influenced the present patterns.

Population Densities and Distribution

Table 9

POPULATION DENSITY RANGES FOR THE CONTINENT OF AFRICA, MID-1967

Density per square mile	Pop. %	Area %	Cumulative Pop. %	Cumulative Area %	Reverse cumulative Pop. %	Reverse cumulative Area %
−10	5.86	59.21			100.01	99.99
10−	6.72	13.75	12.58	72.96	94.15	40.78
20−	5.20	6.55	17.78	79.51	87.43	27.03
30−	6.62	5.41	24.40	84.92	82.23	20.48
40−	5.23	3.33	29.63	88.25	75.61	15.07
50−	4.32	2.22	33.95	90.47	70.38	11.74
60−	4.00	1.73	37.95	92.20	66.06	9.52
70−	1.75	0.68	39.70	92.88	62.06	7.79
80−	2.73	0.89	42.43	93.77	60.31	7.11
90−	4.56	1.30	46.99	95.07	57.58	6.22
100−	14.59	3.00	61.58	98.07	53.02	4.92
200−	7.54	0.86	69.12	98.93	38.43	1.92
300−	3.21	0.27	72.33	99.20	30.89	1.06
400−	3.51	0.23	75.84	99.43	27.68	0.79
500−	2.50	0.13	78.34	99.56	24.17	0.56
600−	1.47	0.06	79.81	99.62	21.67	0.43
700−	1.29	0.05	81.10	99.67	20.20	0.37
800−	0.58	0.02	81.68	99.69	18.91	0.32
900−	1.44	0.04	83.12	99.73	18.33	0.30
1000+	16.89	0.26	100.01	99.99		

Crude density—28.4

Fourth, the existent distributions are usually the result of the interaction of a multiplicity of factors and, while it is sometimes possible to distinguish a predominant correlate, the relations are often so complex as to defy accurate assessment of the relative importance of individual factors. Fifth, the significance of the several factors may change with the scale of observation. The distribution of population in the Sudd region of southern Sudan, for example, cannot be depicted accurately without using large-scale topographic maps with very small contour intervals. As a second example, the soil factor often takes on greater weight when a region is examined in closer detail.

Finally, the amount of attention that has been devoted to analyzing and describing African population distributions is still quite limited. References to a number of excellent studies are included in the footnotes and in the bibliography, but there are many countries and many more subregions for which far more attention to population studies is required.

DISTRIBUTIONAL PATTERNS

Table 10 presents data regarding the distribution of Africa's population by major region. The table is valuable only as a first-step introduction to the distribution of population on the continent. Its limited utility is revealed by noting that the population density for northern Africa is given as 23.7 per square mile, whereas only about an eighth of the total population lives at or below that density.

The figures of mid-1967 national-population estimates included on Map 2 provide a second step. A brief perusal will reveal, for example, the high proportion of the West African population concentrated toward the east of that region, with Nigeria, Dahomey, Togo, and Ghana having 71.0 percent of the regional population total on 21.7 percent of its area. Similarly in eastern Africa the use of regional or national totals conceals marked differences within the area, which is characterized by a considerable number of "islands" of high population concentration set in vast expanses of low density, plus a number of real islands of very

Table 10

DISTRIBUTION AND DENSITY OF AFRICA'S ESTIMATED POPULATION BY MAJOR REGION, SELECTED YEARS, 1930-1967

Region	Population in millions					Percent of pop. mid-1967	Crude[a] density per sq. mi. mid-1967	Area	
	1930	1940	1950	1960	mid-1967			Million sq. mi.	%
Northern Africa	39	44	53	66	79	24.1	23.7	3.29	28.1
Western Africa	48	58	67	88	104	31.7	43.8	2.37	20.3
Middle Africa	21	23	25	29	34	10.4	13.8	2.55	21.8
Eastern Africa	46	54	63	77	90	27.4	40.8	2.44	20.9
Southern Africa	10	12	14	18	21	6.4	20.1	1.04	8.9
Total	164	191	222	278	328	100.0	28.4	11.69	100.0

SOURCE: U.N., *Demographic Yearbook 1967* (New York, 1968), p. 97; Table 6 and other calculations.

[a] Density figures from Table 8; these figures may differ from U.N. data because of differing national areas, exclusion of lakes, etc.

Population Densities and Distribution   63

high density such as Mauritius and Réunion. The eight most densely populated political units in the eastern region, in fact, account for 14.0 percent of the estimated population but only 2.5 percent of the total area.

One special type of distribution of great significance is that distinguishing between rural and urban populations. Africa is estimated to have about 19.8 percent of its population in urban communities and 15.8 percent in urban centers above 20,000. The degree of urbanization varies from country to country as does the definition of urban centers. These and other topics related to urbanization are discussed more fully in Chapter 4.

The patterns of population distribution in Africa are not as readily described as those of South America, most of whose high-density nodes are peripheral, either extending inland from the coasts or situated in intermountain basis of the Andean cordilleria paralleling the Pacific coast, and where several moving frontiers can be delineated if successive periods are mapped. Indeed the patterns in Africa appear more inchoate than for any of the other inhabited continents.

Maps 3 and 4 are useful in attempting to describe such gross patterns as are discernible. Quite obviously, however, large-scale maps are more necessary than for most parts of the world. The most striking single feature is the vast emptiness of the Sahara, paralleled to a much less notable degree by the empty or sparsely populated regions of the southwest associated with the Namib Desert, the Kalahari Steppe, and contiguous semiarid regions. Beyond this major point there are few generalizations of value on a continental scale and hence it becomes necessary to turn to a discussion of distributional patterns in individual countries or broad subregions. In several of the major regions, however, only selected countries will be examined.

POPULATION DISTRIBUTION IN NORTHERN AFRICA

Along the northern fringe of Africa is a belt of high population density which is, in fact, less continuous than it appears on most small-scale maps and which is most conspicuous in the coastal lowlands and adjacent mountains of the Maghreb. Even if the urban populations were extracted the general pattern would not be destroyed.

The Maghreb. In Morocco (Map 5) the most densely populated belt runs rather continuously along the Atlantic seaboard from Safi to Kenitra; the *cercles* fronting on the sea in this stretch accounted for

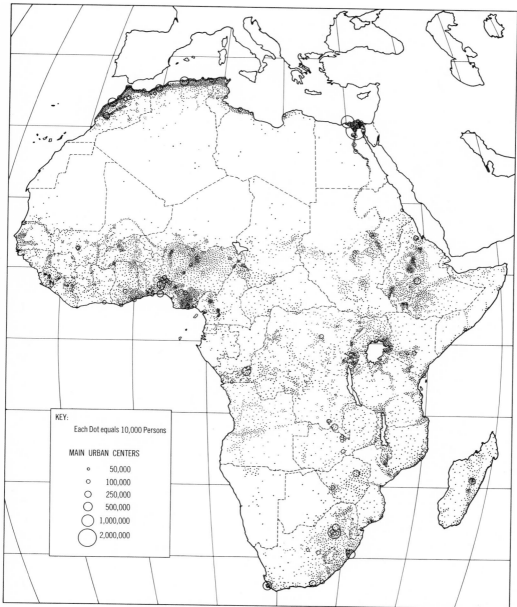

KEY:

Each Dot equals 10,000 Persons

MAIN URBAN CENTERS

∘	50,000
∘	100,000
○	250,000
○	500,000
○	1,000,000
○	2,000,000

MAP 3. Population distribution of Africa about 1960. (Courtesy University of Stellenbosch.)

Population Densities and Distribution

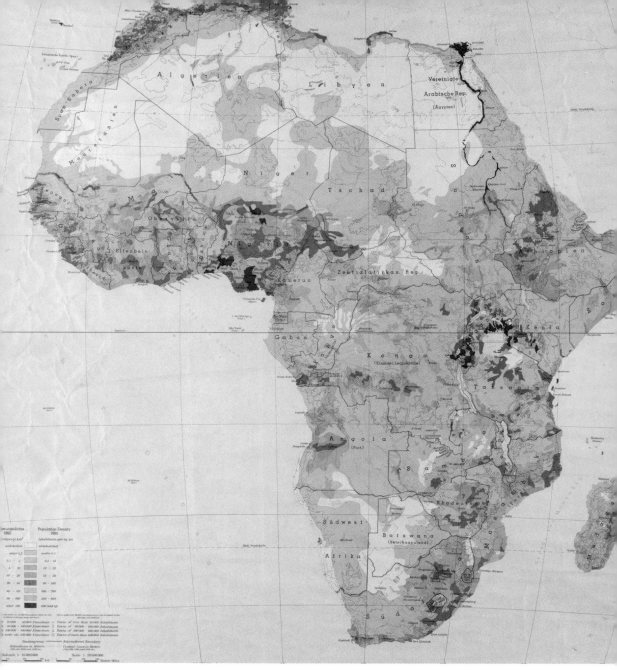

MAP 4. Population densities of Africa about 1960. (Copyright © by
Denoyer-Geppert Company, Chicago, used by permission.)

Population Densities and Distribution

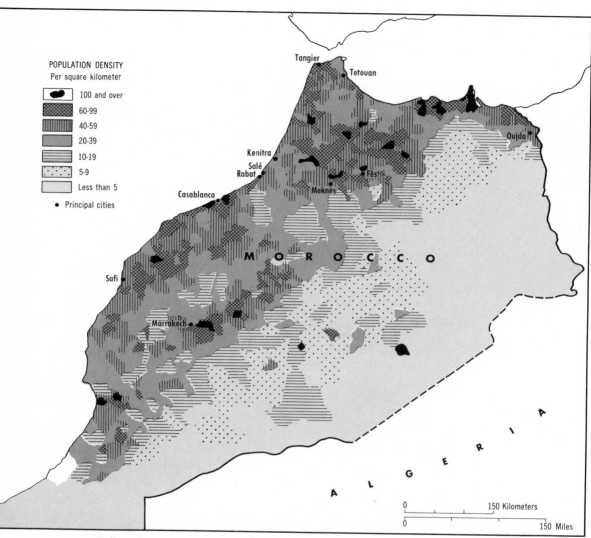

POPULATION DENSITY
Per square kilometer

	100 and over
	60-99
	40-59
	20-39
	10-19
	5-9
	Less than 5
•	Principal cities

Tangier
Tetouan
Oujda
Kenitra
Salé
Rabat
Fès
Meknès
Casablanca
M O R O C C O
Safi
Marrakech
A L G E R I A

0	150 Kilometers
0	150 Miles

MAP 5. Population densities of Morocco, 1960. (After *Atlas du Maroc.*)

22.8 percent of the total 1960 population on 4.9 percent of the kingdom's area. All of the coastal *cercles* from Agadir to the Algerian border had 35.1 percent of the population on 11.4 percent of the area. A second zone of high density includes the low-lying Rharb and the regions inland

Population Densities and Distribution   67

from that around Meknès and Fès. The overall pattern displays declining densities toward the south and southeast with the thirteen driest *cercles* of the country, all in pre-Saharan and Saharan Morocco, accounting for 13.1 percent of the population on about 53.3 percent of the total area.

The availability of water appears to be the single most important factor influencing population distribution; the level of precipitation influences the broadest pattern, while the availability of surface or underground water for irrigation has permitted the dense occupance of a few sizeable areas and many smaller irregular zones and ribbons, such as those along the main rivers or in the oases along the wadis flowing from the Atlas Mountains toward the Sahara.

Some unusually high densities occurring in difficult mountainous terrain may also be correlated with the higher and more certain precipitation due to the orographic influence, though the attraction of defensible sites was also important. The heavy migration in more recent decades to the Atlantic cities, and especially to the Casablanca-Rabat zone, reflects the increasing commercialization and industrialization of the economy as well as the selection of Rabat as the capital of the Cherifien Empire.

A village in the Rif Mountains of Morocco. Many topographically difficult areas of the Maghreb have very high population densities.

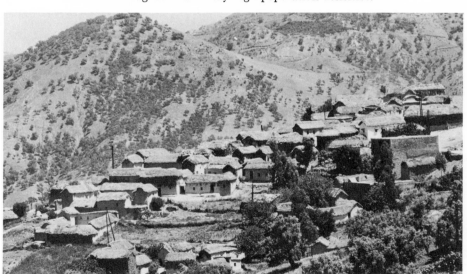

Population Densities and Distribution

In Algeria the coastal lowlands occur as a series of pockets or valleys between the echeloned prongs of the maritime Atlas. The most important lowlands—the plains around Oran and along the Sig River, the Chelif Valley, the plains of Algiers and the Metidja, and those near Bejaia (Bougie) and Annaba (Bône)—comprise only 2 to 3 percent of the area of Algeria but contain over half its population. The explanations for these heavy concentrations include the high percentage of lands that are irrigable and can be farmed with great intensity, the excellent piedmont alluvial soils, the ecological suitability of the regions for high value crops such as citrus fruit, vegetables, and wine, the advantage that the area has in the European market in the provision of winter and early vegetables, and the high portion of the country's industries and urban centers found in the coastal lowlands.

Tunisia has its most notable concentrations on the Tunis plain and in the Medjerda Valley which enjoy many of the same advantages as the lowlands of Algeria. Important nodes are also found along the Gulf of Gabès centered on Sousse and Sfax. Some of the mountainous areas of the Maghreb have surprisingly high densities as, for example, in the Rif Atlas, Kabylia, and the Aurès Massif. While average densities decrease inland and towards the Sahara in all three countries many of the people in the dryer zones live at very high densities in oasis communities.

Libya. In Libya the vast bulk of the population is concentrated in two regions: the coastal plain and Jefara in the northwest and the Cyrenaican plateau running along the coast east of Benghazi. Ninety-five percent of the country is desert; permanent cultivation is confined to less than ½ of 1 percent of the total area and could probably not be more than doubled. Probably 60 percent of the population occupies less than 1 percent of the area and over 85 percent lives on perhaps 5 to 7 percent of the land, nearly all on or close to the coast. The oases of the south are, like those of the Algerian Sahara, very densely populated. The relatively high percentages of seminomadic and nomadic peoples (22 and 20 percent of the total population respectively in 1965) still do not extend the utilized area of the country much beyond the Mediterranean fringes, while those engaged in oil prospecting and production are too few to alter the pattern significantly.

Egypt. The United Arab Republic (Egypt) has one of the most astonishing of population patterns in the world (Map 6), with almost all

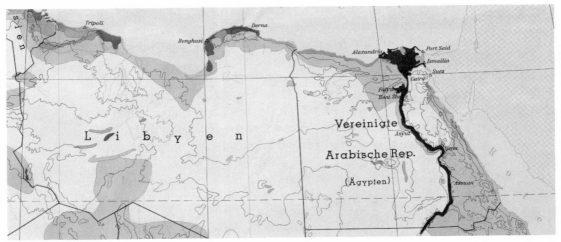

MAP 6. Population densities of northeast Africa, about 1960. (Copyright © by Denoyer-Geppert Company, Chicago, used by permission.)

of the population concentrated along the constricted flood plains of the Nile in Upper Egypt and on the delta, with lesser but still important appendages in the Faiyum Oasis (also watered by the Nile) and along the Sweetwater and Suez Canals. The utilized area of Egypt is about 15,000 square miles or only about 3.9 percent of the total. This gave an overall density in the inhabited areas of 2,060 per square mile in 1967. The farmed area of the Nile Valley will be about 7.3 million acres upon completion of the irrigation works associated with the High Dam, giving a man-to-arable land ratio of about 2,918 per square mile or one person per 0.22 acres. The actual average density experienced in Egypt is over 31 times the crude density for the country.

The tremendous rural densities of Egypt are a tribute to the splendid combination of physical factors found on the lands along the Nile: high average temperatures, exceptionally rich alluvial soils, and water now available on a year-round basis for irrigation. Double and triple cropping is common; yields which are relatively very high, are further increased by heavy applications of fertilizer and by the intensive care applied by its teeming population. But even the great productivity possible on these lands is not enough to meet the needs of the large and rapidly growing population. Nor are the developments in the urban centers, which contain an estimated 41.0 percent of the total popula-

Population Densities and Distribution

tion including 4,500,000 people in Cairo, by all odds the continent's largest city, adequate to provide a satisfactory standard of living.

The Sudan. The course of the Nile may also be readily traced on a population map of the Sudan (Map 7), particularly in the drier, northern half of the country. The Gezira-Managil Scheme between the Blue and White Niles south of Khartoum and the Three Towns conurbation at the juncture of these two great streams continue the band of high population associated with the Nile. The other main belt of higher density in the Sudan runs at right angles to the Nile system and is part

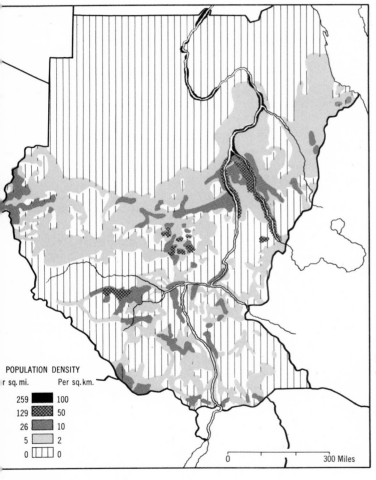

POPULATION DENSITY

r sq. mi.	Per sq. km.
259	100
129	50
26	10
5	2
0	0

0   300 Miles

MAP 7. Population densities of Sudan, 1956. (After K. M. Barbour, *The Republic of the Sudan,* London, University of London Press, 1961, used by permission.)

Population Densities and Distribution   71

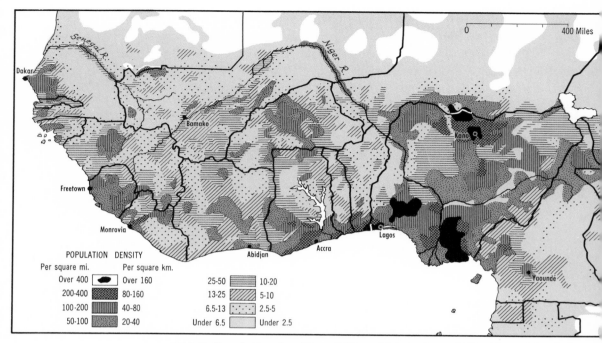

POPULATION DENSITY

Per square mi.	Per square km.		
Over 400	Over 160	25-50	10-20
200-400	80-160	13-25	5-10
100-200	40-80	6.5-13	2.5-5
50-100	20-40	Under 6.5	Under 2.5

MAP 8. Population densities of West Africa, about 1960. (Copyright © by Denoyer-Geppert Company, Chicago, used by permission.)

of the acacia grass steppe and savanna zone of central Sudan; densities in this belt are not particularly high, however, ranging as they do from about 15 to 40 per square mile as compared to over 400 in the Gezira and over 500 in discontinuous bands along the Nile. North and south of the east-west belt are vast areas of nomadism with densities below 3 per square mile.[5] Regions of high density away from the Nile in central Sudan include the Nuba Mountains and the homeland of the Aweil Dinka. In southern Sudan belts of higher density may be delineated along the junction of the clay plain and the ironstone plateau, in the extreme southwest, and around the southern hill masses.

POPULATION DISTRIBUTIONS IN WESTERN AFRICA

In Western Africa (Map 8), observers have sought to distinguish two discontinuous belts of high population density separated by a middle

[5] K. M. Barbour, "The Nile Basin: Social and Economic Revolution" in R. M. Prothero, ed., A Geography of Africa (London, Routledge and Kegan Paul, 1969).

Population Densities and Distribution

belt of lower densities. The northern high-density belt is the best delineated of the three; it runs, with interruptions, mainly but not entirely
along the east-west dry savannah and steppe vegetation-climatic zone,
with nodes of concentrated population in Senegal and Gambia, in the
Fouta Djalon Mountains of Guinea, around Bamako in Mali, in the
Mossi country of Upper Volta, around Korhogo in northern Ivory Coast,
the Kabrai country of northern Togo, the Atakora region of northern Dahomey, and regions centered on the emirate cities of northern
Nigeria, particularly those in the north-central portion whose major
node is Kano. Northward from this belt population densities decrease

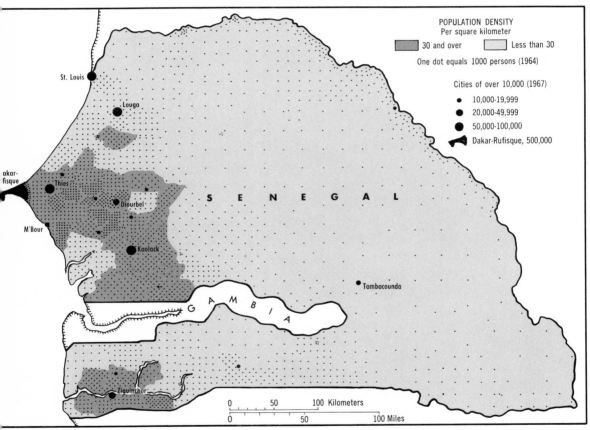

MAP 9. Population map of Senegal, 1964 rural, 1967 urban. (After
République du Sénégal. Ministère du Plan. *Cartes pour Servir a
l'Aménagement du Territoire*, Dakar, August, 1965.)

Population Densities and Distribution   73

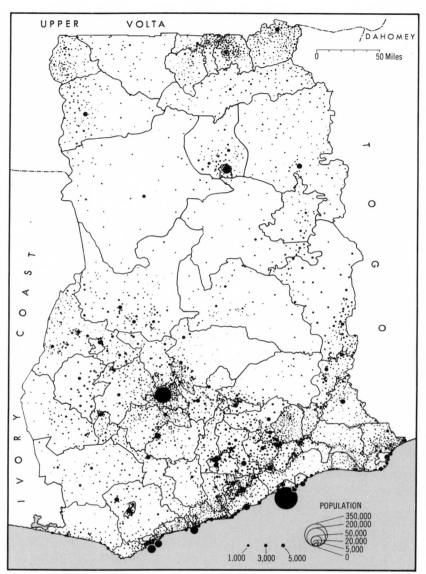

MAP 10. Population density map of Ghana, 1960. (After *1960 Population Census of Ghana: Atlas of Population Characteristics,* 1964, Ghana Survey Department, by permission.)

Population Densities and Distribution

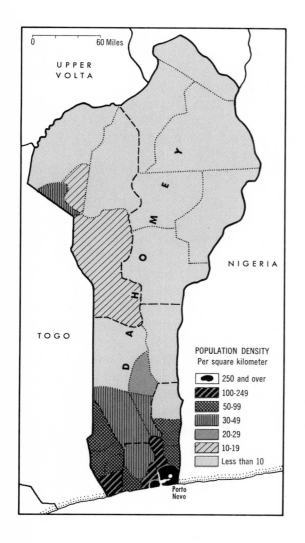

POPULATION DENSITY
Per square kilometer

⬤	250 and over
▨	100-249
▨	50-99
▥	30-49
▤	20-29
▨	10-19
░	Less than 10

MAP 11. Population density
map of Dahomey, 1961.
(After I.N.S.E.E.)

rapidly at first and then more gradually into the empty wastes of the Sahara.

Two notable exceptions to this pattern are found north of the series of high density nodes in the otherwise sparsely populated dry steppe and desert zone; these are the ribbons of high density associated with the Senegal River and with the middle course and interior delta of the Niger River. The presence of the Senegal River helps to account

Population Densities and Distribution   **75**

for the concentration of about 80 percent of Mauritania's population in the southern seventh of that country. In Senegal (Map 9) the population concentrations along the river form an exception to the generalization that densities tend to decrease from south to north and from west to east in that country. The central delta of the Niger River in Mali similarly attracted densities about four times the average for the country and considerably higher than the non-Saharan portions of the country dependent on rain-growing or pastoralism.

The second or middle belt, occurring between the two discontinuous north and south belts, is one of relatively low densities with minor exceptions such as the concentrations in Upper Guinea, around Bouaké in the Ivory Coast, centered on Kaduna in northern Nigeria, and an extension of the high-density Yoruba node into the belt in Ilorin Province.

The southern belt is actually only half a belt since it cannot be clearly delineated west of Ghana. It includes zones of particularly high density as follows: (1) Ashanti, Ghana; (2) a coastal region running from southern Ghana through Togo, Dahomey, and western Nigeria; (3) coalescing with this belt, Yorubaland in western Nigeria; and (4) Iboland in eastern Nigeria. Ashanti contained 16.5 percent of the population of Ghana in 1960; the coastal belt in that country accounted for 53.9 percent (Map 10). The four circonscriptions in the maritime region of Togo had an estimated 34 percent of that country's 1968 population on 12.5 percent of its area. In coastal Dahomey 48 percent of the population in 1961 was concentrated in 15 percent of the country with densities exceeding 750 per square mile around Porto Novo (Map 11). According to the suspect 1963 Nigerian census the Eastern and Western Regions plus Lagos, which contained the two large southern nodes of high density, accounted for 41.9 percent of the total Nigerian population on 16.8 percent of the country's area. Belts of somewhat lower densities separate the major nodes of the southern belt but it can reasonably be considered as a continuous region of population concentration.

West of Ghana the coastal-belt pattern disintegrates. There are a series of small nodes from Dakar to Monrovia; from there to Abidjan in the Ivory Coast is a broad region of generally very sparse population. The pattern of Sierra Leone is atypical (Map 12). As Professor Clarke

Population Densities and Distribution

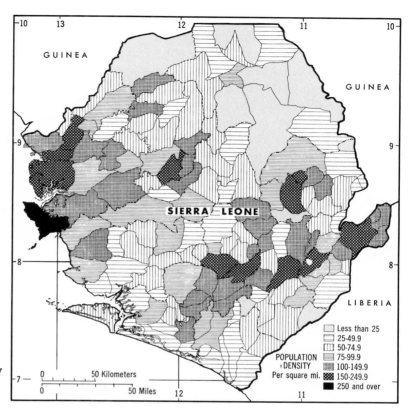

10 13 12 11 10

GUINEA

GUINEA

9

9

SIERRA LEONE

8

8

LIBERIA

MAP 12. Population
density map of Sierra
Leone, 1963. (After John
I. Clarke, *Sierra Leone in
Maps*, London, University
of London Press, 1966,
used by permission.)

POPULATION
DENSITY
Per square mi.

	Less than 25
	25-49.9
	50-74.9
	75-99.9
	100-149.9
	150-249.9
	250 and over

0   50 Kilometers
0   50 Miles

7

12   11

has noted,[6] it is difficult to delineate any simplifying population pattern
for Sierra Leone. In addition to the node of high density centered on Freetown and the Peninsula, which had an average density of 762 per square
mile according to the 1963 census, two zones or belts of higher density
may be indicated. Clarke calls these the two "lungs" of the country and
notes that they are also the main areas of population growth. The first
belt, north and northeast of Freetown, is associated with the most intensive zone of swamp-rice farming, important fishery operations, iron-
ore mining around Marampa, and market gardening in the Bullom
Peninsula. The second, in the south and southeast, has been influenced
by diamond mining and cash-crop farming areas and shows a marked

[6] Clarke. *Sierra Leone in Maps*, pp. 41-42.

Population Densities and Distribution   77

correlation with the belt of towns situated along the line of rail inland from Bo. Zones of lower density in the country, which has one of the higher crude densities in Africa (78.7 per square mile in 1967), are associated in the south with swamp zones, forest reserves, and mountain massifs, and in the northeastern Koinadagu districts, the largest area of relatively low population, with accidented relief, remoteness, dependence on grazing, and lack of modern amenities and transport.

While it is possible to present satisfactory explanations for portions of the population in western Africa, the available data on demographic, physical, cultural, and historical factors do not permit more than tentative suggestions regarding either the broad three-belt pattern or the intimate local patterns.

The Southern Belt. Explanations that are cited for the southern high-density belt, many of which would apply to the more scattered nodes west of Ghana, include the following:

1. The rainfall and temperature regimes which permit year-round production.

2. The comparative ease of producing such subsistence crops as manioc, corn, yams, plantains, and palm oil.

3. The existence of some relatively favorable soils, as for example, those of Yorubaland which are comparatively rich and which have good water relations. One of the major problems in the demographic pattern is, however, explaining the exceptionally high densities found on some poor soils, as in portions of Iboland.

4. The ecological suitability of this zone for such important exports as cocoa, coffee, palm products, rubber, and timber products, and such lesser crops as coconuts, bananas, pineapples, cashew nuts, and cola nuts.[7]

5. The particular attractiveness of the rainforest/savanna border zone, in part because of the more ready use of fire as an aid to field clearance.

6. The occurrence of several important mineral deposits whose exploitation has been fairly labor intensive, leading to the formation of minor nodes within the southern belt.

7. The existence of several early kingdoms or powerful tribal groupings whose political stability permitted the development of substantial concentrations. These agglomerations, in turn, stimulated the rise of

[7] See F.A.O., *Crop Ecologic Survey in West Africa*, Vol. II, *Atlas*, (Rome, 1965).

Population Densities and Distribution

more intensive and diversified economies. Notable among these were the kingdoms of Ashanti, Abomey, Porto Novo, Yorubaland, and Benin. The high densities associated with the Ibo of Eastern Nigeria, who did not have comparable political organizations, appear to provide an exception, but their centralized economic structure and cultural cohesiveness may have provided similar strength and security.

8. Longer and more intensive contacts with the outside world during the modern period when technological advances have been most dramatic and levels of trade have been far higher than in the earlier period of contact across the Sahara enjoyed by the sudanic groups of the north. Important contributions from these contacts which may have influenced the population patterns of the southern belt include the introduction of high yielding subsistence crops, earlier and more intensive development of cash-crop production, the construction of generally better and denser transport systems, the reversal of early trade directions of the inland areas away from the trans-Saharan route and to the Guinea Gulf ports to the advantage of the economies of the southern belt, and the concentration of modern political bureaucracies at coastal points for all of the states fronting on the sea. Modern, consumer-oriented manufacturing has tended to concentrate in these port-administrative cities which are also often the financial, educational, and cultural capitals of their countries. The relatively high percentage of such non-Africans as there are in western Africa in the southern belt and coastal cities has also contributed to the present population patterns, though the influence of non-Africans in this respect has been considerably less notable than in eastern and southern Africa.

9. Migrations, which have contributed to the present patterns in a variety of ways. Early migrations, in which chance may have played a major role, are probably one of the major explanations for existing distributions, but relatively little is known regarding such movements. It has been suggested that the considerable difficulty of moving through rainforest zones may help to explain the greater diversity in the population pattern of the south as contrasted with the sudan belt. Later historical migrations affected the concentrations of various ethnic groups, as, for example, the generally southward movement of various Yoruba clans partially to occupy more readily defensible sites. The impact of emigration associated with the slave trade on population patterns is very poorly understood, although it is likely that the desire to defend against slaving

as against intertribal war helps to explain the tendency of various groups to agglomerate rather than to reside in a more dispersed pattern. It should be noted that slaving was conducted both by coastal groups providing slaves for transatlantic destinations and by interior groups who sought slaves for themselves and for other Muslim tribes. More recently, the southern belt has received very large numbers of migrants from inland areas, the most important receiving zones being southern Ghana and southeastern Ivory Coast.

A good many of the "explanations" for the population patterns of the southern belt, as for other parts of Africa, are obviously permissive, not causal, factors. The strong contrasts between the rainforest zone in the southwest and that in the east indicate that historical and cultural factors are of greater significance in understanding the total pattern than are physical factors, which can only be permissive or restrictive.

The Northern Belt. Turning to the northern belt it may first be noted that the zone between it and the Sahara reflects the strongly limiting influence of declining precipitation. The most striking contrast is between the region of cultivation along the south and the grazing areas to the north. Immediately beyond the last fields the densities drop to about ten per square mile in a narrow band and then very rapidly to under three per square mile. The exceptions to this pattern occur in belts or zones where the regional limitation—inadequate precipitation—is offset either by the presence of through-flowing streams such as the Senegal and the Niger, the existence of mountainous areas which receive greater-than-average rainfall because of the orographic influence as in the massifs of Air and Adrar des Iforas, or by the occurrence of water points which attract usually quite small semipermanent settlements.

Explanations for the high-density nodes of the northern belt include the following:

1. The receipt of enough rainfall to permit annual cultivation of such subsistence crops as sorghum, millet, and peanuts, and of such cash crops as peanuts and cotton.

2. The existence of some through-flowing streams which provide water for small garden areas or whose floodplains may be tilled after the waters have receded following the annual inundations of the rainy season. The Sokoto Rice Scheme in northwestern Nigeria was based

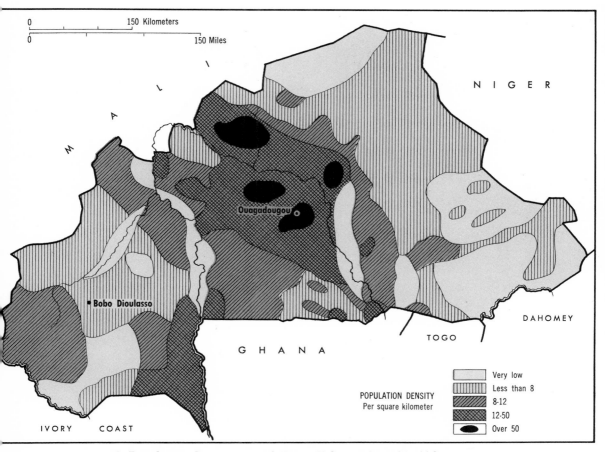

MAP 13. Population density map of Upper Volta, 1960-1961. (After I.N.S.E.E.)

upon the mechanical cultivation of such lands to permit increasing the per capita productivity of the local inhabitants. Urvoy notes, however, that areas subject to inundation have no common population pattern.[8] Stretches where river borders are submerged a good part of the year are often deserted except for a few fishermen, whereas hummocky areas cut by serrated streams are eminently suitable to sustain high densities. Elsewhere, the borders of certain rivers, as along some of the main

[8] Urvoy, *Petit Atlas Ethno-Démographique du Soudan entre Sénégal et Tchad*, Mémoirs de l'Institut Français d'Afrique Noire, No. 5, p. 11.

Population Densities and Distribution   *81*

A boy guiding two victims of onchocerciasis or river blindness in Gombélédougou, Upper Volta. Twenty million people in the continent have been estimated to suffer from onchocerciasis, carried by the black fly (simulium damnosum). Prevalence of the fly accounts for several belts of sparse population settlement, especially in West Africa.

streams of Upper Volta (Map 13), are virtually uninhabited because of the danger of contracting onchocerciasis or "river blindness."

3. The presence, in some portions of the belt, of unusually shallow water tables which could be tapped with the primitive techniques available. Examples are found in the closed-settled zone around Kano and to the east of Niamey, Niger.

4. The availability of some relatively good soils, such as the easily worked drift soils of the Kano area.

5. The use of relatively advanced agricultural techniques to permit greater output per acre and more continuous use of the land. An excellent example is the heavy application of fertilizing agents, including human excrement, in the lands around the major centers such as Kano. Also important is the seasonal grazing of nomadic cattle on the stubbled fields of the sedentary farmers who encourage this practice to secure the enriching manures.

6. The early development of commercial nodes at termini of caravan routes and at the intersections of routeways. This region had contacts with peoples across the Sahara much before those which affected the south. The build-up of very substantial artisan activity contributed to the relatively large urban nodes of the emirate cities.

Population Densities and Distribution

A village and fields in the close-settled zone around Kano, Nigeria.

7. The relative ease of movement across the sudan and sahel belts. This may help to explain the existence of nodes along the entire belt, which contrasts with the more diverse pattern of the southern belt. The movement is well illustrated by the historical migrations of the Fulani, who, during the past 500 years, moved eastward from Senegal as far as northern Cameroon.

8. The more recent provision of improved transport to specific centers in the belt has undoubtedly contributed to the growth of individual nodes. Thus the completion of the rail line to Kano in 1912 undoubtedly strengthened the Kano region in relation to Sokoto or Bornu Provinces. The westernmost nodes of the northern belt, in Senegal and Gambia, have the additional and atypical advantage of having easy access to port cities so their contact with the West improved during the period when that of other sudanese centers was decreasing. The selection of Dakar as the capital of the vast French West African federation and the concentration there of a large percentage of the Europeans in that grouping also contributed to its build-up as a population node.

9. The formation and consolidation from the eighth to the sixteenth centuries of a series of extensive and powerful states provided, as did the kingdoms of the south, sufficient security to permit and sustain

Population Densities and Distribution

the economic development required to support a dense population. Some of these kingdoms have disappeared or disintegrated over the years, but others are represented today by the nodes associated with the Mossi kingdom and with the former Hausa states of Katsina, Zaria, and Kano. The node around Sokoto was more recently formed under Fulani rule. Separating a number of the population agglomerations are zones of low population density, some of which were apparently depopulated by wars and slave raids, as, for example, the empty area in southwestern Upper Volta in the Comoé River area between the Sénoufo and Lobi peoples. Urvoy notes that several tribes of the sudan belt seem to have densities proper to themselves even when following similar ways of life and having comparable environments; he suggests that this pattern must be explained by socio-political structures and little-known historical chance.[9]

10. Somewhat antithetical to the previous point is the explanation for some population concentrations in more or less difficult hill or mountain regions, namely the movement to these areas by weaker ethnic groups seeking refuge from more powerful tribes. Such enclaves are found sporadically across the entire sudan belt from the Atlantic to the Nile; in western Africa they include the Dogon occupation of mesas east of Bandiagara in Mali, the Somba concentration in the Atakora Massif of northern Dahomey, and the densely settled Fouta Djalon in Guinea. Pressure of population in these refuges has frequently stimulated the adoption of intensive practices, including the construction of more or less elaborate terracing systems. While their hilly regions often suffer from difficult topography and isolation, they are advantaged by somewhat higher precipitation than surrouning plains and, sometimes, by better-than-average soils or by an abundance of springs.

The Middle Belt.   The population pattern of the middle belt is considerably more difficult to explain than those of the other belts. Physical factors which might help in understanding the relative sparseness of most of the belt and which have been noted by various observers include: the presence of the tsetse fly in the belt which would preclude the movement of northern pastoralists into the area at least seasonally; a greater incidence of human trypanosomiasis than in the coastal belt; the widespread absence of permanent and dependable water supplies; the greater need to practice shifting agriculture than in either of the

[9] Urvoy, *Petit Atlas Ethno-Démographique du Soudan entre Sénégal et Tchad,* pp. 13-14.

The Dogon village of Irély on the slopes of the Bandiagara Cliffs in Mali, an example of settlement in a refuge area.

other belts because of poorer soils as compared to the north and the more restricted suitability of tree crops, particularly palm oil, as compared to the south; and the lower suitability of the area for a variety of crops commonly grown either to the north or to the south.

But no one or combination of these physical deterrents, even accepting the validity of the claims, provides adequate explanation for the observed population pattern of the middle belt. First, because the zone can produce a great variety of crops. Second, because the several exceptions to the general pattern do not appear to be occasioned by superior physical conditions. These exceptions assume two patterns: (1) "islands" of high density such as those around Bouaké, Ivory Coast, and Tiv country in eastern Nigeria, and (2) a generally considerably higher average density in the east of the middle belt as contrasted with the west.

Economically there is no doubt that the middle belt is relatively poorly developed. Little is produced for export except beniseed and ginger; yams are the only major cash crop sold to other domestic regions. Local transport is not well developed, though parts of the region are advantaged by through roads and rail lines catering mainly to the northern belt. Urbanization is also far less significant than in the other belts. The cause-and-effect relations between these economic measures and population are, however, not at all clear.

Slaving from both north and south has also been given as a major explanation for the relatively low densities of the middle belt. Specific examples include the southwestern section of the former Northern Region of Nigeria which is said never to have recovered from the depredations of the Emir of Kontagora, and middle Dahomey, which is occupied by Yoruba peoples who did not develop powerful kingdoms comparable to those of several Yoruba groups in Nigeria and who were repeatedly attacked by the latter and by the Fon of southern Dahomey. It has also been suggested that depopulation of middle-belt areas by slaving would have caused abandoned farm land to revert to bush which would in turn increase the prevalence of tsetse and contribute to further out-migration.

Historical evidence suggests, however, that many more slaves were taken from regions which are now densely populated than from the middle belt, which suggests that this explanation may have been exaggerated. Nonetheless, the presence of agglomerations of powerful political groups both to the north and to the south and the general absence of such groups in the middle belt leads one to suspect that historical factors are probably of greatest importance in explaining the population pattern of the middle belt.

Conclusion. Gleave and White have recently questioned the validity of the whole concept of a middle belt, suggesting that

until more is known, we accept the Middle Belt as a fact in the physical environment that has important consequences for economic development. But it is also a geographer's fiction, introduced to aid comprehension of a complex continental area, and it may eventually prove to be misleading.[10]

[10] M. B. Gleave and H. P. White, "The West African Middle Belt: Environmental Fact or Geographer's Fiction?" *The Geographical Review,* LIX, No. 1 (January, 1969), 139.

They further suggest that "Hance's concept of economically developed 'islands' may prove more fruitful in the analysis of human geography in West Africa." One measure which might make the triple belt concept somewhat more useful, and it can probably be little more than a device for generalizing part of a very complex and somewhat inchoate pattern, would be the acceptance of population belts with discontinuous nodes of population concentration with no attempt made to force the belts to conform to physical belts such as climatic and vegetation zones, which do follow reasonably parallel belts running east and west across much of western Africa.

POPULATION DISTRIBUTION IN MIDDLE AFRICA

The population patterns of Middle Africa are more haphazard than for the other major regions of the continent, though satisfactory correlations can be made with physical and cultural factors for some of its subregions.

Chad. The gross distribution in Chad, for example, may be related primarily to the precipitation levels of its several zones. The northern zone is the desert and sahel region of the "B.E.T." (Borkou-Ennedi-Tibesti) covering about 300,000 square miles or three-fifths of the country north of the 14° N parallel. The desert portion, consisting mainly of vast, denuded plateaus, with a very meager vegetation cover, has some population concentrations in scattered palm oases and below the Ennedi and Tibesti massifs; the sahel portion is a thornbush steppe with rainfall increasing to about 12 inches in the south. An estimated 400,000 people occupy this vast zone, very unevenly distributed but with increasing densities toward the wetter south, where cattle and goat grazing is the essential activity as contrasted with the "grand nomadism" of the desert which occupies perhaps 50,000 of the total population.

South of this zone is a sahel-sudan belt of thorn steppe and gum acacia; rainfall ranges from about 16 to 35 inches and falls during a longer period of the year. About 1.65 million people occupy this zone at an average density of about 9.8 per square mile, compared to 1.3 per square mile for the B.E.T. Stock raising remains the main activity, but millet, corn, and peanuts are grown along the wadis. The third zone is in southern Chad, in the basins of the Chari and Logone, and is the richest and most densely peopled belt. Precipitation ranges from about

27 to 50 inches south of a line from Moundou to Fort Archambault; it is a wooded savanna zone with vast, grassy plains in the regions subject to annual inundation. Southern Chad is the main cotton-producing area of the country, cotton providing over 80 percent of the total value of exports; rice, peanuts, and a variety of foodstuffs are also grown, while fishing is an important activity. With about 10.4 percent of the country's total area, the south has about 46 percent of the total population; its average density is about 31.1 per square mile, though concentrations are considerably heavier toward the west. Settlements in the south are generally agglomerated along the roads and rivers, partly because this pattern was encouraged if not enforced by the colonial administration, as it was in most of the former French territories. Towards the north, villages are characteristically situated at water points; they are smaller than in the south and more likely to be dispersed in hamlets which sometimes become itinerant in the dry season.

The Central African Republic. Covered mainly with grass or wooded savanna but possessing some tropical rainforests in the south-west and gallery forests along the numerous watercourses, the C.A.R. has large areas in the north and east which are essentially empty. Only around Bangui, which appears to be far larger than the economy of the country would justify, is there a zone of high density. None of the political regions has a density above 10 per square mile, but if data were available for lower-order administrative districts they would doubtless reflect the higher densities achieved in the numerous and sometimes lengthy street-villages positioned at varying intervals along the rather sparse main roads. Historical factors, and particularly the chance migrations of various ethnic groups, must be cited as the main explanation for the population distribution, including the generally low densities, of the C.A.R. While the country does have several repulsing factors such as presence of the tsetse fly, poor soils, difficulty of overland movement, and, in the modern context, remoteness from world markets, its physical attributes, including rainfall averages of 40-80 inches per annum and its great size (larger than France and the Low Countries combined) would surely have permitted a substantially larger population than the estimated 1.5 million in 1968.

Cameroon. This country has one of the higher densities of Middle African countries, about 31 per square mile in 1967 as compared with about 5–18 per square mile for the other mainland states of this major region. These overall figures, however, conceal marked disparities within each nation. Two-thirds of Cameroon's population is concentrated in three areas: in the north, from the Benue to Lake Chad; in the western highlands; and in the southwest, which includes the two cities of Douala and Yaoundé. Much of the northern region is comparable to adjacent areas of Chad and the Central African Republic, with which it shares the Chari and Logone basins, in tribal, physical, and economic conditions. The Massif of Mandara along the Nigerian frontier is another example of densely populated mountain refuges. Separating the north and south of Cameroon is the east-west trending Adamoua Massif which ranges in elevation from 2,500 to 4,500 feet and which occupies about a seventh of the country. While this massif is favored by relatively heavy precipitation and absence of the tsetse fly, rinderpest, and foot-and-mouth disease, it has densities usually under 10 per square mile. Again, historical considerations must largely explain the low densities, particularly the relatively late arrival of Mboums and of Peuls and Bororos related respectively to the Hausas and the Fulani of Nigeria.

The western highlands of southern Cameroon consist of a chain of volcanic peaks culminating in Mt. Cameroon (13,370 feet) and sometimes rising from and bounded by high crystalline and volcanic plateaus. Many of the volcanic soils are superior in quality, while much of the area is ecologically suitable for a great variety of tropical crops including those of higher value usually associated with tropical highlands. The region is also advantaged by nearness to the sea and relatively good road and rail connections. West Cameroon, formerly part of the British Cameroons, is not as densely settled as that portion of the highlands within East Cameroon, the former French trusteeship territory. But the southern quarter, where oil palm, rubber, banana, tea, coffee, cashew nuts, and pepper are produced on the Cameroons Development Corporation plantations, on private plantations, and on smallholdings, and which has a sizable export of tropical timber, probably has an average density around 150–180 per square mile. West Cameroon does, however, have considerable unused land, some of good to excellent potential.

The adjoining highlands in East Cameroon comprise less than 4.5 percent of the area of Cameroon but have over 17 percent of the total population. The Bamiléké, the major ethnic group, occupy about a quarter of the area with average densities of about 275 per square mile, though some portions have densities up to 850 per square mile. While the region has superior resources for agriculture and accounts for most of Cameroon's coffee exports (the first-ranking export by value) plus shipments of palm products, cinchona, cola, bananas, and tea, it faces a serious problem of protecting the equilibrium of the people and the land, to which a relatively large out-migration has provided only partial relief.

The rest of southern Cameroon is essentially a vast plateau with average elevations of about 1,950 feet rising progressively toward the Adamoua Massif, bordered on the west by a series of narrow coastal plains and to the southeast by the Sanaga and Ngoko Valleys. The western part of the plateau has a markedly accidented relief with deep valleys and mountains over 3000 feet, though the only notable depression is that of the Sanaga Valley. The climate is equatorial with heavy rainfall in two seasons and even temperatures; the vegetation grades from a broad belt of tropical rain forest across the south to wooded and grass savannas toward the north. Cocoa, the second export of Cameroon, is the predominant cash crop, while subsistence crops such as manioc, taro, and the plantain are well developed throughout. The population on the plateau is heaviest around Yaoundé where it ranges from about 90 to 130 per square mile. Three concentric zones surround this core area, each with decreasing densities; the first has densities ranging from about 25–45 per square mile, the second from 8–20, and the third, primarily to the north and east, has densities under 3 per square mile. The population tends to be particularly dispersed in the dense forest zones. Douala, on the coast, is the most important urban node in Cameroon; it is the country's main port, commercial, and industrial center. Rubber, timber, and some palm products produced in the coastal belt contribute to its wealth.

Congo (Brazzaville) and Gabon. These two countries of former French Equatorial Africa have among the lowest crude population densities of Africa, 6.5 and 4.6 per square mile respectively in 1967. The major characteristics of the population distributions in these two countries are: great unevenness; an unusually high percent urbanized, particularly in Congo (38.5 percent), but also in Gabon (23.0 percent); generally lower densities as one goes northward; and concentration of settlements along the major roads and waterways of the region (Map

Population Densities and Distribution

14).[11] According to Sautter's exhaustive study, one-third of the area had densities under 0.5 per square mile, 54–60 percent had under 2.6 and only 4 percent had over 15.5 per square mile in 1955. He concluded that the distributions were not well correlated with physical factors (climate, relief, geology, vegetation, soils, water supply, and "natural regions").

Slave raids, ethnic conflict, and migration to Brazzaville appeared to account in part for the very low densities of the Congo Basin portion of Congo (Brazzaville). In that country, about 70 percent of the population resides in the southern portion. The Brazzaville district, with about 200,000 in the city itself or 22.9 percent of the total 1968 population of the country, and with a relatively dense peripheral zone, is the most heavily populated part of the country. This concentration reflects both the attraction of the urban agglomeration and repulsive forces in the zones of out-migration. A relatively intensive agriculture extends outward from the city, producing crops primarily for the urban populace, while a large number of people are engaged in fishing in the Stanley Pool. But Brazzaville, like Kinshasa across the river, owes its importance primarily to its location at the base of navigation on the inland Congo system which provided the major external routeway for Chad, the Central African Republic, and large parts of Congo itself under French rule, and to its selection as the capital of the former Federation of Equatorial Africa. But the city has attracted many more people than the level of its economic activities has justified; unemployment has existed since at least 1952 and now affects a considerable portion of the population.

The overall low densities of Gabon and Congo (Brazzaville) are, according to the hypothesis preferred by Sautter,[12] explained mainly by the tropical rain forest acting as a screen to slow down historical migratory flows from other regions into the area. He feels that the forest also worked against the survival of migrant groups through the difficulty of social communications and demographic segregation.

THE POPULATION PATTERNS OF EASTERN AFRICA

The Horn. The population patterns of Ethiopia (Map 15) must remain somewhat obscure until a national census has been held. We do know that the lowlying portions of the country, which are excessively

[11] Gilles Sautter, *De l'Atlantique au Fleuve Congo: une Géographie de Sous-peuplement: République du Congo, République Gabonaise* (Paris, Mouton, 1966), I, 114.
[12] *Ibid.*, II, 997.

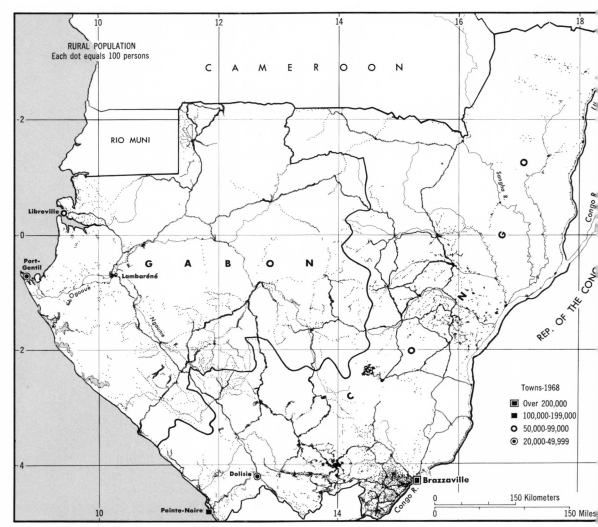

MAP 14. Population distribution in Gabon and Congo (Brazzaville). (After Gilles Sautter, *De l'Atlantique au Fleuve Congo: une Géographie du Sous-peuplement, République du Congo, République Gabonaise*, Paris, Mouton, 1966, v. 1.)

Population Densities and Distribution

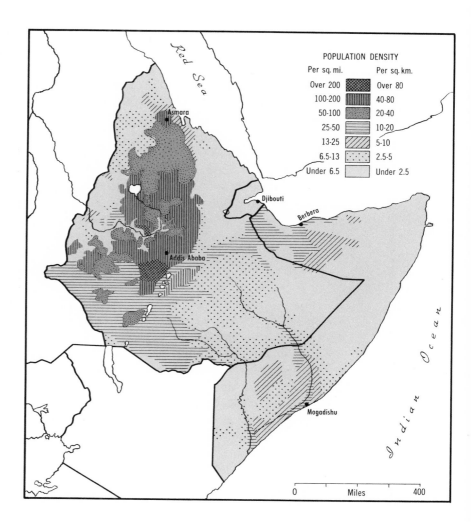

POPULATION DENSITY

Per sq. mi.		Per sq. km.
Over 200		Over 80
100-200		40-80
50-100		20-40
25-50		10-20
13-25		5-10
6.5-13		2.5-5
Under 6.5		Under 2.5

MAP 15. Population density map of the Horn, about 1960. (Copyright © by Denoyer-Geppert Company, Chicago, used by permission.)

hot and dry, are sparsely populated by nomadic peoples. The highlands, which are usually bounded by steep scarps, split by the great rift, and broken up into subregions by hills, mountains, peaks, cliffs, and some extremely deep canyons, have the bulk of the population of the country, with generally heavier densities on the richer "plateau basalts" as compared with the soils formed from crystalline rocks.

Somalia, too, has not yet had a population census. About 45 per-

Population Densities and Distribution   93

cent of the total are nomads and seminomads, generally ranging over vast areas at very low average densities, but sometimes gathering in large numbers at watering points or at winter settlements. Most of the country has too little and too erratic precipitation to permit more intensive use; about two-fifths of it is practically useless. Aside from the cities, which may have about 13 percent of the total population, the more densely populated parts of the country are along the two main rivers, the Giuba and the Webi Scebeli. Less than a third of 1 percent of the total area is cultivated, and irrigated lands total only about 0.035 percent of the total; nonetheless they account for the bulk of Somalia's exports and, on the sugar and banana plantations, a high percentage of the country's wage employees. Most of the population of the French Territory of Afars and Issas is resident in Djibouti which is supported by its handling of transit trade for Ethiopia and by some bunkering traffic.

East Africa. The population pattern of East Africa (Map 16) displays some of the sharpest contrasts of any tropical African area. "Islands" of sometimes exceptionally high density abut regions of very low density. Crude densities for the individual countries or for the region as a whole are, therefore, next to meaningless. While historical factors are of considerable importance in explaining the present pattern, the correlations between population levels and physical factors are striking, suggesting that the pattern may be somewhat more mature than in many other tropical regions.

The most important "control" is precipitation, which is, in turn, related to location with respect to the Indian Ocean, Lake Victoria, tropical air masses, and to topographic features. "Intensive arable cultivation requires a minimum rainfall of about 30 ins. per year, and . . . nearly all the populous parts of East Africa receive at least this rainfall in four years out of five."[13] This rule applies to the coastal belt, the highlands and elevated peaks of all three countries, and the Lake Victoria basin. The only exceptions are portions of Sukumaland in Tanzania and Machakos in Kenya. There are, however, some regions which have a fair to good chance of receiving 30 inches of rain which are not densely populated, particularly in southern Tanzania and in parts of Uganda. Presence of the tsetse fly may help to explain the low population of such districts, but, since the fly can be forced to retreat if there is close settle-

[13] Blacker, "The Demography of East Africa," in Russell, ed., *The Natural Resources of East Africa*, p. 27.

Population Densities and Distribution

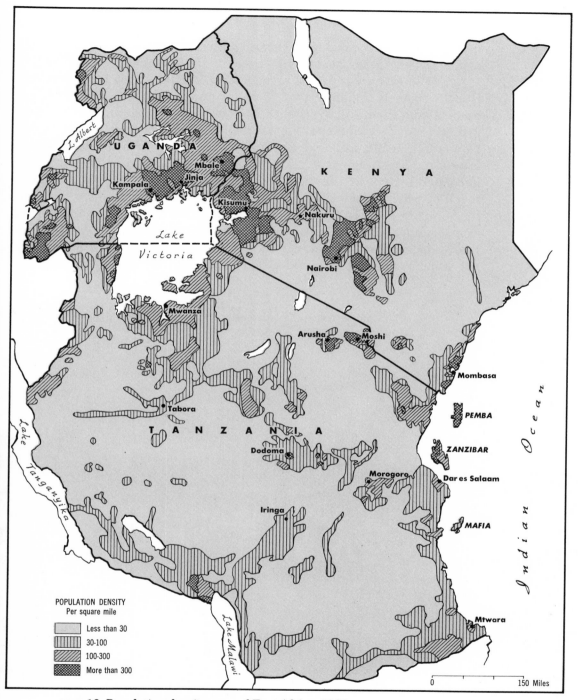

MAP 16. Population density map of East Africa, 1962.

Within the image:

L. Albert

UGANDA

Mbale
Jinja
Kampala
Kisumu

Lake
Victoria

KENYA

Nakuru

Nairobi

Mwanza

Arusha   Moshi

Mombasa

Tabora

PEMBA

TANZANIA

ZANZIBAR

Dodoma

Morogoro   Dar es Salaam

MAFIA

Iringa

Lake
Tanganyika

Indian Ocean

POPULATION DENSITY
Per square mile

Less than 30
30-100
100-300
More than 300

Lake Malawi

Mtwara

0   150 Miles

ment, it must be concluded that historical factors are significant in explaining the lower densities in some of these regions.

The vast areas of low precipitation in East Africa are characteristically sparsely inhabited by pastoralist groups, except where the tsetse fly precludes their keeping cattle. Thus only about a sixth of Kenya has favorable precipitation and this portion has about 88 percent of the total population; at the other extreme 76 percent of the area has only 8 percent of the population at densities below 10 per square mile. The sparsely populated regions of that country include the huge Northern Province, much of which receives under 10 inches per annum, and large parts of the Southern Province, especially the Masai Districts. Karamoja in northeast Uganda is similar to the adjacent regions in Kenya. Acholi and Lango districts in that country have somewhat more plentiful rainfall, but water supplies are not adequate to justify intensive agriculture or

Small livestock in the rift valley near Ol Longonot, Kenya. Much of Kenya is arid and suitable only for very extensive grazing.

Population Densities and Distribution

dense populations. Huge, low-populated regions with low precipitation in Tanzania include the Masai Plain, the Serengeti Plain, large parts of the low eastern plateau, and the Central Plateau. The tsetse fly, which is present in about 60 percent of mainland Tanzania, may have been more important than precipitation in inhibiting settlement in some large zones.

Intensive farming in Kikuyuland, Kenya. Farms in this and other high-density areas of the country have been regrouped and rationalized under the Swynnerton Plan.

Factors other than precipitation which help to explain the high population concentrations in the highland zones of East Africa include: the existence of some good to excellent soils, including the rich volcanic soils of Mounts Kilimanjaro, Meru, Kenya, and Elgon and the red and chocolate loams of the fertile crescent of Uganda; the possibility of producing a very wide range of both subsistence and cash crops; the high value and labor-intensive character of several of the important cash crops such as tea, coffee, and pyrethrum; the attractiveness of the climate for European settlement, especially the moderate temperature and humidity and relatively high percentage of sunshine; and the lower inci-

Population Densities and Distribution   *97*

Tea pickers on an estate at Limuru, Kenya. High-value intensively produced crops such as tea, coffee, and pyrethrum can support dense populations in favored areas.

Chagga farming on the windward slopes of Mt. Kilimanjaro, Tanzania. High precipitation, some supplementary irrigation, and fertile volcanic soils permit very high densities.

Population Densities and Distribution

dence of several tropical diseases and of the tsetse fly at the higher elevations of the highlands. These basic attributes permitted many of the highland regions to develop rapidly as "economic islands" and this led to the rise of urban communities, most notably in the former White Highlands of Kenya which is the most urbanized portion of East Africa. Government, services, and industry provided increasing employment opportunities and hence contributed their own impetus to the urbanizing surge, which has, in fact, tended to outpace the effective demands in the last decade. There are contrasts of some significance between the more accessible highlands and the more remote ones such as those of the West Nile Province and Kigezi in Uganda or the Ufipa Plateau and portions of the Southern Highlands in Tanganyika; the latter are pre-

The leeward slopes of Mt. Meru, Tanzania, another volcanic mountain. Contrast the land use with that on Mt. Kilimanjaro.

Population Densities and Distribution   99

dominantly concerned with subsistence production and are likely to be source areas of migrants to the more commercial economic regions. An asset of increasing importance to several of the highland areas is their great touristic interest and the ease of reaching many of the prime areas of attraction from Nairobi.

The non-highland agglomerations share certain of the attractions attributed to the highland nodes and have certain advantages of their own. The coastal belt produces sisal, coconuts, cashew nuts, wattle, cotton, and a variety of fruits and vegetables for the major centers. The sisal plantations in mainland Tanzania are concentrated close to shipping points partly because of the necessity to reduce transport costs for this relatively low-value and bulky commodity, and these plantations attracted a substantial number of permanent and temporary workers to the zone. The significance of Mombasa as the gateway for almost all overseas shipments of Kenya and Uganda and of Dar es Salaam as the major gateway and capital of Tanzania gave the coastal belt the third- and second-ranking cities of East Africa, Mombasa having been larger than Nairobi in the first half of the colonial period. The longer contact of the coastal communities with outside forces also helps to explain their present importance. Zanzibar shared many of the physical attributes of the coastal belt, though its port and entrepôt functions were diminished with the cessation of slaving and the decline of Arab influence on the mainland. The concentration of the economy on high-value cloves, whose planting was required by the sultan beginning in 1818, and on coconut products has helped to sustain a very high density on the two main islands of Zanzibar and Pemba.

The assets of the Lake Victoria Basin regions of high density are most notable in the fertile crescent north of the lake. In addition to the heavy and reliable precipitation, most of the soils are quite productive. Contributing to the average high densities is the concentration of coffee production and the use of much migrant labor on the indigenous farms, no less than 45 percent of the total population of Buganda being from outside that province. The elevations provide some moderation but the region is more hot and humid than the highlands of East Africa. Important in the explanation of several populous regions in Uganda is the early existence of several military kingdoms such as those of the Ganda and the Nyoro.

European settlement had relatively little impact on the popula-

Population Densities and Distribution

tion patterns of Uganda and Tanzania, but a very significant effect in portions of the Kenya highlands. Alienated land in Kenya totaled 12,850 square miles, of which 91.5 percent was in the former "White Highlands," now called the Scheduled Areas; the latter was 5.15 percent of the total area of Kenya but included 18.2 percent of the high potential land of that country. The European farms did give employment to over 250,000 Africans and about a million Africans were resident on them, but densities on the high quality lands were markedly lower than on comparable African areas. The introduction of the African Settlement Program in 1961 and the concomitant purchase of European farms has resulted in the closer settlement of the areas thus far involved and it is likely that the process will continue for some time, though at a slower pace. Company-owned tea and coffee estates have not yet been affected but their take-over would not have a major effect on population since they already employ a large number of workers, about one per acre in the case of tea plantations.

   Just as there were contrasts in population densities between the European and African areas so there are contrasts among the lands held by the numerous tribal groups in East Africa. The East Africa Royal Commission wrote that "it was constantly brought to our notice . . . how local sub-tribal or tribal boundaries have become crystallized in such a manner as to exclude the landless of neighbouring communities" and "at present it is often difficult for appreciable numbers of persons to migrate from areas where land is relatively scarce to settle in areas where it is relatively abundant owing to tribal jealousies and restrictions."[14]

   The effect of modern economic development on population concentrations is apparent in several ways in East Africa, although it is not always possible to tell which came first, the chicken or the egg. Certainly most of the important productive zones are high-density areas but not all high-density areas participate proportionately in the modern economy. Examples of the attraction to more developed regions of greater populations than would have otherwise been expected include southern Uganda and the sisal-producing areas of Tanzania, a few generally small "islands" associated with mineral exploitation, and the burgeoning urban centers, most of which were initiated under colonial rule. Soja has shown with respect to Kikuyuland how proximity to the most rapidly develop-

[14] *East Africa Royal Commission 1953-1955 Report* (London, H.M.S.O., Cmd. 9475, June, 1955), pp. 13, 34.

ing European areas in Kenya was positively correlated with participation by the Kikuyu in the modernizing process, both by development in their own areas and by their predominant share in internal migration to urban areas, European farms, and now to settlement in the Scheduled Areas.[15]

Rwanda and Burundi. These countries share with many of the highland areas of eastern Africa some exceptionally high densities. The populations on the heavily dissected uplands are rather uniformly dispersed, although elevations above 5,000 feet have the highest densities. One of the unusual features for such densely populated areas has been the very low percentage of urban residents. Only Bujumbura in Burundi is a large city and its growth is a postwar phenomenon; in 1944 its population was estimated to be 10,000, while it is now about 100,000. Many of its residents are, however, of non-Rundi origin.

Factors which have permitted the great densities of Rwanda and Burundi include: a generally excellent climate with high average precipitation; ecological suitability for a variety of crops, including high value cash crops such as coffee and tea; absence of the tsetse fly; a generally lower incidence of tropical diseases, although malaria is prevalent near interior marshes, intestinal diseases are common on the plains, and tuberculosis is a serious threat; the existence of some relatively good soils; the development of some intensive agricultural practices including the extensive terracing of many steep slopes; and the relative security against slaving and tribal invasions due to the social and political solidity of the population under the feudal rule of the Tutsi kings.

Zambia. Kay suggests that five unequal sectors running approximately north–south may conveniently be used to delineate the population pattern of this country (Map 17).[16] Starting with the westernmost the sectors are:

1. A moderately or sparsely populated area paralleling the upper Zambezi, within which are two "islands" of dense population: the heart of Barotse Province centered on the Barotse floodplain, and the heart of Luvale country astride the Zambezi upstream of Balovale. This is a tsetse-free area and the Barotse floodplain is considered the richest part of the province, though it is not capable under present land-use systems

---

[15] Edward W. Soja, *The Geography of Modernization in Kenya: A Spatial Analysis of Social, Economic, and Political Change,* pp. 24-26.

[16] George Kay, *A Social Geography of Zambia,* pp. 47 f. Much of the following analysis is derived from Kay.

of providing all of the food needs of its densely peopled portions. In the seasonally flooded sections, "minor relief features such as terrace remnants, sand-banks, and former levees are of the utmost significance in the life of local peoples,"[17] although most of them are vacated prior to the height of the flood when there is an annual ceremonial procession to the border villages. Fishery resources provide support for a portion of the population. The remoteness of the region from the line of rail and the Copperbelt and a relatively sparse road system are negative factors. Very important in explaining the population concentration along the floodplain is the heritage of the powerful Barotse kingdom, one of the several military states which existed in Zambia in the second half of the nineteenth century. On the other hand, the special status of Barotseland as a protectorate within the country, a status which existed from the 1890 treaty with Lewanika, Paramount Chief of the Barotse nation, until independence, affected the developing population pattern because of the lesser attention accorded to the region. This resulted in greater isolation than might otherwise have existed, which was reflected in a variety of ways including a lower migration to the country's more developed regions.

2. A belt of very low densities paralleling the north-south section of the Kafue River and centered on the Kafue National Park. The Kafue Flats, once the floor of a lake, comprise over a million acres of grass-covered clay lands much of which is flooded annually. Presence of the tsetse fly repulsed cattle-keeping tribes, while seasonal inundations of extensive stretches made tillage agriculture precarious. More recently it has been realized that there are potentialities for fisheries development and that portions of the flats could be polderized for intensive production of commercial crops. The National Park will presumably continue to limit the population on a large portion of this belt, although it should attract a rapidly increasing number of tourists and hence contribute to the national income far more importantly than in the past.

3. A belt of moderate to dense population bisected by the Katanga pedicle. The southern portion of this belt contains the main economically developed portions of the country: first and foremost, the Copperbelt, which ranks as the single most important "island" of export production in tropical Africa; but also the best-developed commercial agricultural

[17] *Ibid.*, p. 19.

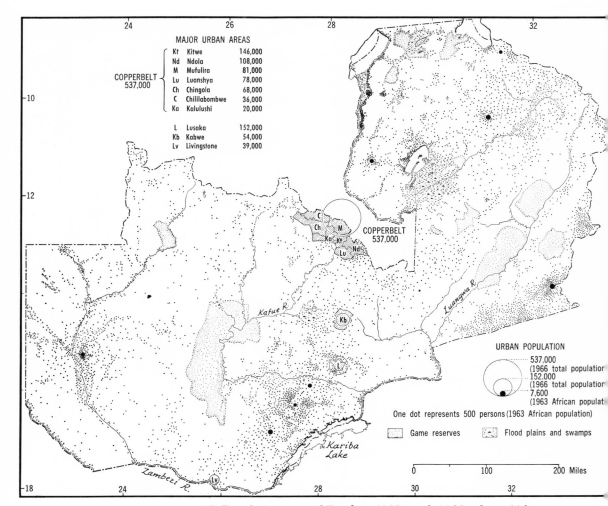

MAP 17. Population map of Zambia, 1963 rural, 1966 urban. (After George Kay, *A Social Geography of Zambia,* London, University of London Press, 1967, used by permission.)

zone in Zambia; and the capital city, Lusaka, which has been growing more rapidly than any other urban community in the country in recent years. The attraction of the line of rail, paralleled by the country's main road, is a striking feature of the population pattern.

The Copperbelt, whose notable population concentrations are

The Luanshya mine on the Zambian Copperbelt. The huge mineral resources of this area and adjacent Katanga in Congo (Kinshasa) have attracted one of the most important urban agglomerations in tropical Africa. A high-density township is seen beyond the plant area.

obviously related to the discovery, exploitation, and processing of the vast reserves of ore in the area, alone contained 12.9 percent of the total African population of Zambia according to the 1963 census, with the seven major urban centers accounting for 92.9 percent of the total. The Copperbelt accounted in that year for two-fifths of all male employees in the country. Its cities also held 57.3 percent of the European population of Zambia (74,640) at the time of the 1961 census of non-Africans. The total population of the seven centers in 1966 was estimated to be 537,000 or 14.0 percent of the total for the country.

The line of rail south of the Copperbelt includes the three remain-

Population Densities and Distribution

ing large towns and sixteen small towns; these contained 6.0 percent of the African population of the country in 1963 and 30.3 percent of the European population in 1961. The line of rail accounted for 30 percent of male employees in 1963. Lusaka grew from about 58,000 in 1956 to 152,000 in 1966.

The northern portion of the third population belt, running northward of the Katanga pedicle, contains several "islands" of dense population. First are those associated with the fishing grounds of the lower Luapula, Lake Mweru, and the Bangweulu lakes and swamps. The average density on seven of the actual islands in the Bangweulu basin was about 260 per square mile in 1963. Concentration upon cassava as a major subsistence crop has permitted higher densities than would otherwise have been possible in some areas. Second are a number of "islands" associated with important local settlements such as Kasama, Fort Rosebery, and Mpulungu. Existence of the Bemba kingdom contributed to some population concentration, though the Bemba characteristically did not settle as densely as peoples of the other major Zambian military states.

4. A northeast-southwest belt of sparsely populated or empty lands associated with the Luangwa trough and the middle Zambezi Valley. These low-lying areas are not attractive for settlement, being tsetse-ridden, hot and humid, having lower precipitation than the surrounding plateaus, and having a general scarcity of permanent surface-water supplies because most streams flowing from the north and west drain only the escarpment zones. Only in a few places are there small, densely populated nuclei; these are usually associated with excellent alluvial soils.

5. A densely populated belt more or less paralleling the Luangwa trough and lying on the eastern plateau. Two population cores are associated with Ngoni states, one of them centered on Fort Jameson, the other on Lundazi. The present high densities of these core areas is explained in part by the inhibitions on expansion imposed by Pax Britannica, while the former no man's land surrounding them has gradually been filled in by neighboring groups who no longer needed to seek refuge from the Ngoni. Both this and the previous belt suffer considerably from remoteness and poorly developed surface transport.

Kay concludes that the most important physical factors influencing the population distribution of Zambia are the location of dry-season water supplies, presence or absence of the tsetse fly, types of land use including a changing emphasis to cassava, the availability of fish in lakes, rivers, and swamps, and the existence of mineral resources. He states that soils are not closely correlated with population on a national basis, though areas where current dissection provides a balance between soil formation and soil removal have soils that are among the most valuable and are, therefore, capable of supporting relatively dense populations. Such areas, however, require special precautions to avoid excessive erosion.

Rhodesia. The single most significant correlation with the population distribution of Rhodesia (Map 18) is the amount of precipitation. The most attractive parts of the country from this standpoint are the high veld, a belt of generally level land above 4,000 feet running nearly across the country from northeast to southwest, and the mountain country along the eastern border. The high veld, comprising about a fifth of the country, contains most of the European population and the densest African settlement as well. It is the physical and economic backbone of the country, containing the most important commercial agricultural zones, the bulk of the varied mining operations, a predominant share of the relatively well-developed manufacturing sector, and most of the cities and towns of the country, including Salisbury and Bulawayo which had an estimated 13.9 percent of the total population in 1968. Obviously, a variety of factors have attracted population to this favored zone, but the continued dominance of the rural population suggests that precipitation is the leading one. The eastern mountains, which have the highest precipitation, have the particular advantage of ecological suitability for high-value tea and for wattle plantations. The moderate temperature pattern of the high veld and mountains has also been attractive to Europeans.

The land of Rhodesia drops off from the high veld through a broad, more heavily eroded, middle veld to the low veld of the Zambezi Valley in the northwest and of the Limpopo and Sabi basins in the southeast. The Zambezi trough is deep and rather narrow, with very steep sides; it has only a few population concentrations such as those associated with the Kariba hydroelectric installation and with a sugar plantation near Chirundu. In addition to being dry, hot, and humid it suffers from presence of the tsetse fly and limited level land, part of which is now devoted to a game park. The southeastern low veld is

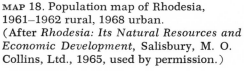

MAP 18. Population map of Rhodesia,
1961–1962 rural, 1968 urban.
(After *Rhodesia: Its Natural Resources and
Economic Development*, Salisbury, M. O.
Collins, Ltd., 1965, used by permission.)

Towns with populations
of over 5,000 (1968 estimate)

Sa	Salisbury	380,000
Bu	Bulawayo	271,000
Um	Umtali	54,000
Gw	Gwelo	41,000
Wa	Wankie	23,000
Qu	Que Que	21,000
Ga	Gatooma	17,000
Sh	Shabani	17,000
FV	Fort Victoria	13,000
Se	Selukwe	10,000
Re	Redcliff	9,000
Ha	Hartley	8,000
Mar	Marandellas	8,000
Bi	Bindura	7,000
Ka	Kariba	7,000
Ma	Mangula	7,000

Livingstone

Wa

B O T S W A N A

RURAL POPULATIONS

• Each dot equals 1,000 Africans (1962)
▲ 100 non-Africans (1961)
▢ Tribal Trust Land
++++ Railways

0   50   100 Miles

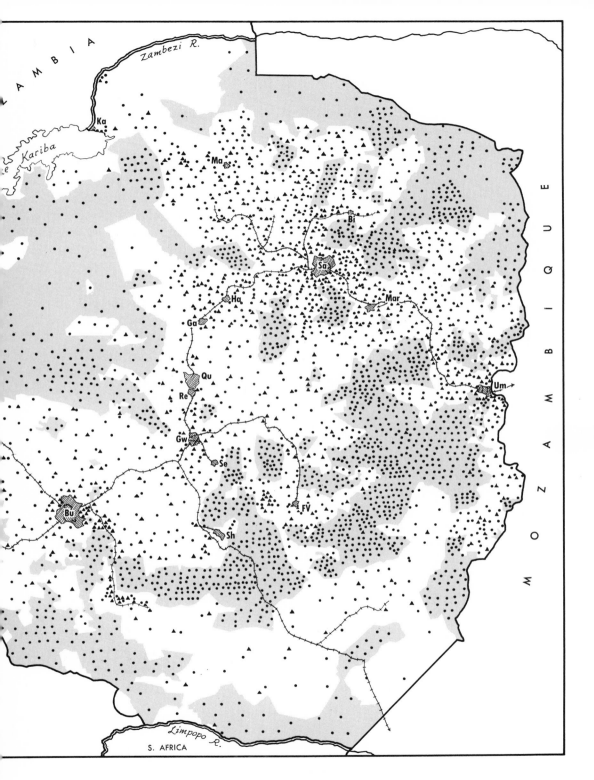

ZAMBIA

Zambezi R.

Kariba

Ka

Ma

Bi

Sa

Ha

Mar

Ga

Qu

Re

Um

Gw

Se

FV

Bu

Sh

Limpopo R.

S. AFRICA

M O Z A M B I Q U E

considerably broader and, while it is still largely undeveloped and sparsely populated, it has very attractive potential for intensive irrigation schemes, of which the Hippo Valley and Triangle Estates represent present examples. The ambitious Sabi-Lundi scheme calls for the staged development of sizable acreages in the southeast which could support large numbers of persons.

The population pattern of Rhodesia is strongly affected and to some extent distorted by racial land apportionment. Europeans have allotted themselves about 37 percent of the area of the country and Africans have about 46 percent, while 6 percent is unreserved, and about 11 percent is in game reserves, parks, and forest reserves. A disproportionate share of the better lands are found in the European areas which were estimated in 1961 to have 47.0 percent of their total area ecologically suitable for semi-intensive or intensive farming as compared to 25.3 percent for the African lands. The European farms, however, housed about 836,000 Africans in 1961 while an additional 653,000 were resident in urban centers outside the reserves. Not all of these were Rhodesian Africans, since 47 percent of the 576,000 male employees in 1961 were immigrants, mainly from Malawi and Mozambique. Thus 42 percent of the total African population of Rhodesia but only 35 percent of the indigenous African population lived in the European areas. Their absence from the reserve areas doubtlessly reduced the pressure there and contributed, through remittances, to support of families in the reserves, but it is considered that the African lands, which carry an average density more than double that of the European farming area, are not capable of sustaining their present population under prevailing land use systems. Most of the tribal lands fall between 2,000 and 4,000 feet above sea level and much is in the less favored and more heavily dissected middle veld. Densities vary from 8 to over 100 per square mile as compared to densities on European lands ranging from under 5 per square mile in the cattle-ranching areas of the low veld to about 40 in the good farming areas of Mashonaland.

The impact of the sanctions imposed on Rhodesia after its unilateral declaration of independence in 1965 on population distribution is not clear, though two trends of importance have been a reduction in immigration of foreign African males from an average of 33,290 per year in the period 1963–1965 to about 15,000 per year in 1966–1968;

and a lowered use of farm workers as tobacco production is replaced by less labor-intensive farm emphases.

The relatively well-developed urbanization of Rhodesia, which had an estimated 18.1 percent in towns above 10,000 in 1968, is another reflection of the dual economy of the country. Over three-quarters of the non-African population of the country (237,000 Europeans and 23,300 others in 1968) live in the nine main urban areas, with 61 percent residing in the two major cities, Salisbury and Bulawayo.

Malawi. This country contains some of the notable "islands" of high density in eastern Africa. Extracting its water surface, Malawi had an average density of 111.3 in 1966, but 38.7 percent had densities above 300 per square mile. The regional figures used conceal some of the highest densities, which exceed 800 per square mile in parts of the Cholo and Mlanje Highlands. As in other countries of eastern Africa, the higher densities are frequently associated with highland zones which have ample precipitation and good to excellent soils. However, pressure on the land and lack of alternate opportunities result in a heavy migration of workers, particularly to Rhodesia and South Africa. As many as a third of the able-bodied men in Malawi work outside the country at any one time. Northern Malawi, which has densities well below the national average, is disadvantaged by greater isolation, a low development of commercial agriculture, and large areas of stony, shallow soils.

Madagascar. The population pattern of this great island reveals unusual variation, not all of which is readily comprehensible (Map 19). When its crude density was 19.9 per square mile only 31.4 percent of the population lived at or below this density but occupied 77.2 percent of the total area. Nearly 20 percent of the area, in fact, was empty, notably the limestone regions of the west. Areas with very low densities included the region west of the central massif and the steep and highest parts of the Ankaratra and Antongil massifs and of the eastern scarps. Explanations for the low densities over much of Madagascar include: the generally poor quality of much of the soils, the high percent of the central massif which is in slope, the aridity and precarious nature of precipitation in much of the west and south of the country, and historical factors which are poorly understood but which include the migration of the seafaring immigrants from Indonesia to the interior of the island.

At the other extreme about 27.0 percent of the population lives at densities in excess of 100 per square mile on 2.3 percent of the area.

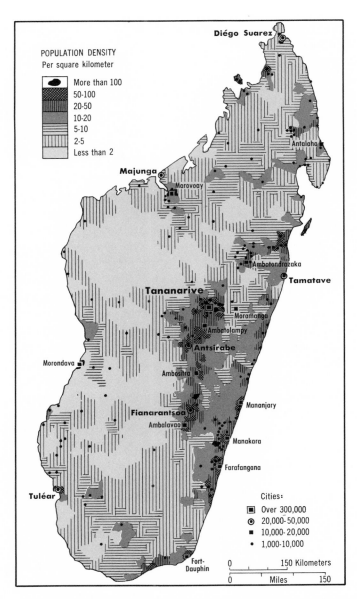

MAP 19. Population density map of Madagascar, 1960 rural, 1965 urban.
(After Institut Géographique National, Annexe de Tananarive, 1965.)

Population Densities and Distribution

The most populated portions of the island include the east-central part of the interior massif, the borders of Lac Alaotra, and a series of nodes along the east coast from Andapa to Fort Dauphin.

Explanations for the concentration on the massif, which contained 29.6 percent of the population on 5.6 percent of the country's area, include: the presence there of the powerful Merina kingdom which gained hegemony over the entire island before it was subjected to French rule; the selection of their capital as the colonial capital and the concentration there prior to and following independence of a very high proportion of the country's administrative, commercial, financial, industrial, and cultural establishments and employees; the existence of the best areas on the island for the development of the rice culture favored by the Merina and the Betsileo, including the large Betsimitatatra Plain adjacent to Tananarive, and the ability of this intensive culture to support dense rural populations; and the ecological suitability of this tropical highland area for a great variety of tropical, subtropical, and middle latitude crops. The Lac Alaotra Basin has some of the same advantages enjoyed by the region around Tananarive, especially the existence of a large area which could be converted to paddy rice production.

The east coast nodes and alignments may be explained by a variety of factors including: the tropical rainy climate which permits the production of a variety of commercial crops including robusta coffee, cloves, vanilla, bananas, and sugar; the possibility of producing rice in paddies along the numerous streams flowing down the scarp and onto the narrow coastal plain; and the development of Tamatave as the main port of the country and its rail link servicing the most productive section of the highlands which is by all odds the leading consumption area of the island. Minor areas of high density include portions of the northwest, including the island of Nossi-Bé, which has very favorable conditions for high-value tropical crops, and urban nodes associated with Majunga, the island's second-ranking port, and Diégo-Suarez, the third port and naval base situated near the northern tip of the island on one of the world's finest natural harbors.

Indian Ocean Island Groups. Eastern Africa as defined by the U.N. includes four island groups in the Indian Ocean, one of which, the British Indian Ocean Territory, is not considered in this book. The other three—the Comoros, the Seychelles, and the Mascarenes—now rank among the most densely populated portions of Africa and its appurtenances. The last two were void of human inhabitants when first visited by European explorers. The Seychelles and Réunion, one of the two main islands of the Mascarenes, are selected for discussion at this point.

The Seychelles are composed of two distinct collections of islands, the Mahé group, 32 granitic islands with peaks rising to 2,971 feet, and the Outlying Islands, 60 mostly coralline islands which are waterless and have no permanent population, though 18 of the 60 recorded about 3 percent of the total population at the time of the 1960 census. The granitic islands total 85 square miles or 54.5 percent of the total; Mahé, the main island, has an area of 55 square miles. The coralline islands do contribute to the total carrying capacity in that they are worked by labor under government contract for copra, guano, bird's eggs, turtles, and salted fish.

The Seychelles were uninhabited until 1770 when the French colonized Mahé primarily to deny to Britain a port of call on the route to India. In 1814 it became a British possession and was inhabited by an estimated 5,000 persons including at least 4,000 slaves from East Africa and Madagascar. Slavery was abolished in 1834, which destroyed the existing economy. Since then the islands have not fed themselves, their chief dependence being on copra and expenditures of people who have chosen the islands for retirement. French, Africans, British, Malagasy, Indians, Chinese, and others have contributed genes to the common pool, but, unlike the Mascarenes, Indians and Chinese comprise little more than 1 percent of the total population. The present population thus mainly represents natural growth from the early settlers plus an influx of 2,409 Africans liberated by the British Navy from slave dhows between 1861 and 1872. Population growth was interrupted in the period 1831–1841 when a group of settlers and their slaves totaling about 4,000 emigrated, ostensibly because they objected to efforts to convert them to the Church of England (the Seychelles remain today, despite their long association with Britain, heavily influenced by French culture, language, and outlook, and by Catholicism). Population growth was slowed again by a serious outbreak of smallpox beginning in 1883, heavy losses to Spanish flu in the intercensal period 1911–1921, and by the absence of a large number of young men serving in the Seychelles Pioneers in the 1941–1951 period. Today, the density for Mahé, which has about 83 percent of the total population, is about 800 per square mile. While the concentration of population on Mahé may be understood in relation to the lack of opportunities on the other islands, it would be difficult to claim that Mahé can adequately support its present density let alone what may be projected for the years ahead. Two factors which help to explain the existent situation are the absence of certain tropical diseases, including malaria, and the difficulty of escaping from the finite size of a small island. The lessons for larger areas scarcely need emphasis.

Réunion. This island provides an interesting example of the effects that physical factors on a confined and densely populated space have on population distribution (Map 20). It is composed essentially of two volcanic masses, one of which is active on an average of once every two years. The older massif rises to 10,067 feet in Piton des Neiges and covers about three-quarters of the island. It is extremely dissected by erosion, three great, pear-shaped cirques edged by sheer cliffs as much as 3,000 to 4,500 feet cut deeply into the mass. The active volcano, Piton de Fournaise, occupies the southeast quarter of the island and is joined to the main body by a broad, high saddle.

The greatest part of the population is grouped along the main roads running along three bands on the mountain slopes: a coastal strip, almost uninterrupted except where recent lava flowed from Fournaise into the sea at the east end of the island and in the driest part of the west; a band at about the 800–1,000 foot line; and one at about 1,600–2,000 feet. A coastal strip about 4⅓ miles wide contains about 85 percent of the total population; settlements are characteristically on the undissected slopes rather than in the valleys. The upper levels began to be used in the nineteenth century primarily due to rising pressure nearer the coasts; migration to the towns has more recently reversed this move to some extent.

Réunion received its first permanent settlement in 1671, when its population was 76. In 1711 the populace had grown to 1,500 and it was about 46,000 in 1788. Following the abolishment of slavery in 1848 most black workers quit and migrated to "Les Hauts" where they opened smallholdings, largely subsistent in character. A good many "petits blancs" had previously moved to the higher levels and their descendants continue to operate essentially self-subsistence units. Indians, Pakistanis, and Chinese were brought to the islands to replace the

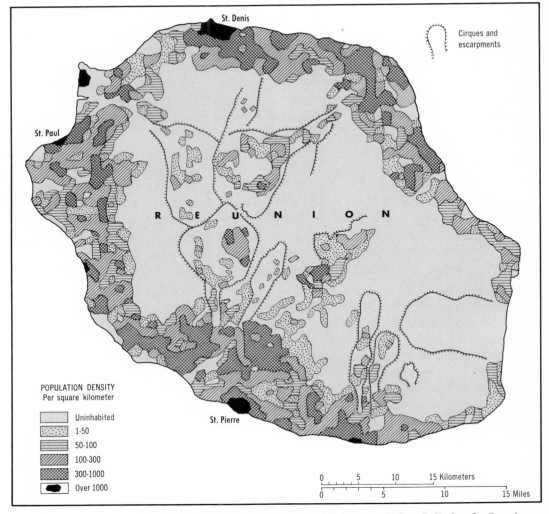

St. Denis

Cirques and
escarpments

St. Paul

R E U N I O N

POPULATION DENSITY
Per square kilometer

	Uninhabited
	1-50
	50-100
	100-300
	300-1000
	Over 1000

St. Pierre

0   5   10   15 Kilometers
0   5   10   15 Miles

MAP 20. Population density map of Réunion. (After J. Defos du Rau.)

former slaves, mainly by the larger sugar-plantation holders, who were the only ones who could afford to pay their passage.

Today the island is occupied by about 418,000 people divided approximately as follows: whites 20 percent, Indians 20 percent, Chinese 1+ percent, Pakistanis 0.5 percent, with the remaining percentage being of African or Malagasy origin, frequently mixed. The crude density of population was about 432 in 1967, but actual densities were roughly twice as great, and the man-to-arable-land ratio was only about 1 to 0.36 acres.

The major factors in restricting the population to little more than half of the island's surface are: the high percentage of the area in lands which are too high or too steep for effective use, exclusion from lands affected by recent volcanic flows, excessive precipitation on the windward side (with rainfall averaging over 150 inches and occurring on about 230 days a year); excessive aridity on small portions of the leeward side, and the desire to avoid maximum exposure to the large number of cyclones which hit the island (from the seventeenth century to 1967, 190 cyclones of damaging proportions have affected the island).

The island supports its present population at a precarious level and does so only because of the richness of its volcanic soils and the ability to export sugar at higher than world prices.

THE POPULATION DISTRIBUTION OF SOUTHERN AFRICA

South Africa. The population distribution of South Africa (Map 21) is most closely correlated with climate, mineral occurrences, modern urbanization, and reservation of land for specific racial groups.

The rural population of South Africa comprises half of the total, estimated to be 20.1 million in mid-1970. Non-African groups, accounting for about 30 percent of the total population, are much more strongly urbanized, about 85 percent of whites and Asians and perhaps three-quarters of the Coloured (persons of mixed race) population residing in urban centers. About 65 percent of the African population is resident in rural areas. The rural distribution shown in Map 21 for 1960 is most closely associated with amount of precipitation, with the densest populations found in the humid subtropical region along the southeast. Lesser concentrations are visible in the dry subtropical section of the Cape and on portions of the plateau, particularly in northern Transvaal. The huge arid to semiarid west, suitable only for the extensive grazing of livestock, is sparsely populated.

Very important in explaining the relative densities is the division of land by race. Europeans have allotted themselves about 87 percent of the total area of the Republic; Africans are supposed to consider the remaining portion, which is split into about 238 distinct and usually separated blocks, as their "homelands." Some sections of the European

Farming in one of the large cirques on Réunion. A large part of this island is unused because of topographic and climatic handicaps, giving a density on utilized areas of about 900 per square mile.

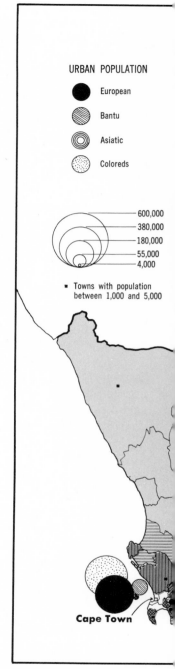

URBAN POPULATION

European

Bantu

Asiatic

Coloreds

600,000
380,000
180,000
55,000
4,000

Towns with population between 1,000 and 5,000

Cape Town

MAP 21. Population density of South Africa, 1960.

Population Densities and Distribution

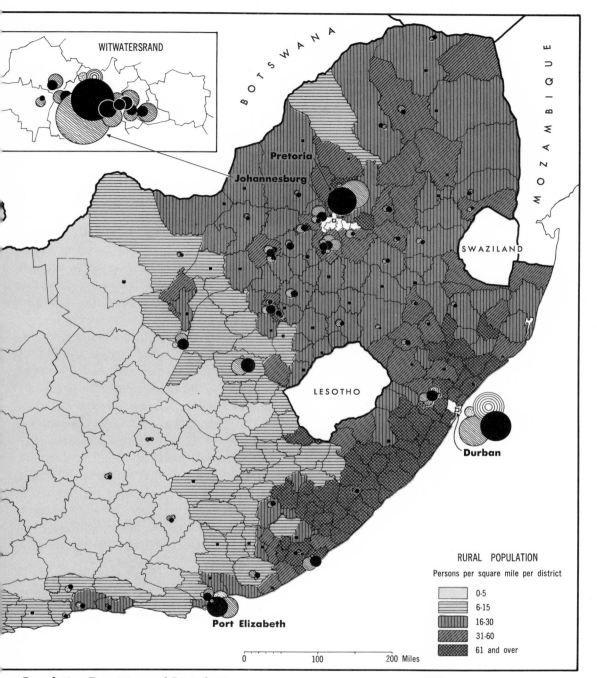

WITWATERSRAND

B O T S W A N A

M O Z A M B I Q U E

Pretoria

Johannesburg

SWAZILAND

LESOTHO

Durban

Port Elizabeth

RURAL POPULATION

Persons per square mile per district

	0-5
	6-15
	16-30
	31-60
	61 and over

0   100   200 Miles

Population Densities and Distribution   *119*

portion have Asian and Coloured farms, but these groups are found primarily in portions of urban areas assigned to them. The present distribution of African areas is explained by a variety of factors including: the early and historic migration of various ethnic groups; the existence of several well-integrated tribes which could not readily be ousted from their core areas, the retreat of some groups to refuge areas following upon white penetration and subjugation of the hinterland, and the subsequent delineation of specific areas for particular groups.

Of the 65 percent of the total African population residing in rural areas about 54 percent live in the reserves and 46 percent on white farms, many of the latter on at least a *de facto* permanent basis and the others as migrant workers. Comparison of Maps 21 and 39 clearly reveals the substantially higher densities in the African areas, estimated to be about 78 per square mile in 1970 as compared to about 13 per square mile in the European rural areas. The fact that African lands have a higher percentage of potentially productive lands than the average for the country only partially diminishes the artificiality of the population distribution as between African and white rural areas.

South Africa obviously has an unusually high percentage of its total population urbanized for African countries, which reflects its status as the economically most developed country on the continent, its great mineral wealth and the concentration of several important minerals in relatively confined and well-delineated zones, and the unattractiveness of much of it for nonurban pursuits. The urban population would be considerably higher than it is, however, if it were not for restrictions on the movement of Africans to the urban areas and particularly if migrant workers were permitted to bring their families instead of being restricted to the so-called "bachelor" migration. The urbanized region of the Rand, whose major center is Johannesburg, is the most important of South Africa's urban belts. While it owes its origin and much of its continuing significance to exploitation of the gold-uranium ores of the Witwatersrand, the zone has also developed as the leading manufacturing region of the country. It is further strengthened by development in adjacent communities, especially at Pretoria, the administrative capital of the Republic and an important industrial center in its own right, and the group of cities to the south including Vereeniging, Vanderbijl Park, and Sasolburg with their heavy industrial and chemical enterprises. The other major urban regions are those focused on the country's ports, particularly Cape Town and Durban.

The government of South Africa continues to add to its already formidable powers to engineer the population distribution of the country. Concerned to maintain and even to increase the present ratio of whites to non-whites in the European areas, and particularly in the cities, it has passed a whole series of laws which are designed to restrict or indeed reverse the movement of Africans to urban centers and to retain

A view in the Transkei, South Africa. Like other reserve areas, the Transkei is characterized by overgrazing, archaic farming practices, and pressure on the land.

and increase the reserve and Bantustan populations by development of these areas or by the build-up of industry in border locations. These laws and programs fly in the face of powerful economic and demographic forces and are not at all likely to achieve their stated goals.[18] Indeed, it may be predicted with some confidence that the African ratio will increase in all major subdivisions of the country in the years ahead.

Of the remaining countries in Southern Africa attention will be focused on Lesotho and Swaziland; Botswana and South–West Africa,

[18] See the discussion in William A. Hance *et al.*, *Southern Africa and the United States* (New York, Columbia University Press, 1968), pp. 141-56.

Population Densities and Distribution

two huge, largely arid, and sparsely populated countries, will not be examined.

Lesotho. With an estimated 964,000 people on only 11,720 square miles, this is another of Africa's "islands" of high density. Its population pattern is correlated with the major topographic zones which, in turn, have specific climatic and soil conditions which are important factors in the physical complex affecting the distribution. The Maluti mountain zone, mostly above 8,000 feet, is traditional "cattle post" country, formerly used exclusively for summer grazing. Despite the physical limitations, population pressure has led to increasing settlement and arable farming in this zone. Lying between the Maluti Mountains and the western lowlands is a foothill zone, rolling country between 6,000 and 7,000 feet interrupted by mountain spurs and river valleys. In the valleys and on the flatter areas, rich volcanic soils provide some of the best soils of the country and erosion is not so advanced as in the west. Accounting for about 15.6 percent of the total area the foothill zone contains about 21.3 percent of the total population at densities roughly 2.3 times the average density of the mountain zone.

The three lowland zones of Lesotho comprise 26.6 percent of the country and contain 49.1 percent of the population. On the west the narrow border lowland is characterized by severe erosion and soil impoverishment but has a considerable portion of the arable lands, in places below 5,000 feet. Extending east of this zone is a more fertile, less arid, lowland belt between about 5,000 and 6,000 feet elevation interrupted by sandstone plateaus. This is the most important zone of the country, containing over a quarter of the population at densities now about 250 per square mile. The third lowland area is a long wedge along the Orange Valley in the southern part of the mountain zone.

Lesotho is predominantly an agricultural country, but only about an eighth of the area is used for tillage, including about 40 percent of the lowlands and 6 percent of the highlands. Land-form, soil, and climatic conditions are highly restrictive and the country usually does not produce enough food to meet local requirements. It is heavily dependent on the earnings of its migrants working in South Africa; about 53 percent of the adult males are absent at any given time. While the high densities of Lesotho reflect the cohesiveness of the Sotho and their desire to retain their independence and integrity, it is not unlikely that many more Sotho would have migrated permanently if controls had not been instituted in the Republic. Thus Lesotho represents a kind of extreme example of a high density "island" whose population concentrations have been intensified by both voluntary choice and political restrictions on movements of its peoples.

Swaziland. The two major correlations of the population distribution of Swaziland are with the tenurial pattern and the physiographic regions of the country. In this respect it bears resemblance to Rhodesia. Swaziland may be divided topographically into four distinct regions, each running north to south. On the west the high veld, a north-easterly extension of the Drakensberg, occupies about 29 percent of the country at elevations usually between 3,500 and 5,000 feet but with maxima of 6,000 feet. Its precipitation is relatively high but its elevations bring winter cold and occasionally severe frosts, and it is further disadvantaged by relatively poor soils, sour grasses, and heavy dissection. It is used for grazing, dryland farming, and forest plantations which have been introduced in postwar years and which represent a more intensive use than the former grazing of trek sheep in the summer. While the high veld had a population density in 1966 close to the average for the country (55.9 per square mile), it carried only 22 percent of the rural African population. Several concentrations associated with urban, mining, and plantation operations increase the importance of the zone. These include Mbabane, the burgeoning capital, the iron mine at Ngwenya, the asbestos mine at Havelock, and the forest plantations centered on Pigg's Peak and Bhunya. These activities help to explain the presence in the high veld of 45 percent of the total of 7,987 Europeans in Swaziland and for 55 percent of the urbanized Europeans in the country at the time of the 1966 census.

The middle veld, with elevations ranging from about 2,700 to 1,300 feet, comprises about 26 percent of the country and contains about 42 percent of its population. Its surface is undulating to more markedly rolling and is dissected at intervals by eastward-flowing streams. The grazing potential of this zone is superior to that of the high veld and the Swazi have always considered the region best suited to their primarily pastoral economy. Rain-grown crops give fair to good yields in favorable seasons and a great range of crops can be grown under irrigation. It is not surprising, therefore, to find 48 percent of the rural African population in the middle veld. Concentrations within the zone are associated with the occurrence of better soils, the town of

Manzini situated near the royal homesteads of the Swazi nation, and the Malkerns irrigation scheme along the Usutu River. The early selection of the area near Manzini by the Swazi is explained in part by its easy defensibility against neighboring tribes.

The low veld, with average elevations of about 500 to 1,000 feet, covers about 37 percent of the country and has about 25 percent of its population. Most of it is gently undulating acacia bushveld; rainfall will support only grazing or the production of hardy annuals such as sorghum and millet; temperatures are high. Malaria was formerly endemic and presence of the tsetse fly inhibited cattle-keeping. Until relatively recently it was shunned by most Swazi except as a hunting reserve, although a royal cattle post was established as early as 1880 and the northern part was used to settle refugee clans. The presence of some very good alluvial and basaltic soils and a number of transverse streams has stimulated the development of several important irrigation schemes which have in turn attracted population concentrations.

The Lebombo Plateau accounts for 7.6 percent of the area and 5.2 percent of the population. It is an undulating to rolling, narrow plateau with elevations of 1,500 to 2,700 feet dropping off abruptly on the west into the low veld and deeply dissected by the three main streams leaving the country. Precipitation of about 33 inches permits the support of good pastures, but there are only a few areas of reasonable topography and soils which support dense concentrations, the remainder being too rugged.

Affecting the general pattern of population distribution in a very significant way is the complex landownership superimposed on the relatively simple topographic base. In the period 1880–1890 the Swazi king Mbandzeni granted numerous concessions which were sorted out in the Proclamation of 1907; with some later additions to the Swazi area the present situation is that 52.4 percent of the total area of the country is held by Swazi, 47.1 percent is held in freehold farms, mainly by Europeans, and 0.5 percent is in urban areas. The Swazi and European holdings are intricately interwoven, which results in population and land-use patterns showing sharp contrasts from section to section within each major zone. The density of population on the African-held land was 74.7 per square mile in 1966 as compared to 27.1 per square mile on the freehold farms. About 22 percent of the African population was enumerated on the freehold farms, some as owners, some as farm dwellers, but most as farm workers, 72 percent in African rural areas, and 6 percent in urban areas. About 55 percent of the European population lives in the freehold areas, 41 percent in urban areas and the remaining 4 percent in the African rural areas. In the African rural areas of the middle veld densities average 106.5 per square mile but reach locally over 500 per square mile. The Swazi characteristically live in dispersed settlements, which contrast with the villages and compounds associated with recent agricultural, industrial, and mining developments.

Bibliography

Baker, S. J. K. "The Population Geography of East Africa," *The East African Geographical Review*, No. 1 (April, 1963), pp. 1-6.

Barbour, Kenneth M. *Population in Africa: A Geographer's Approach*. Ibadan, Ibadan University Press, 1963.

——. *The Republic of the Sudan: A Regional Geography*. London, University of London Press, 1961.

——, and R. Mansell Prothero, eds. *Essays on African Population*. London, Routledge and Kegan Paul, 1961.

Barlet, Paul. *"La Haute-Volta* (Essai de presentation géographique)," *Etudes Voltaiques*, No. 3, 1962, pp. 5-77.

Birmingham, Walter, I. Neustadt, and E. N. Omaboe. *A Study of Contemporary Ghana.* Vol. I: *The Economy of Ghana.* Vol. II: *Some Aspects of Social Structure.* London, George Allen and Unwin, 1966.

Blacker, J. G. C. "The Demography of East Africa," in E. W. Russell, ed. *The Natural Resources of East Africa.* Nairobi, D. A. Hawkins, 1962.

Boateng, E. A. "Some Geographical Aspects of the 1960 Population Census of Ghana," *Bulletin of the Ghana Geographical Association*, V, No. 2 (July, 1960), 2-8.

——. *A Geography of Ghana.* New York, Cambridge University Press, 1966.

Brookfield, H. C. "Population Distribution in Mauritius," *The Journal of Tropical Geography*, XIII (December, 1959), 1-22.

Buchanan, K. M., and J. C. Pugh. *Land and People in Nigeria.* London, University of London Press, 1955.

Church, R. J. Harrison. *West Africa: A Study of the Environment and of Man's Use of It.* 6th ed. London, Longmans, Green and Co., 1968.

——, et al. *Africa and the Islands.* London, Longmans, Green and Co., 1964.

Clarke, J. I. *Sierra Leone in Maps.* London, University of London Press, 1966.

Defos du Rau, Jean. *L'Ile de la Réunion.* Bordeaux, Institut de Géographie, 1960.

Fair, T. J. D. *The Distribution of Population in Natal.* Natal Regional Survey, No. 3. Cape Town, Oxford University Press, 1955.

——, and N. Manfred Shaffer. "Population Patterns and Policies in South Africa, 1951-1960," *Economic Geography*, XL, No. 3 (July, 1964), 261-74.

Federation of Rhodesia and Nyasaland. *Atlas of the Federation of Rhodesia and Nyasaland.* Salisbury, Federal Surveys Department, 1960-65.

Forde, Enid R. *The Population of*

Ghana: A Study of the Spatial Relationships of Its Sociocultural and Economic Characteristics. Evanston, Northwestern University Press, Studies in Geography, No. 15, 1968.

Fortems, G. *La Densité de Population dans le Bas-Fleuve et le Mayombe.* Brussels, 1960.

Gleave, M. B., and H. P. White. "The West African Middle Belt: Environmental Fact or Geographer's Fiction?" *The Geographical Review,* LIX, No. 1 (January, 1969), 123-39.

Gourou, Pierre. *Madagascar: Carte de Densité et de Localisation de la Population.* Brussels, CEMUBAC/ORSTOM, 1967.

Great Britain. *East Africa Royal Commission 1953-1955 Report.* London, H.M.S.O., Cmd. 9475, June, 1955.

——. Colonial Office. *Land and Population in East Africa.* Colonial No. 290. London, H.M.S.O., 1952.

Grove, D. *Population Patterns—Their Impact on Regional Planning.* Kumasi, Kwame Nkrumah University, 1963.

Hance, William A. *The Geography of Modern Africa.* New York, Columbia University Press, 1964.

Hilton, T. E. *Ghana Population Atlas.* Edinburgh, T. Nelson, 1960.

Hodder, B. W., and D. R. Harris, eds. *Africa in Transition: Geographical Essays.* London, Methuen and Co., 1967.

Holleman, J. F., ed. *Experiment in Swaziland.* London, Oxford University Press, 1964.

Institut Scientifique Chérifien. *Atlas du Maroc.* Rabat, 1960.

Jarrett, H. Reginald. "Population and Settlement in the Gambia," *The Geographical Review,* XXXVIII, No. 4 (October, 1948), 633-36.

Jensen, S. *Regional Economic Atlas, Mainland Tanzania.* Dar es Salaam, University College, Bureau of Resource Assessment and Land Use Planning, Research Paper No. 1, 1968.

Karmon, Yehuda. *A Geography of Settlement in Eastern Africa.* Jerusalem, Hebrew University, 1966.

Kay, George. "The Distribution of African Population in Southern Rhodesia: Some Preliminary Notes," *Rhodes-Livingstone Communication,* No. 28. Lusaka, 1964.

——. "A Population Map of the Luapula-Bangweulu Region of Northern Rhodesia with Notes on the Population," *Rhodes-Livingstone Communication,* No. 26. Lusaka, 1962.

——. *A Social Geography of Zambia.* London, University of London Press, 1967.

Kenya. *Atlas of Kenya.* Nairobi, The Survey of Kenya, 1959.

Lebon, J. H. G. "Population Distribution and Land Use in Sudan," in *The Population of the Sudan.* Report on the Sixth Annual Conference of the Philosophical Society of Sudan. Khartoum, 1958.

Mountjoy, Alan B., and Clifford Embleton. *Africa: A New Geographical Survey.* New York, Praeger, 1967.

Noin, D. *Atlas du Maroc, Notices Explicatives: Population (1960).* Rabat, Comité National de Géographie du Maroc, 1963.

Prescott, J. R. V. "Population Distribution in Southern Rhodesia," *The Geographical Review,* LII, No. 4 (October, 1962), 559-65.

Russell, E. W., ed. *The Natural Resources of East Africa.* Nairobi, D. A. Hawkins, 1962.

Sabbagh, M. Ernest. "Some Geographical Characteristics of a Plural Society: Apartheid in South Africa," *The Geographical Review,* LVIII, No. 1 (January, 1968), 1-28.

Sautter, Gilles. *De l'Atlantique au Fleuve Congo, Une Géographie du Sous-peuplement, République du Congo, Ré-*

publique Gabonaise, Vol. I, Paris, Mouton & Cie., 1966.

Smits, Lucas G. A. "The Distribution of the Population in Lesotho and Some Implications for Economic Development," *Lesotho (Basutoland Notes and Records),* No. 7, 1968, pp. 3-19.

Soja, Edward W. *The Geography of Modernization in Kenya: A Spatial Analysis of Social, Economic, and Political Change.* Syracuse, Syracuse University Press, 1968.

Steel, R. W. "Land and Population in British Tropical Africa," *Geography,* XL, No. 187 (January, 1955), 1-17.

——. "The Population of Ashanti: A Geographical Analysis," *The Geographical Journal,* CXII, Nos. 1-3 (July-September, 1948), 64-77.

——, and R. Mansell Prothero, eds. *Geographers and the Tropics: Liverpool Essays.* London, Longmans, Green and Co., 1964.

Talbot, A. M., and W. J. Talbot. *Atlas of the Union of South Africa.* Pretoria, The Government Printer, 1960.

Tanganyika. *Atlas of Tanganyika.* 3d ed. Dar es Salaam, Department of Lands and Surveys, 1956.

Thomas, I. D. "Geographical Aspects of the Tanzania Population Census 1967," *The East African Geographical Review,* No. 6 (April, 1968), pp. 1-12.

Thompson, Virginia, and Richard Adloff. *French West Africa.* London, George Allen and Unwin, 1958.

Trewartha, Glenn T., and Wilbur Zelin-

sky. "The Population Geography of Belgian Africa," *Annals of the Association of American Geographers,* XLIV, No. 2 (June, 1954), 163-93.

——. "Population Patterns in Tropical Africa," *Annals of the Association of American Geographers,* XLIV, No. 2 (June, 1954), 135-62.

Uganda. *Atlas of Uganda.* Entebbe, Department of Lands and Surveys, 1962.

United Nations. *The Population of Ruanda-Urundi.* New York, 1953.

——. *The Population of Tanganyika.* New York, 1949.

——, F.A.O. *Crop Ecologic Survey in West Africa.* Vol. II: *Atlas.* Rome, 1965.

Urquhart, Alvin. "Settlement Patterns in Southwestern Angola," *African Studies Bulletin,* IV, No. 4 (1961), 38-39.

Urvoy, Y. *Petit Atlas Ethno-Démographique du Soudan entre Sénégal et Tchad.* Mémoires de l'Institut Français d'Afrique Noire, No. 5. Paris, Librairie Larose, 1942.

Verrière, Louis. *La Population du Sénégal.* Dakar, Université de Dakar, July, 1965.

Williams, Stuart. "The Distribution of the African Population of Northern Rhodesia," *Rhodes-Livingstone Communication,* No. 24. Lusaka, 1962.

Wilson, A. S. B. "A Regional Comparison of Human and Some Meat Animal Populations in Africa," *Economic Bulletin of Ghana,* VII, No. 1 (1963), 3-16.

CHAPTER 3

POPULATION MOVEMENTS
IN AFRICA

MIGRATION is one of the demographic dynamics of particular
concern to geographers, though interest in population movements is also
strong among anthropologists and sociologists. In this chapter, the subject is introduced by a brief historical survey of migration in Africa
followed by a discussion of the methods of measuring migration, none
of which are readily applied in Africa. Next, the major and some of the
minor migratory patterns are summarized on a regional basis. An examination of the possible ways of classifying migration follows, which
prefaces a discussion of the causes and motivations of migrations. An
assessment is then made of the impact of migration on source and
destination areas and on the migrants and their families. Finally, an
effort is made to delineate the factors which are likely to affect the
future trends of migration in Africa.

Population movements on a substantial scale have been going on in Africa from time immemorial. Sometimes they have occurred in the form of conquests by warrior tribes or peoples; sometimes they have taken place by a slow and largely peaceful shift to new lands. A good deal is known about some of the early migrant waves in the north; much less is known regarding the origins and migratory routes of various ethnic groups south of the Sahara. Research by many disciplines will be required to shed light on these movements—research on the origin and dispersal of plant foods and domesticated animals, on blood-type distributions, on paleogeographical conditions, on archeological sites, and so on.

Included among the important early migrations were: those from India, which began before recorded history; the trans-Indian Ocean movements of the forebears of the present Merina, Betsileo, and other ethnic groups of Madagascar, the first probably dating from several centuries B.C. and the most recent from the fifteenth century; the succession of Arab invasions across North Africa and into the Sudan; the general southward movements of various Nilotic and Hamitic groups in eastern Africa and of Bantu groups in eastern and southern Africa, the latter forcing the earlier-arriving Khoisan and Hottentots into less productive areas; migrations across the sudan belt and from equatorial Africa into West Africa, and an opposing trend of nomads coming southward along the Atlantic periphery and then eastward along the steppes south of the Sahara.

PRE-COLONIAL MIGRATION

Even in more recent times, that is in the period before European partition of the continent, there were numerous migrations going on. Particular note must be made of movements associated with slaving, which included not only the export of slaves to the western world but large-scale transport of slaves to North Africa and the Middle East, the subjugation of one group by another within tropical Africa, and tribal migrations which were undertaken to avoid seizure by others. Estimates vary regarding the numbers involved in slaving; for the slave trade with the West figures range from 8 to 20 million with 14 million emigrant slaves being a widely used estimate. Large-scale movements from certain areas left scars of depopulation still discernible, and elsewhere a heritage of hatred and fear is yet to be fully expunged.

But many movements not associated with slaving were transpiring in all parts of the continent in the pre-colonial period; these affected larger numbers than were involved in slaving and have undoubtedly influenced the present distribution of population and of ethnic groups more significantly. As examples may be cited: the Fulani expansion in West Africa, Luba-Lunda extensions into what are now Congo and Angola; Indian migration to Zanzibar and coastal regions of East Africa under Seyyid Saïd and his successors from 1830–1890, leading to their domination of commerce in these areas as early as 1860; migration of the Ndebele from Zululand to Rhodesia in 1837; movement of Sotho groups to Botswana and the western Transvaal; migration of several Ngoni groups from the south to Congo and Tanganyika between 1800 and 1884; a considerable internal migration on Madagascar following formation of the Merina kingdom; and the almost incessant migrations of the Fang in present Cameroon and Gabon beginning in the latter half of the eighteenth century and lasting for about 150 years.[1]

In places, particularly in West Africa, movement of laborers and of cash-crop farmers had begun well before the partition of the continent. In Ghana, for example, Akwapem farmers were migrating by the middle of the nineteenth century to empty lands where they could grow oil palms and subsistence crops, palm products then ranking as the leading cash crop of the area.[2] And elsewhere "many Africans had early developed an adjustment to a form of labor migration."[3]

MIGRATION IN THE COLONIAL PERIOD

The colonial era did alter the patterns of migration significantly, although the changes must be seen, as has been suggested, as part of a much longer history of population movements. Such movements "have been a feature of Africa in the past and are one of its most important demographic features at the present day. There is no phase of African

[1] Virginia Thompson and Richard Adloff, *The Emerging States of French Equatorial Africa* (Stanford, Stanford University Press, 1960), p. 343.

[2] E. V. T. Engmann, "Population Movements in Ghana," *Bulletin of the Ghana Geographical Association*, X, No. 2 (January, 1965), 44-45.

[3] Elizabeth Colson, "Migration in Africa: Trends and Possibilities," in F. Lorimer and M. Karp, eds., *Population in Africa*, (Boston, Boston University Press, 1960), p. 60.

history which can be understood without reference to the movements of people both before and during it."[4]

Contrasts of colonial-period migration with pre-colonial migrations include the following:

1. The greater significance of economically motivated movements.

2. The much greater importance of individual migration as opposed to group movements, which accounted for the vast bulk of pre-colonial migrations. The security introduced under colonial rule was a permissive factor in this change; the nature of labor needs and of stimuli for migration were positive elements affecting the contrast. Group movements were by no means eliminated, but colonial rule, by outlawing intertribal wars and defining tribal reserves, impeded previous types of migrations of both peaceful and warlike nature. Indeed Colson argues that

what is new is the attempt to stabilize population. . . . an attempt to tie people to given areas of land as their permanent homes and thus to perpetuate the population distribution that existed when the European Powers took over. Permanent migration was discouraged by administrative regulations and by a freezing of the rules under which land was held.[5]

Once the reserves and boundaries were demarcated, village clans and tribal authorities became possessive of their interests and frequently made it difficult for outside groups to migrate as they might previously have done. There is little doubt, however, that colonialism brought much more intermingling of tribes than existed before, particularly in the present cash-cropping, mining, and urban areas. In some cases it also permitted the settlement of new areas by certain tribes, as was the case in the so-called "down hill" migration of small pagan tribes in Nigeria, Sudan, and elsewhere from the sites to which they had fled to permit defense against slaving by adjacent tribes.[6] In some areas this led to abandonment of intensive farming systems, including intricate terracing in favor of less onerous farming on lower levels and plains. Another example of the greater freedom of movement under colonial rule

[4] R. Mansell Prothero, *Migrants and Malaria*, p. 25.
[5] Colson, "Migration in Africa: Trends and Possibilities," pp. 60-1.
[6] See Michael B. Gleave, "The Changing Frontiers of Settlement in the Uplands of Northern Nigeria," *Nigerian Geographical Journal*, VIII, No. 2 (December, 1965), 127-41.

enjoyed by previously weak tribes is seen in Northern Rhodesia (now Zambia) where neighbors of the Ngoni withdrew from the marshes and thickets where they had sought refuge and built in open country right up to the borders of Ngoniland. Similarly, the movement of numerous tribal groups in East Africa into areas formerly held or threatened by the Masai, who were confined by the colonial authorities to a fraction of the lands they previously ranged, was one of the most important population dispersions occurring in eastern Africa in the colonial period.[7]

3. The great significance of European-initiated developments in stimulating population movements. De Kiewiet writes

the development of Africa . . . can be more easily understood if it is seen as the result of two movements of migration. The first is the migration of European traders, officials and settlers into Africa together with their skills, investments, equipment and governmental organization. The second is the migration of the African tribesman into the new world created by European enterprise.[8]

A somewhat different example of the effect of European rule on population movements is seen in the effort in numerous countries to agglomerate dispersed groups and to require scattered peoples to align themselves along routeways in order to permit more effective control, to allocate responsibility for the maintenance of these routes, and to assist in the introduction of education and other social services. The French followed this policy, particularly in such sparsely populated countries as Gabon and the Central African Republic. Kay writes regarding a region in the former Northern Rhodesia that the

dispersion of relatively small groups of people was characteristic of the nineteenth century, but was opposed by the governments of the British South Africa Company and the Colonial Office because of difficulties it presented to their administrative machinery. Settlements were amalgamated into larger units which were officially registered and known as "villages"; each was under an officially recognized headman.[9]

[7] A. W. Southall, "Population Movements in East Africa," in K. Michael Barbour and R. Mansell Prothero, *Essays on African Population* (London, Routledge and Kegan Paul, 1961), p. 161.

[8] C. W. de Kiewiet, *The Anatomy of African Misery* (London, Oxford University Press, 1956), p. 25.

[9] George Kay, "Chief Kalaba's Village," *Rhodes-Livingstone Papers*, No. 35 (Manchester, Manchester University Press, 1964), p. 18.

Kay further notes, however, that from about the mid '30s these regulations were relaxed as roads were extended and a process of decentralization reasserted itself. Elsewhere there was a spontaneous move by many peoples to areas with better developed route-ways.

4. A substantial increase in the intercontinental movement of persons to Africa, particularly from the metropoles and from south Asia. The largest such movements were to the Maghreb, particularly Algeria, South Africa, Rhodesia, East Africa, and some of the islands in the Indian Ocean. Indo-Pakistanis and some Chinese were brought to eastern and southern Africa and to the islands to provide a more reliable, amenable, and skilled supply of laborers, in some early cases to replace Africans who refused to work after slavery was abolished. In Tanganyika, for example, Indian immigration was accelerated in the German period and the Indian population quadrupled under the British in the period 1919–1939, at the end of which they dominated retail trade and artisan activity throughout the country, held a large number of clerical posts in government and business plus semiskilled and skilled jobs in industry, dominated the sisal and cotton industries, and had entered the professions in increasing numbers. In West Africa, Syrians, Lebanese, and West Indians performed some of the functions handled by Indians in East Africa, though their total numbers were much smaller and the share of commerce they controlled in most countries was much lower, with the exceptions of Sierra Leone and Liberia. Those Europeans going to Africa presumably on a permanent basis went predominantly to urban centers. Various incentives were provided to encourage such movement in a number of countries, these incentives often continuing well into the period when the long-term presence of European settlers became subject to serious question. While study of the impact of non-African migration is not attempted in this chapter, it is obvious that political, social, and economic repercussions were complex and powerful and out of all proportion to the numbers involved.

The Evolution from Forced to Voluntary Migration. In the early years of colonial rule a degree of compulsion was often required to secure an adequate supply of labor. Many governments resorted to forced labor under one or another term ("compulsory service," the "corvée," "prestations") in order to get Africans to the areas where they were needed. In some cases, a certain number of days of work each month or year were required of each adult male (and, sometimes, female) and work on roads and other public projects was occasionally required without compensation. Such systems of obligatory labor lasted until after World War II in a number of countries.

Forced labor led on occasion to gross malpractices both on the part of administrators and of African chiefs who were given responsibility for recruitment. In constructing the Congo-Ocean Railway in Congo (Brazzaville) in the period 1923–1934, for example, a total of 120,000 workers were conscripted and an estimated 15–18,000 perished, in part due to improper care and nourishment, in part because workers recruited in the north were subjected to diseases for which they had no immunity. A large-scale flight to adjacent countries resulted from local residents seeking to escape recruitment for work on the line.[10]

The British used forced labor, called the "Kasanvu System," in trying to get manpower for public works in Uganda; the system worked poorly, however, and was abolished in 1922. Belgian, Liberian, and Portuguese use of forced labor led to repeated objections and several cases were brought to the International Court of Justice to seek redress for the conscripted workers.

More common was the imposition of a head or hut tax which had to be paid in cash. Such taxes provided an enormous impetus to migration, though some authors have stressed their role in the rationalization of compulsion in labor recruitment. In any case, the imposition of taxes tended to institutionalize migration and to spread the source areas over a whole country rather than to the areas immediately adjacent to the places requiring labor.

In cases where there was a substantial labor shortage government or private interests often organized recruitment, sometimes on a rather elaborate basis. Incentives included free or assisted travel either by public transport or by truck, bus, boat, and plane provided by the agency itself, construction of camps and shelters along the major routes, provision of food and minimal clothing, preselection examinations, and health provisions. Most common has been the provision of free transport, as, for example, for the holders of "navétane cards" proceeding to work in Senegal, for recruits of SIAMO, an association of employers in the Ivory Coast which worked in liaison with the Labor Inspectorate and the Government Employment Office, and for laborers recruited by

[10] Thompson and Adloff, *The Emerging States of French Equatorial Africa*, pp. 141-42, 256.

Population Movements in Africa

two British and two semiofficial Belgian companies which handled early recruitment for the UMHK mines in Upper Katanga, the Ulere or Free Migrant Labour Service of Rhodesia, the Witwatersrand Native Labour Association (WNLA or Wenela), established to provide the very large number of laborers needed by the gold mines of the Rand and by the supporting coal mines, and the South West Africa Native Labour Association (SWANLA), the counterpart of Wenela for the mines of South West Africa. Wenela recruited far afield, not only in South Africa but in the former High Commission Territories, Central Africa, and Mozambique. It developed and still maintains a net of recruiting stations from which it provides transport by land, water, and air to the Rand. It has, indeed, sometimes built its own roads to facilitate recruitment.

In some early cases, particularly where transport had not been developed, workers were assembled in groups and walked to the destination area along specially marked routes, using camps at convenient stages where they were provided with food for the following day. This system was used for Northern Rhodesian workers going to Katanga.[11] And in the 1920s, when active recruitment of Rwandans was begun for indigenous farms and for expatriate-owned plantations in the fertile crescent of Uganda, camps were provided for migrant workers along the main roads.

In some countries the method of recruitment was confined to a system of chiefly payments whereby the local chief was paid a stipulated sum in return for his providing a given number of men to the employer. In Liberia, for example,

with the willing assistance of the Liberian government, Firestone has been using labor recruitment since 1926. . . . Paramount chiefs are assigned quotas and are paid for sending laborers to the plantations. The quotas established in the late 1920's have not been revised appreciably since that time.[12]

LAMCO, the large iron-mining company with mines at Mt. Nimba and processing and port facilities at Buchanan, used this system at first but by 1962 had reduced the number so recruited to only 10 percent of the total employed.

[11] Bruce Fetter, "Elisabethville," *African Urban Notes*, III, No. 2 (August, 1968), 23.
[12] Robert W. Clower *et al.*, *Growth without Development: An Economic Survey of Liberia* (Evanston, Northwestern University Press, 1966), p. 158.

When recruitment involved movement from one country to another efforts were soon made to control it both by local regulations and international agreements. The motives for such regulations and agreements varied from concern to protect the health and wellbeing of the migrant both enroute and on the job, through concern to maintain what was considered a satisfactory demographic position in the source area, to the desires simply to assure a continuing supply of workers and to protect the source country's earnings from migrant workers. In a number of cases consuls or other officials of the source country are stationed in the destination area to ensure the proper working of these regulations.

As an example of regulations respecting migrant workers the Malawian laws are designed primarily to reduce the undesirable effects on the source communities. They require that the worker, unless he is accompanied by his family, must return after two years, that monthly deductions be made from his wages, a part of which is remitted by the employer to a specified dependent in his home district and the remainder of which is paid to him upon his return, and that he secure an identity certificate before leaving, the certificate only being issued if the individual has paid his taxes and, in many cases, if he has planted gardens and made provision for his dependents.

One of the more elaborate of international agreements is the Mozambique Convention, the first version of which dates from 1897 and the latest from 1938. The Convention provides for the recruitment by South Africa of a minimum of 65,000 and a maximum of 100,000 Mozambique Africans from that part of the country south of the 22°S parallel in return for routing via Lourenço Marques ("L. M.") of a guaranteed minimum of 47.5 percent of overseas traffic for a defined area in the Transvaal. Wenela has handled the recruitment from Mozambique since 1900. Until recent years there was no difficulty in meeting the shipping quota via "L. M." because that port is closer to the Rand than Durban and large tonnages of petroleum products moved upline while mineral exports were railed downline. Construction of two refineries at Durban and of a pipeline from Durban to Johannesburg has greatly reduced the share of imports using "L. M.," and South Africa has had to pay compensation to Mozambique amounting to $910,000 in 1965–1966 and an estimated $2,285,000 in 1968–1969. Not all migrants from Mozambique follow the procedures stipulated by the

Provincial Government, there being a substantial number who move clandestinely to seek jobs in the Republic, particularly on European farms.

Some large employers, preferring not to rely on the somewhat precarious recruiting systems or to be so heavily dependent on migrant workers whose terms precluded training for more skilled work, chose to attract workers on a permanent basis. UMHK, for example, plagued by a serious labor shortage, began about 1925 to encourage workers to bring their wives and children to the Katanga mines and offered reliable employees three-year instead of six- to twelve-month contracts. This resulted in a sharp reduction of migrants from Northern Rhodesia, who were required to return before contracting additional terms, in the rise of the Lomami region as the main source of recruits, and to an increased reliance on settled and permanent employees.

As the years passed in the colonial period the element of compulsion in labor service was generally removed, in part because it was no longer necessary, since African labor more and more presented itself freely and in adequate numbers to obviate either requisitioning or elaborate recruitment methods. As Skinner put it, "a complex of factors permitted the abolition of forced labour in most African areas, . . . [and] many African workers, who formerly had to be forced to do extratribal labour, began to migrate voluntarily to centers of European employment."[13]

The French in Senegal, for example, who had provided free rail transport for *navétanes* since 1937 and accommodation centers from 1943, cut down their aid in 1954 and today most labor used in the peanut harvest comes from within Senegal. The example of LAMCO's reduced use of chiefly payments has already been cited. In Uganda the problem became one of a surplus of migrants and in many other areas no special provisions were required to obtain an adequate number of workers.

Berg places the crossover point from coercive to voluntary migration at about 1930, though it varied considerably from one region to another, occurring relatively early in western and southern Africa and Angola.[14]

[13] Elliott P. Skinner, "Labour Migration and Its Relationship to Socio-Cultural Change in Mossi Society," *Africa*, XXX, No. 4 (October, 1960), 377.
[14] Elliot J. Berg, "Backward-Sloping Labor Supply Functions in Dual Economies," *The Quarterly Journal of Economics*, LXXV, No. 3 (August, 1961), 478.

Navétanes or migrant workers participating in collecting peanuts in Senegal. The use of such workers has tended to decline in recent years.

Among the factors which led to the changed attitudes toward migration the most important was the increasing desire of Africans to acquire the material and cultural accoutrements of modern life. Cultural motivations, such as initiation to adult life or the rîte de passage, now appear to be of much less significance than had been thought (or rationalized).

POST-COLONIAL CHANGES IN MIGRATION

The end of colonialism brought several significant changes in migration, though the basic patterns have not been fundamentally altered. First was the exodus of European settlers, on a very large scale in the case of Algeria and on a lesser scale from Morocco, Tunisia, Senegal, Congo, Kenya, and Tanzania. Many countries, however, have as large or larger expatriate populations than before independence, but these are now more predominantly temporary residents. Second, there has been a sharp increase in the number of refugees moving across international boundaries (see p. 183). Third, there has been an increasing number of restrictions placed on foreign Africans by independent governments, and expulsion of foreigners in a number of cases. Wallerstein noted in this respect that

free movement of Africans fits in with the ideology of Pan-Africanism. But government struggling with severe economic problems tend to be restrictionist in their reactions, and, under population pressure, to be even more so.[15]

Indicative of the trend toward restricting foreign African migrants is a recent statement of Ghanaian population policy which calls for controlling the entry and activities of noncitizens to reduce loss of currency and "to insure that services that can adequately be performed by Ghanaians are reserved exclusively for Ghanaians."[16]

Fourth, there have been some migrations resulting from the unleashing of tribal tensions which had been suppressed under colonial rule. Fifth, there has been an increase in the international migration of elites both to and from Africa. And, lastly, there has been an accelerated migration to urban centers, particularly to the primate cities of many countries.

South Africa, which has not, of course, changed from colonial to independent status as have most of the countries on the continent, has seen the introduction of increasingly powerful and restrictive legislation designed to promote separate development which would reduce migration from African reserves; nonetheless there has been a continuing heavy migration to urban centers, the economic forces thus appearing to outweigh the political considerations of the Nationalist Government. Migration of foreign Africans to the Republic has continued unabated though more regulated by legal prescriptions, except from Zambia which forbade movement of its citizens to take jobs in South Africa. The independent governments of Lesotho, Botswana, Swaziland, and Malawi have felt that the heavy dependence of their economies on South Africa precluded any measures to reduce migration of laborers to South Africa, and the Zambian government has been subject to protestations from Barotse Province that edicts forbidding movement of workers to South

15 Immanuel Wallerstein, "Migration in West Africa: The Political Perspective," in Hilda Kuper, ed., *Urbanization and Migration in West Africa* (Berkeley and Los Angeles, University of California Press, 1965), p. 159.
16 Republic of Ghana, *Population Planning for National Progress and Prosperity: Ghana Population Policy*, p. 23.

Africa were depriving the inhabitants of necessary income and that alternate opportunities did not exist in Zambia. The labor system in South Africa does, in fact, tend to perpetuate reciprocal migration in that separate development makes it increasingly difficult to move to European areas, and especially to urban centers, on a permanent basis.

The Measurement of Migration

While many of the major migratory patterns of Africa have been recognized, there are great difficulties in securing adequate data on most movements. A continuous register of migrants at both source and destination areas would be required to provide the desired information, but this is not to be expected. One can derive partial estimates from employment records in some major areas such as the Copperbelt or the gold-mining regions of the Transvaal and the Orange Free State, from records at border posts, from counts taken at strategic bottlenecks such as ferries or bridges, or from records of recruiting agencies. But these sources provide very incomplete and often only discontinuous data. Information on intercontinental movements are usually reasonably reliable, but do not necessarily distinguish between migrants and persons who might more accurately be called visitors.

One major difficulty is distinguishing between one-way and reciprocal migration. More recent censuses have frequently included questions which permit measuring net migration of an enumeration area at the time of the census by recording the place of birth of the informant. The following E.C.A. example, taking data from the 1948 Ghana census, illustrates the use of such data:

(1) Population present in Ashanti territory	817,782
(2) Population born and present there	675,841
(3) = (1) − (2) In-migrants from other territories and foreign countries	141,941
(4) Population born in Ashanti territory	713,231
(5) = (4) − (2) Out-migrants from Ashanti territory	37,390
(6) = (3) − (5) Net migration	+104,551[17]

[17] Economic Commission for Africa, Seminar on Population Problems in Africa, *Population Distribution, Internal Migration and Urbanization in Africa* (E/CN.14/ASPP/L.3, 16 October 1962), p. 32.

Data such as the above from only one census obviously do not tell anything regarding the period of time during which the migratory movements took place. Data from successive censuses can provide information on trends and permit at least rough estimates of annual migration rates.

Net migration can be calculated very approximately by utilizing the total counts of the population of an area in two censuses and then comparing the rate of growth of the area with the rate of natural growth of the whole country. To be valid at all this method requires that: there be no significant international migration or that the population can be divided between local and foreign-born, that the rate of natural increase be roughly the same for the entire country, and that the two censuses are accurate.

The use of the balancing equation—net migration equals population at second census minus population at the earlier census minus births plus deaths—is very rarely possible in Africa, primarily because of the absence of vital registers but also because of the unequal validity of successive censuses. Furthermore, the formula, by giving only net migration, may reveal very little regarding the actual levels of in- and out-migration. Use of the "cohort" method is also very difficult in Africa. Under this method, each age-group enumerated in a given area at the first census is aged, by applying probabilities of survival to the date of the second census; the results, when compared with the population actually enumerated, give an estimate of the volume of net migration.

It is likely that future censuses will include questions which will be more helpful in estimating migration, such as "how long have you lived here?" and "where were you living x years ago?" Until such questions are asked, the picture must remain cloudy, and information on the size, length of stay, and source and destination areas will be only approximate. But the one-time character of some migrations, the ever-changing dimensions of others, the element of personal decision involved, and many other factors mean that we will never be able to keep up very accurately with many movements.

More information has been provided on migrations by special surveys than by regular censuses. The one-time character of most of such surveys reduces their value, but they nonetheless provide a wealth of qualitative and some quantitative data. The goals and methodologies of such surveys vary greatly, but two significant trends have been the effort to quantify the reasons for undertaking to migrate and the correlation of migration with numerous socioeconomic variables, both of which can be of considerable value in understanding migrant flows. Reference is made to a number of migration studies in the following sections and others are included in the bibliography. An excellent example of a study of correlatives is the recent report by Caldwell of sample surveys conducted in Ghana from 1962 to 1964, which revealed that the volume of migration tended to be inversely related to the distance from the source area to the nearest large center and directly correlated to: the size of the source village; the economic condition of the households, with the wealthier ones having a greater representation among migrants; the presence of relatives in the destination areas, "chain migration" being an important mechanism in rural-urban migration in Ghana; the amount of education, literacy, and the ability to speak English; and family size and birth rank, the younger children in a family being more likely to migrate. The flows were also predominantly of young adults, with the pattern of male dominance changing insofar as the propensity for females to migrate was rising faster than that of the males. It was also apparent that migrant streams were being increasingly diverted from rural destinations to towns.[18]

Migratory Movements

In this section an attempt is made to outline some of the major migratory patterns in Africa, with particular reference to the main reciprocal links. Map 22 gives these movements in a very approximate way both as regards relative size and location of source and destination areas. Certain types of migration such as nomadism, transhumance, and refugee flights are not depicted, though these are briefly discussed in a later section. For each major region, discussion is confined to selected and representative countries or to an outline of the major regional patterns.

NORTHERN AFRICA

Morocco. The major migratory movement in Morocco in recent decades has been a massive rural-urban shift which began about 1912 with the establishment of the protectorate, although large urban centers

[18] J. C. Caldwell, "Determinants of Rural-Urban Migration in Ghana," *Population Studies: A Journal of Demography*, XXII, No. 3 (November, 1968), 361-77.

Seminomads near Ksar es Souk in Saharan Morocco. The less favored parts of Morocco have seen large-scale out-migration particularly to the modern cities of the country.

had long existed in Morocco. The coastal zone from Safi to Kenitra received the largest share of the influx; indeed in the 1936–1952 period Casablanca alone received three-fifths of the total rural exodus or one-third of the total population increase of the country. There has also been a very substantial rural-rural shift to the west and northward from the drier and less advantaged portions of the country.

The years following independence witnessed a large-scale exodus of Europeans and Moroccan Jews, the non-Moroccan population declining from 535,000 in 1952 to 396,000 in 1960 and to an estimated 200,000 in 1965, while the number of Moroccan Jews decreased from 215,000 in 1952 to an estimated 100,000 in 1965. The 1960 non-Moroccan population included about 300,000 Europeans and 96,000 Algerian refugees, most of whom have presumably since returned to that country. Recent years have also seen a considerable migration of Moroccan laborers to France; some 20,000 were reported leaving in 1965 with the trend steadily upward.

Seasonal migrations affect practically all labor-force categories in the Kingdom: farmers moving between harvests to construction, handicraft, and other types of employment; some urban-rural movement to participate in harvesting on large farms and plantations, and migration to food-processing plants which operate on a largely seasonal basis.

Algeria. This country has experienced many of the migratory trends noted for Morocco, often on a substantially larger scale. Emigration of Muslim Algerians to Europe (almost entirely to France) started during World War I to replace men who were mobilized; the movement slowed in the interwar years, especially during the depression, but became heavy again after 1945, reflecting a rising rate of unemployment in Algeria. The densely populated and resource-poor Kabylia accounted for about half of the migrants in the early postwar years. The number of Algerians in France rose from about 50,000 in 1946[19] to 160,000 in 1948, 300,000 in 1955, and about 600,000 in 1967. Most of the migrants are males, there being very few Algerian families permanently installed in France. The total of 600,000 for 1967 compared to a total of about a million heads of family employed in Algeria, revealing in a crude way the very great significance of foreign employment to the Algerian economy.

The 1962 Evian agreements called for unrestricted migration from Algeria to France but restrictions were placed on such movements in 1964 and again in 1967, when the numbers allowed to migrate for work were limited to 250 per week, though the limits set in these years were regularly exceeded. In October 1968 a new agreement stipulated that 35,000 Algerian workers would be admitted to France in the following three years.

The years of the Algerian war for independence saw large-scale population movements and dislocations within the country and the flight of some 250–300 thousand refugees to Morocco and Tunisia. Whole villages migrated from the mountains to seek protection near the military camps. The French also set up a large number of regroupment centers, partly to give them greater control over the population; these involved as many as 1.5 million Algerians.

One of the largest movements of postwar years in all of Africa was the exodus of Europeans from Algeria, at the end of the Franco-Algerian war. The European population of Algeria exceeded one million

[19] James R. McDonald, "Labor Migration in France, 1946-1956," *Annals of the Association of American Geographers*, LIX, No. 1 (March, 1969), 125.

in 1960–1961; no fewer than 650,000 left in 1962, and in 1967 the number of French residing in Algeria was only 80,000, of whom half were in Algiers and a quarter were in Oran. Following the mass emigration of Europeans in 1962, Muslims flooded into the cities and towns to replace them, sometimes occupying the vacated dwellings at several times the previous density.

Sudan. In the Sudan there has been a substantial attraction to the Three Towns and to the various large irrigation areas, including the Gezira-Managil, Khashm el Girba, and pump schemes along the Blue and White Niles. The Khashm el Girba area was used to accommodate people evacuated from the Wadi Halfa district, which was to be flooded by Lake Nasser; Egypt paid $42 million compensation to Sudan for this flooding, but this did not cover the full cost of the dam, village construction, and the sugar factory at Khashm el Girba, which totaled about $84 million.

An important if somewhat unusual type of migrant in the Sudan is the "Westerner," described by McLoughlin as

typically a Mecca-bound pilgrim whose original home is either in the extreme west of the Republic . . . or West Africa. One person in six in Sudan is a Westerner, both because the trip normally takes five to seven years and the pilgrim must earn income *en route* and because Westerners have formed farming colonies across the Qoz Sands and Central Clay Plains and are in every major northern city and town. The Westerner is over half of the nation's *wage* labor force—without him economic development, particularly in Gezira, would collapse. He is, in economic as well as social fact, the slaves' descendant.[20]

According to the 1955–1956 census, Westerners, largely from Darfur, comprised 43 percent of the Three Towns population and this represented 6.3 percent of the total Westerner population in Sudan. Assimilation of the Westerners, who tend to retain their own customs and languages, is a problem of some dimensions in Sudan.

Several examples of "down-hill" movements have occurred in Sudan where Nuba and Fur tribes from the Nuba, Marra, and lesser massifs have migrated from earlier refuge sites to more accessible areas and to the surrounding plains.

[20] Peter F. M. McLoughlin, "The Sudan's Three Towns: A Demographic and Economic Profile of an African Urban Complex," *Economic Development and Cultural Change*, XII, No. 1 (October, 1963), footnote 10, 76.

Several major migratory patterns may be delineated in West Africa (see Map 22). Most important are the movements from more remote areas to economically developed regions closer to the coast. These include:

1. Permanent and seasonal migrations to Dakar and to the peanut-growing areas of Senegal and Gambia. Berg estimated that about 75,000 *navétanes,* "strange farmers," and laborers move to these areas annually.[21] The *navétane* is a migrant who contracts to cultivate a part of his employer's lands under various arrangements whereby he gets some time and area for his own cultivation and works the rest of the time for the proprietor. The *navétanat* system in Senegal reached a peak in 1938 when 70,000 migrants were involved; in 1949, when there were 57,000 *navétanes,* about 39 percent originated in Mali, 33 percent in Guinea, and 26 percent in Senegal. The system has since continued to decline and may, according to Lombard, be on its way to disappearing.[22]

2. Migration, largely within the two countries, to the cities, mines, and cash-cropping areas of Sierra Leone and Liberia. In Liberia about four-fifths of the labor force of about 100,000 are thought to be migrants. In Sierra Leone one of the larger movements was associated with the search for diamonds, which assumed the dimensions of a veritable rush in the 1950s when as many as 75,000 Africans were engaged in pot-holing, or surface digging, of diamonds. In the winter of 1956–1957 the colonial authorities evicted a number of foreign diggers variously estimated at 16,000–20,000. Liberia halted an incipient diamond rush about the same time, partly to protect the labor supply for the Firestone Plantations and the Bomi Hills iron mines which then provided the bulk of government revenues. Despite a decline in output of diamonds in Sierra Leone the diamond area had an excess of 46,309 males at the time of the 1963 census; this took place in April, the height of the mining season, when five of the nineteen chiefdoms involved had sex ratios less than 70 females/100 males, while several towns in the

[21] Elliot J. Berg, "The Economics of the Migrant Labor System," in Kuper, ed., *Urbanization and Migration in West Africa,* p. 161.

[22] République du Sénégal, Ministère de l'Éducation Nationale, *Études Sénégalaises,* No. 9, *Connaissance du Sénégal,* Fasc. 5, *Géographie Humaine,* by J. Lombard (Saint-Louis, 1963), pp. 28-32.

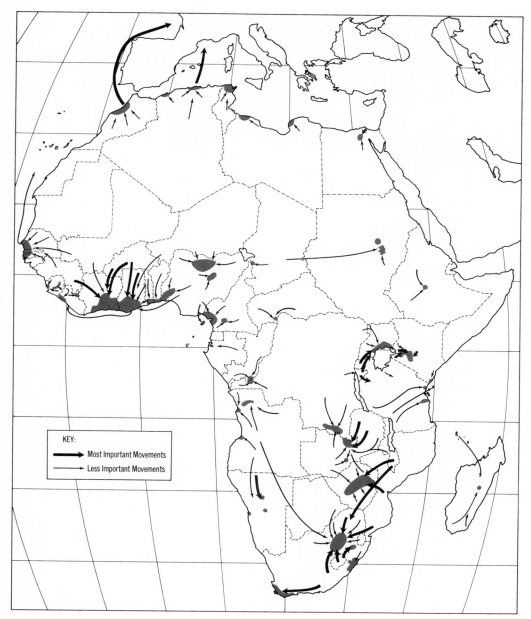

MAP 22. Population movements in Africa.

Population Movements in Africa

Potholers digging for diamonds in Sierra Leone. As many as 75,000 Africans, mostly migrants from other regions, have been engaged in such surface digging.

areas had ratios below 50 females/100 males, indicating a continuing high involvement of migrant workers.

3. Migration, both from internal and extra-national sources, to southern Ivory Coast and Ghana, two of the most important destination areas on the continent. The largest number of migrants go to work on indigenous farms (mainly coffee and cocoa farms in the Ivory Coast and cocoa farms in Ghana), but also to European-owned plantations, especially in Ivory Coast, to the mining areas of southwest Ghana, and to the major urban centers. Some migrants work on a share-cropping basis. Estimates of the numbers which have been involved in these movements vary considerably and it is very difficult to distinguish between temporary and semi-permanent or permanent migrants. For the Ivory coast it was estimated in 1965 that 17 percent of the total population of about 3.84 million had been born outside the country; this included about 220,000 from Mali, 200,000 from Upper Volta, and 150,000 from other African countries. In 1966 it was estimated that 200,000 Guineans were in the Ivory Coast. The exceptionally heavy dependence of the Ivory Coast on migrant workers is indicated by the 1965 estimates that only 46 percent of unskilled workers in the private industrial sector were

Population Movements in Africa

natives of that country and that commerce employed foreign workers almost exclusively.

According to the 1960 Ghana census 12.3 percent of the 6,726,-800 recorded were persons of foreign origin, of whom about 67.6 percent had been born in other countries. The largest number by origin were from Togo (280,670), Upper Volta (194,570), and Nigeria (190,780), which together accounted for about 80.5 percent of the total residents of foreign origin. A substantial number of women and children accompany migrants in Ghana, as is suggested by the ratio of 68 females/100 males among the foreigners, and of 59 females/100 males among the foreign born. There is also a substantial migration from northern Ghana to the south; Hilton notes that 115,670 members of the eleven main Upper Region tribes were counted outside northern Ghana in the 1960 census, a 3.6 fold increase over the number recorded in the 1948 census.[23] That movement within Ghana was both substantial and increasing is suggested by the estimates that 42 percent of the total 1960 population were enumerated outside their place of birth as compared to 26 percent in 1948 and that 12 percent were counted in a different region. However, in late 1969 a decree was issued requiring the repatriation of as many as one million foreign Africans and Ghana may be expected to limit immigration rigidly in the years ahead, primarily as a response to increasing unemployment of its own citizens.

4. Movements on a smaller scale to southern Togo, southern Dahomey, and to Lagos and Yorubaland in Nigeria. The total number of Togolese migrants was estimated at about 100,000 in 1960, of whom 30,000 were in Ghana; the major internal movement was from north to south, particularly of Moba from Dapango cercle and of Kabré, who also compose the bulk of migrants to Ghana. In Dahomey about 20 percent of the population was recorded outside their place of birth in 1961; the percentage of males recorded as absent at the time of the survey ranged from only 10 percent in the southwest to 16.7 percent in the center to from 33.3 to 50 percent in some cantons of the northwest.

Berg has estimated that about a million people are involved in West African labor migration, of whom about half go to Ghana and the Ivory Coast. These make up a very substantial portion of the labor

[23] T. E. Hilton, "Population growth and distribution in the Upper Region of Ghana," in Caldwell and Okonjo, eds., *The Population of Tropical Africa*, p. 288.

force in the money sector, including almost all those in paid agricultural employment and possibly half of those in non-agricultural employment.[24] Destination areas of some importance outside the southern and coastal zones already noted include the mines of the Jos Plateau in Nigeria, the peanut-producing regions of Mali and north-central Nigeria, and, in nearby parts of middle Africa, the southern Cameroons and Fernando Po, where Nigerians provide a high percentage of the labor force.

Other West African migrations of varying significance include: the movement of fishermen along various coastal stretches; migration of Kru stevedores on vessels planning to load lumber and logs, particularly in Gabon and Congo (Brazzaville); the migration of African elite, usually from southern areas, to northern regions within their own countries or to the landlocked nations; and migration to Europe. In 1966 there were an estimated 50,000 black African workers in France, about 70 percent of whom were believed to be from the region of the Senegal Valley shared by Senegal, Mauritania, and Mali.

MIDDLE AFRICA

Former French Areas. The relatively low level of economic achievement in much of former French territories of Middle Africa is reflected in a generally less important development of migration. The major cities—Brazzaville, Douala, Yaoundé, Bangui, Ft. Lamy, and Pointe Noire—have attracted migrants in increasing numbers, despite high rates of unemployment. There is also a steady migration to the forest and plantation enterprises of West Cameroon; the Bamiléké migrate from their densely populated part of the highlands in East Cameroon to adjacent lowlands and to Douala and Yaoundé; and Gabonese move to the forestry operations and to the manganese mine at Franceville.

The Central African Republic, when it was a province of French Equatorial Africa, was both an importer and an exporter of labor. Workers were recruited for employment in Gabon and Congo (Brazzaville) in the 1940s while the diamond-mining companies in the C.A.R. had considerable difficulty in securing labor and were forced to recruit

[24] Berg, "The Economics of the Migrant Labor System," in Kuper, ed., *Urbanization and Migration in West Africa*, pp. 161-62.

Population Movements in Africa

in Chad and Cameroon. Concerned that the shortage of labor would inhibit development and that returning emigrants would refuse to accept the lower wages prevailing at home, the C.A.R. government instituted several regulations in 1948 designed to control the flow, and banned completely recruitment for work outside the province in four regions. From 1950 on, however, the concern dissipated, particularly after unemployment in Bangui became acute.[25]

Congo (Kinshasa). Migrations in Congo (Kinshasa) before independence involved movements to: the mining areas, not only the major copper mines of Katanga but also the more scattered tin-tantalo-columbite-beryl zone of northern Katanga and Kivu, the northeastern gold area, and the diamond mines of Kasai; the plantations scattered about the Congo Basin and the estates concentrated in the eastern highlands; and to the several large cities, all of which were European creations.

The 1958 census revealed the summation of movements near the end of the colonial period; 22.7 percent of the total population were listed as *"extra-coutoumier"* or living outside their native areas. Of the 1.1 million Africans under "workman's contracts," 40.2 percent were employed outside their native territories, and 5.5 percent originated in foreign countries. Of the 443,000 employed outside their native territories, however, only about a fifth were employed outside their native province.

As noted earlier, Belgian policy came to focus upon settling workers at the employment sites rather than depending on migratory workers. UMHK, for example, recruited far afield in the early years, principally in the Rhodesias. Its labor position improved with completion of the rail line to Bukama in 1918 and again after 1926 with the more effective establishment of a stabilization program. The company employed 14,000 workers in 1925, when only 2,500 women and 770 children were resident at their installations; by 1955 the numbers had increased to 21,000 workers, 17,500 women, and 40,000 children, and in 1963 the totals were 20,000 workers, 16,000 women, and 60,000 children. The annual turnover rate among UMHK workers declined from 165 percent in 1921 to 78 percent in 1931, 12 percent in 1939,

[25] Thompson and Adloff, *The Emerging States of French Equatorial Africa*, pp. 422-23.

and a low of 7 percent in 1960. Post-independence political difficulties resulted in the rate rising to 27 percent in 1963.[26] The ratio of adult females to adult males among the total *extra-coutoumier* population in 1958 was 77.4 : 100, not as high as in many West African situations, but considerably higher than in many employment centers in eastern and southern Africa.

The disruptions and internal conflicts occurring in Congo after independence led to some massive internal population movements, whose extent can only be very roughly estimated. Tens of thousands of Baluba were forced to migrate while as many as a half-million refugees flooded into Kinshasa mainly to seek security. In addition to the very substantial refugee movements within Congo after independence, there were also flights from the country and, because of difficulties in adjacent countries, several large-scale movements into the country, particularly from Angola, Sudan, and Burundi.

Angola. Numbers of Angolan Africans have migrated for some years to adjacent countries or to South Africa. In addition to cross-boundary nomadic movements from Cuando, Cubango, and Huíla districts, labor migrants go to South-West Africa (c. 12,500 in 1963), South Africa (c. 17,000), and Zambia (c. 5,000), the majority working on the mines. The Bakongo from Uíge and Zaire Districts formerly migrated in large numbers to Leopoldville and to the Kasai and Katanga Provinces of Congo, but the Belgians precluded such movements in the late 1950s. From 1961–1967, however, very large numbers of Bakongo and other Angolans fled to Congo as a result of the revolt in Angola. Within the country, labor migration has been heaviest from the poorer areas of Malange, Huíla, Bié, and Huambo, usually on short-term contracts in the under-manned coffee-producing areas and on the mines. About 82,000 contract workers were employed in 1964, or 22 percent of the total wage labor force in the territory. Migration has, as elsewhere, led to considerable unemployment in the major cities.

EASTERN AFRICA

East Africa. Labor migration again assumes considerable importance in East Africa. The main destination areas are or have been the fertile crescent of Uganda, the scheduled areas of Kenya, the sisal producing zones of Tanganyika, the clove orchards of Zanzibar and Pemba,

[26] Union Minière du Haut Katanga, 1964 monograph, p. 54.

and the major urban centers, particularly Nairobi, Mombasa, Dar es Salaam, Kampala, and Jinja.

Southern Uganda has been one of the most important receiving areas on the continent. While there are a variety of employment opportunities in this region, including mining near Tororo, industries there and at Jinja and Kampala, and a number of large non-African plantations, the largest numbers of migrants are absorbed on indigenous African farms whose main cash crops are coffee and cotton. The receptiveness of migrants by local farmers is related not only to the needs for labor at peak periods but to the desire to avoid physical labor in part for reasons of prestige. A considerable number of migrants have been given access to land which is farmed on a rental or share-cropping basis, while some have been permanently settled in the area.

According to the censuses of the East African countries held in the period 1957-1962 Uganda had 56.7 percent of the total of 206,000 residents of East Africa recorded as present in other East African countries and accounted for only 13.5 percent of those enumerated in other countries. The 1959 Uganda census also recorded the presence of 379,-000 Rwandans and 139,000 Rundi. In addition, there is a substantial reciprocal migration within Uganda, particularly from the remote and densely populated West Nile and Kigezi districts. Uganda has also become a major receiver of refugees in recent years and has taken steps with aid from international agencies to allot areas where some of these groups could provide for themselves.

Kenya has had the largest number of emigrants to other East African countries, about 47.1 percent of the total in the years noted above, but the numbers have been small in relation to those engaged in internal migration. The very high population densities found on much of the high-quality lands in Kenya and the increasing problems of unemployment, underemployment, and land shortage help to explain why the most important economic node in East Africa—the Highland-Rift Valley zone of Kenya—has little importance in international migration and why Kenya is the largest exporter of labor among East African countries. Application of the Swynnerton Plan in Kikuyuland and other African areas has also contributed to an increase in landless people who are likely to migrate in search of work.

The extent of internal migration which had occurred in Kenya was revealed in the 1962 census which enumerated 604,700 persons

The densely populated and overused Kigezi District of southwest Uganda
is one of the important source areas of migrants in that country.

Pyrethrum pickers being paid on a European farm in Kenya. The former
White Highlands, now the Scheduled Area, has been the main receiving
area for migrants in that country.

Population Movements in Africa

or 7.0 percent of the total population outside the province in which they were born. Of this number 44 percent had moved to the Rift Valley Province and 26 percent to the Nairobi district, 83.6 percent of whose African population at the time of the census had been born elsewhere. The former White Highlands, now the Scheduled Area, was the main receiving area; with only about 100,000 in 1900, this area had about 1.2 million in 1962, of whom seven-eighths were Africans. The Central Province, which contained the densely settled Kikuyu country on the eastern flanks of the Aberdare Mountains accounted for the largest percentage (36) of out-migrants. As Soja has noted, the Kikuyu have dominated the African element in the European subsystem of Kenya, the numbers living outside their home districts having increased from 330,000 in 1948 to 715,000 in 1962, or 44 percent of the entire Kikuyu population at the latter date.[27] Nyanza Province, which also experiences very high densities and which has participated only marginally in cash cropping, ranked very close to Central Province as a source area, with 35 percent of those recorded outside their province of birth. The Southern Province, with 17 percent of the total, ranked third among source areas.

A substantial share of the population movements in Kenya was of a semipermanent nature. European plantations and estates provided housing for families, some of whom were given rights to farm small plots and to keep livestock. With the subdivision of European farms and their transfer to African smallholders there will probably be a further trend away from circular migration and toward permanency of occupation. The absorptive capacity of the scheduled areas is limited, however, and increasing pressure may be expected from persons migrating to the urban centers of the country.

Migration has not been as significant in Tanganyika as in the other East African countries, though substantial numbers characteristically move to the economic nodes of Tanga Province and along the central rail from Dar to Morogoro, frequently from the remote and less developed portions of the Southern Highlands. The recent severe decline in sisal prices may have a serious impact on this long-standing movement. Other movements worthy of note include: (1) the annual migration of workers for the clove harvest on Zanzibar and Pemba. This

[27] Soja, *The Geography of Modernization in Kenya: A Spatial Analysis of Social, Economic, and Political Change*, p. 54.

involved a large number of people up to 1958, when there were 41,327 foreign-born on the islands, 57 percent from Tanganyika. The trend in recent years has been toward a greater dependence on local workers. The size of the flow characteristically varied widely from year to year, depending on the size of the harvest and the current wages, which fluctuated very sharply; (2) the receipt of migrants from Burundi and Rwanda, the 1957 census having recorded the presence of 122,000 Rundi and 35,000 Rwandans. With a large-scale migration from these countries to Uganda already noted plus additional movement to Congo it will be obvious that these two small and very densely populated countries comprise a major source area of migrants in eastern Africa; and (3) the migration of some thousands of Tanzanians to Zambia, Rhodesia, and even to South Africa.

Central Africa and Mozambique. The southern part of Eastern Africa contains two of the most important destination areas of the continent and several of the major source regions. Both internal and international migrations are of long-standing significance and contribute to one of the more complex migratory patterns in Africa.

In Zambia the major attracting area is the Copperbelt, but the line of rail, and particularly Lusaka, has also been important. The high percentage of adult males working for wages at least part of the time and the concentration of wage employment in the Copperbelt and on the line of rail suggest the significance of migration; in 1962, when there were an estimated 641,000 adult male Africans, 48.3 percent were wage employees. Only 29.7 percent were working in their own province; 51.4 percent were employed in other provinces, and the remaining 18.9 percent were employed outside the country. About half of the migrants moving to other countries were in Rhodesia, while Congo and South Africa accounted for about 17 percent each. As noted earlier, Zambians are now prohibited from accepting work in South Africa.

According to the 1963 Zambia census 7.2 percent of the total African population had been born outside the country. About a third of the foreign-born Africans were in urban areas and about three-fifths were recorded in African rural areas while only 8 percent were on European farms. The distribution in rural areas was roughly the reverse of the pattern for foreign Africans in Southern Rhodesia.

Reduced turnover rates and increasing productivity on the Copperbelt are tending to diminish the absorptive capacity of that node.

Population Movements in Africa

The booming economy of Zambia has, however, led to increased opportunities in several sectors, such as construction, transport, and services, and Lusaka, the capital, now the most rapidly growing city in the country, has attracted many migrants in recent years.

Rhodesia has sought migrant labor for its farms, mines, industries, and services in what has been one of the best-balanced and developed of tropical African economies. The extent of migration has also reflected, however, the allocation of land by racial group, earlier restrictions on urban residence, and, for some, the desire to secure workers from foreign countries who would work for lower wages and be more docile.

The 1961 African census in Rhodesia indicated that 11.3 percent of the 3.6 million total population had been born outside the country; 54 percent of the 406,000 foreign-born Africans were living in European farming areas and 40 percent in urban areas. Malawi was the country of birth for 49.4 percent of the foreign-born, Portuguese territories (mainly Mozambique) for 29.0 percent, and Zambia for 17.2 percent, these three sources accounting for 95.6 percent of the total.

In 1961 some 277,000 Rhodesian Africans and 278,000 foreign Africans were in paid employment in Rhodesia. Malawi accounted for 42.2 percent of the foreign Africans, Mozambique for 38.5 percent, and Zambia for 16.1 percent. The trends with respect to source country from 1931 to 1961 showed a marked reduction in the share from Zambia and a notable increase in the proportion from Mozambique. The total number of immigrant laborers increased from 103,000 in 1931 to 247,000 in 1951 and 278,000 in 1961.

The impact of UDI (the unilateral declaration of independence taken by Rhodesia on November 11, 1965) and of the subsequent imposition of sanctions on migration to Rhodesia is difficult to assess. An act predating breakup of the Federation of Rhodesia and Nyasaland prohibited non-Federal Africans from working in the main urban centers with the exception of Umtali, close to the Mozambique border. The number of immigrants from Mozambique decreased from 55,000 in 1957 before the act to 12,400 in 1962 and has since declined to 4,300 in 1967 and 3,750 in the first three-quarters of 1968. Immigrants from Malawi and Zambia have also decreased sharply, but the decline began well before UDI. The number of immigrants from Malawi declined from a high of 46,800 in 1956 to 14,800 in 1965 and 9,800 in 1967, while immigration from Zambia decreased from a high of 12,300 in 1955 to 1,400 in 1965 and 325 in 1967. Rhodesia claims in 1969 only a small drop in African employment, which would suggest that such unemployment as does exist has been cushioned by the net-emigration of foreign Africans; nevertheless, reports persist of considerable unemployment among Rhodesian Africans.

It is also apparent that, while sanctions have not affected the economy as strongly as was widely predicted at the outset, they have had a serious impact on tobacco farming, which has been one of the principal employers of foreign migrants. The opportunity cost has also been very substantial, since the economy is not expanding and is no longer providing an increment of jobs needed to absorb the increased numbers entering the market each year.

Malawi and Mozambique are principally important as far as migration is concerned as source areas. For Malawi the dependence on migrant labor is indicated by the estimated number of men absent from the country—115,000 in 1939 and 169,000 in 1958. In 1961 some 200,000 persons born in Malawi were recorded in Rhodesia of whom about 107,000 were in paid employment, while there were an estimated 100,000 Malawians in South Africa in 1968. The heavy dependence of Malawi on migrant earnings, which rank third after exports of tea and tobacco in exchange earnings, has led that country to adopt policies with respect to the Republic and Rhodesia that have been strongly criticized by the O.A.U. That it is becoming more difficult to export labor is indicated by the reduction in number of Identity Certificates (required for working outside the country) issued in recent years, from 61,000 in 1964 to 40,900 in 1966.

While Malawi's dependence on migrant earnings reflects the high population densities over much of the south, the relative poverty of much of the country, and the absence of alternate employment opportunities, the position in Mozambique is more related to the failure of Portugal to exploit the potentialities of the province with sufficient rapidity to create an adequate number of jobs within the country. The numbers of Mozambiqui migrants can only be roughly estimated because many of those born in the province but living abroad have apparently moved permanently to their present places of residence. In 1964 there were about 287,000 Africans registered as working outside the country; they accounted for about $1.1 million in taxes and fees and $3.9 million in voluntary remittances and deferred wages. In 1965, some 87,560 migrants were reported to have gone to South Africa.

SOUTHERN AFRICA

South Africa. It has been estimated that about two million men are involved in migration in South Africa, moving regularly between their "homelands" and the places of employment. It is ironic that the best developed country should have retained such a heavy dependence on migrant labor. The explanations are not far to seek, however, as they reflect both the desire to separate the races and the desire to maintain a supply of low wage labor. The gold mining industry in particular, employing a very large number of Africans, claims that the fixed price of gold has prevented the offering of higher wages; it has an especially heavy dependence on foreign Africans who are housed in compounds and hired on a bachelor basis which in turn is cited as justification for the low wages paid. Wages on the farms, usually paid partly in kind, are also low, and European farms are given certain priorities in the employment of foreign Africans. Industry has found the migrant labor system less satisfactory since it is not conducive to efficiency and high productivity, especially where Africans are needed as semiskilled and skilled operatives. It has, therefore, a distinctly higher percentage of employees who are permanent residents in the European areas, though no African has the right by law to claim such residence.

The reserve and Bantustan areas of South Africa provide a substantial number of the migrants employed in the Republic but it is not always easy to distinguish between those who are temporary and those who are more or less permanently absent from their "homelands." In 1967, for example, it was estimated that about 258,000 Xhosa were employed outside of the Transkei, but about 45 percent of them were thought to be long-term or permanent migrants. The importance to the Transkei of migrant labor is suggested by comparing the number of Xhosa employed outside the area with those gainfully employed within the Bantustan, who totaled 32,700 in 1967, exclusive of teachers. It is estimated for all the reserves that from 50 to 70 percent of the able-bodied males are involved in migration to the European areas.

Nationalist government policy calls for the gradual elimination of Africans from the Western Cape, where Coloureds are supposed to provide the unskilled labor required; about seven-eighths of the total

Coloured population of South Africa is resident in the Cape Province. It has been suggested that this policy reflects the intention of creating a kind of laager to which the Whites might repair in the event of future necessity, but the thesis has doubtful validity. Nor is the policy readily applied, since there is a shortage of labor in the Western Cape and, in 1967, 225,000 Africans remained in the area, of whom 131,000 were contract laborers.

Despite the stated desire to reduce the number of Africans in the European areas there has in fact been an increasing representation, particularly in the cities. Between 1936 and 1960 the percent of the total African population living in cities grew from 38.8 to 46.2 percent. Despite influx-control and job-reservation edicts the flow to the cities has continued and the programs designed to reverse the trend have been entirely inadequate in scope. It is possible, however, that the border industry program could be advanced by requiring the movement of certain industries to selected border sites or that reserve areas could be redefined to make it appear that border industries were more important than they actually were.

Nonetheless it seems clear that the objective of decreasing or even reversing the flow of Africans to white areas is totally unrealistic and that "the government is attempting to move against two forces— economic and demographic—which are probably too powerful to be denied."[28]

South Africa is practically the only African country now experiencing a net immigration of permanent residents from outside the continent, primarily of Europeans who are encouraged by various incentives to move to the Republic. The whites in that country are conscious of their minority position and are eager that their share of the total population remain steady or increase. Since the last war only 1960 saw a net emigration of whites; in the five-year period 1963–1968 net immigration averaged about 33,500 per year; this is probably not a sufficiently high rate to sustain the present percent of whites to the total population, about 19.0 percent in 1967.

Lesotho. Deficient in resources and entirely encased in South Africa, Lesotho has the highest relative involvement in and dependency on external labor migration of any African country. The gold mines of

[28] William A. Hance, "The Case For and Against United States Disengagement from South Africa," in Hance *et al., Southern Africa and the United States,* p. 145.

the Rand were the main employers of Sotho labor, at least until the mid-1940s, the number increasing from 2,300 in 1904 to 20,900 in 1930 and 50,800 in 1940; after that, employment on the mines declined but other jobs became available. The percentage of the *de jure* population of Lesotho resident in South Africa rose from 5.8 percent in 1911 to 19.5 percent in 1956, when it was 155,000; the 1966 census, however, gave a total of 117,000 absentees or only 12.0 percent of the *de jure* population. The exact position is obscured by the absorption of a considerable number of Sotho into the South African population, possibly 80–100,000 by 1946 and another 100,000 by 1960. However it is now increasingly difficult for Lesotho to shift portions of its population to South Africa because of more rigid control in the Republic. With the estimated rate of increase of the *de facto* population now about 3.0 percent per annum it is obvious that Lesotho will face increasing population problems in the years ahead.

Lesotho's dependence on migrant labor may be measured in a variety of ways. Currently about three-fifths of its male labor force plus 20–30,000 women are absent from the country at any one time. Receipts to the workers and their families in the form of voluntary deferred pay, remittances, and money and goods brought back by migrants were estimated to have a total value of $10.2 million in 1966, which compared with the total value of exports of the country in that year of $6.1 million. The government also receives about $250,000 a year from recruiting taxes and fees. Earnings have not, however, been increasing in tune with the needs of the country, while developments within Lesotho have provided only a very limited number of alternate opportunities. The trends now appearing bode ill for a country with "virtually no boot-laces by which to pull itself up."[29]

Migration Theory and Classifications

Some effort has been made to formulate a theory of migration but thus far without very significant results. Lee, for example, classifies factors influencing migration in "A Theory of Migration" as: pull factors or

[29] See Michael Ward, "Economic Independence for Lesotho," *Journal of Modern African Studies*, V, No. 3 (October, 1967), 55-68; see also P. Smit, *Lesotho; A Geographical Study* (Pretoria, Africa Institute, 1967), and G. M. E. Leistner, *Lesotho: Economic Structure and Growth* (Pretoria, Africa Institute, 1966).

those associated with the destination area; push factors or those associated with the area of origin; intervening obstacles, such as ethnic barriers, distance, and cost; and personal factors.[30] Berg's listing of the major factors determining the individual's decision to migrate are not too dissimilar from Lee's "theoretical" factors; they are: the intensity of his preference for money income against "leisure" in the village, the level of his income from village production, the effort-price of income earnable in the village, and the effort-price of income earnable outside the village.[31] Push factors are sometimes labeled centrifugal or impulsive, and pull factors centripetal or attracting. A distinction may also be made between the pull felt by individuals and that motivated by employers, or between a passive and active pull of a destination area.

It is increasingly clear that source and destination areas are both likely to have attracting and repulsing elements, that these will be weighted differently depending on where one is and how long he has been there, and that each element will be valued differently by different individuals. Mitchell states that single factor explanations of migration are totally inadequate and that the listing of all possible motivations is also not very helpful. He sees the need to link together and relate the multiple causes in a logical framework and suggests a classification whose major headings are "the nexus of centrifugal tendencies" and "the nexus of centripetal tendencies" subdivided by "social and psychological" and "economic" factors.[32] It may be wondered whether migrations are not so complex and so changing as to make the development of a unified theory an unrealistic goal.

Migrations may be classified in a variety of ways using differing criteria such as:

1. Source and destination areas. These may be broadly classified as in the system which distinguishes between movements which are internal, intracontinental or intercontinental, or, equally broadly, as rural/rural, rural/urban, urban/rural, and urban/urban, or the areas may be more precisely delineated by naming the source and destination places or regions.

[30] Everett S. Lee, "A Theory of Migration," *Demography*, III, No. 1 (1966), 47-57.

[31] Berg, "Backward-Sloping Labor Supply Functions in Dual Economies," p. 480 ff.

[32] J. Clyde Mitchell, "The Causes of Labour Migration," in C.C.T.A. *Migrant Labour in Africa South of the Sahara* (Publication No. 79, Abidjan, 1961), pp. 262 ff.

2. Method of movement, e.g., whether the migration takes place by land (walking, bicycle, bus, train), sea, or air.

3. Tribal or racial makeup of the migrants. While this is not a very useful classification system, the study of individual tribal movements is valuable in providing information on varying approaches, attitudes, customs, and responses which are needed for comparative analysis. It is also interesting to note that some tribes are attracted and adjust to migration while others migrate only unwillingly and still others apparently scorn it. Southall cites, for example, the early enlistment by the Arabs of the Nyamwezi from the Tabora region of west-central Tanzania who "became the great travellers of Tanganyika and have remained so ever since,"[33] going through all the transformations from porters and leaders in ivory and slave campaigns, to porters and explorers with German and later British officials, and finally to migrant laborers. In Zambia, as a second example, the Bemba-speaking tribes have a long tradition as migrants and have adjusted readily to work on the Copperbelt where Bemba has become the major language, while the more conservative and parochial Barotse have never developed the migratory propensities of the Bemba. In Senegal, as a final example, the Toucouleur and Wolof appear to have great propensities to travel. Degree of contact, cultural differences, and pressure in the source areas appear to be the major influences in explaining these differences by ethnic group.

4. Time of year and duration. Prothero, for example, distinguishes by these criteria four classes of rural/urban mobility, which may, however be broadened to include other types of migration: (a) daily, which might involve moving from the peri-urban fringe to the center of the city; (b) seasonal, which may be related to periods of reduced agricultural activity in the source area or to increased seasonal demand in the destination areas. He notes that such movements are more common in West Africa than elsewhere, particularly from areas with a long dry season. But seasonal migration to participate in harvesting, processing, and other peak work periods in destination areas is common in many parts of the continent and is one of the few types which involves urban-rural movements; (c) short-term movements, normally not exceeding two years; and (d) definitive movements which lead to permanent settlement in the destination area, which may be either

[33] Southall, "Population Movements in East Africa," p. 166.

rural or urban.[34] The permanency of movement is often difficult to measure and to classify. One author distinguishes between temporary town dwellers, stabilized town dwellers who contemplate returning to their village on retirement, and settled town dwellers; others measure the break with the rural source area by using a specific span of years of continuous urban residence, usually five years or more, by considering that the move has become definitive if the town is considered the permanent residence of the wife and children, or at a point where the individual no longer has land rights in the rural area from which he originated.

5. By size, including numbers relative to source or destination area. Such information is valuable, but it is rarely available on a continuing basis and is not very useful as a system for classifying migrations.

6. Wage level and character of destination area. Gulliver, for

[34] R. M. Prothero, "Socio-Economic Aspects of Rural/Urban Migration in Africa South of the Sahara," *Scientia*, November-December 1965, pp. 1-7.

Migrants arriving at the Cameroon Development Corporation plantations near Victoria, West Cameroon.

A mammy-wagon in Ghana. Such buses serve remote areas and are frequently used by migrant workers.

Population Movements in Africa

example, says that there are two general types of labor migration in Africa, to low-wage, rural employment and to higher-wage, industrial employment.[35] Examples of the former would be migration to cocoa farms in Ghana, to sisal estates in Tanganyika, or to European farms in Rhodesia and South Africa; examples of the latter type would be migration to the Copperbelt or to manufacturing plants in urban centers. Gulliver suggests that these types have significant differences which affect the motivation of the migrant and the impact on him and his family. In low-wage, rural employment the workers live under conditions little different from at home, the wages provide limited incentive to remain, little training or learning benefit accrues, experience of the outside world is slight, and low earnings minimize the impact at home. This type of employment is, therefore, attractive mainly to younger men with relatively few opportunities and obligations. The second type usually provides relatively much higher wages, greater chance for

[35] Philip M. Gulliver, "Incentives in Labor Migration," *Human Organization,* XIX, No. 3 (Fall, 1960), 159-63.

Eritrean nomads migrating near Massawa. All of the household possessions, the framework and skins for their huts, and the migrants' two wives and several children are carried on the backs of the camels.

acquiring skills and for advancement, housing well above village standards including lighting and other amenities, better access to education
for himself or his children, and a more intense contact with the modern
world of ideas and technology, all of which are likely to encourage the
migrant to remain for a longer period or to return for additional periods.

7. Historical period. Migrations may, for example, be classified
as pre-colonial, colonial, or post-independence movements. Prothero has
divided population flows on the basis of continuity and change as:
(a) movements that took place in the past but which no longer exist,
but which may help to explain the present distribution of population;
(b) those that have continued from the past into the present, such as
seasonal pastoral migrations, long-term migratory drift, and religious
pilgrimages; and (c) movements that have developed in recent times,
such as downhill or rural-urban migrations.[36]

8. The nature of employment in the destination area.

9. The motivations or causes of migration.

In the following section the motivations for migration are summarized. It must be stressed, however, as was noted earlier, that individual and group migrations are frequently motivated by a multiple of
factors.

Motivations and Causes for Migrations

The listing of single factors motivating population movements is valuable, despite the important reservation noted above, in revealing significant features affecting migrations and thus assisting in an understanding
of very complex phenomena.

MIGRATION RELATED TO ENVIRONMENTAL CONDITIONS

Man in Africa often remains in very close symbiosis with his
physical environment, hence it is not surprising that numerous migrations are caused or motivated by needs to adjust to this environment.
Examples are given in the subheadings that follow.

[36] Prothero, "Migration in Tropical Africa," in Barbour and Prothero, eds., *Essays
on African Population*, p. 250. Prothero has also developed a schematic typology of
all types of population movement in Africa in a paper entitled "Mobility in Africa,"
prepared for a seminar at the Food Research Institute, Stanford University, November 27, 1967.

Population Movements in Africa

The Search for Adequate Grazing. This motivation has existed for millennia in the arid and semiarid areas of Africa and certainly predates the development of settled cropping. It continues to affect hundreds of tribes of nomads and seminomads in the desert fringes and steppes north and south of the Sahara, in the dry lands of the Horn, and in discontinuous zones extending through eastern Africa into southern Africa and across the continent into Angola. Some countries have populations which are predominantly nomadic, such as Spanish Sahara, Mauritania, and Somalia and the majority of African countries have some segment of their population classed as nomadic pastoralists. The nomadism of Mauritania, which involves 73 percent of its population, is divided into three types: grand nomadism, which is the most strenuous, wide-ranging, and continuous, and which is limited to those who travel by camel; middle nomadism, where the relative closeness of watering points permits goat and sheep herding; and light nomadism, which, since it involves the herding of cows, is limited to areas where water is more abundant.[37]

The Fulani, who were originally entirely nomadic, are now classed as town, settled, and nomadic, the last also being referred to as cow or bush Fulani. The nomadic Fulani also vary from those who never spend more than a few days in one place, through those who make seasonal camps in any locality which suits them, to those who have a permanent center where the aged and some other members remain and which is visited from time to time by those who continue to migrate with the herds. Most groups follow a more or less circumscribed seasonal grazing circuit, except when something requires them to seek alternate grounds; some such movements may go on for years.

It is not just grass and herbage which influence the movement of nomads. The availability of potable water both for livestock and humans is obviously a major requirement, but soil, insect, and disease conditions are also important. The example of the Baggara Arabs given by Cunnison is an excellent one:

Movements . . . are dictated by four main needs. These concern the condition of grass, water, and ground underfoot, and the presence or absence of annoying flies. Clearly cattle cannot thrive where water is short or grazing bad. They can exist where there are biting flies, but these are better avoided, since

[37] Alfred G. Gerteiny, *Mauritania* (New York, Frederick A. Praeger, 1967), p. 14.

they disturb the grazing and may be carriers of disease. They can exist on wet clay, but they are inclined to develop hoof trouble, and travel badly with loads through mud.[38]

That these factors may require very frequent moves is suggested by the case of one Baggara cattle owner who changed camp sixty-one times in a year.[39]

Pastoral migrations often assume a regular pattern. Seminomads south of the Sahara, for example, may be based at permanent settling points from which their annual migrations begin as the rains begin to move northward in the Spring. First moving south they then swing around and move northward into the desert as far as the season's grass growth permits, remain there as long as there is adequate pasturage, and finally return to their settlements where forage has been reserved for the winter months.

Regularity is also apparent in the annual east-west or west-east movements along the Nile in the Sudan reflecting the seasonal rise and fall of the river and the availability of riverine pastures after the waters have receded, in the seasonal migration to and from the slightly higher lands in the Sudd, and in the north-south movements in West Africa to take advantage of retreat of the tsetse fly in the dry winter months.

Since precipitation and vegetation growth vary markedly from year to year in steppe and desert areas there are likely to be desperate conflicts over pasture and water rights among pastoral tribes or between clans of the same tribe. Some of these conflicts were never fully suppressed under colonial rule and continue to persist today as, for example, the clan feuds of Somalia or the recurrent raiding between the Turkhana of Kenya and the Karamojong of Uganda. Conditions may become particularly dangerous when the migration involves the crossing of national frontiers.

While the nomadic and seminomadic herdsmen of Africa may appear to follow a life closely dictated by factors beyond their control, which might be expected to create an aversion to the constant hardships and discomforts and a desire to adopt a more settled existence, many nomadic groups in fact look upon their way of life as superior to that of sedentarists, as assuring them independence and freedom from

[38] Cunnison, *Baggara Arabs: Power and Lineage in a Sudanese Nomad Tribe*, p. 19.
[39] *Ibid.*, p. 22.

Population Movements in Africa

A young Karamojong grazier, northeast Uganda. Pastoral tribes inhabiting semiarid and arid areas are perforce engaged in at least seasonal migrations and are represented in most African countries.

what they consider unwelcome interdependence and demeaning supervision by authorities outside the tribe.

Many pastoralists abandon their migrations only when conditions more or less force them to adopt more intensive methods. The great rinderpest epidemics which hit much of Africa in the latter part of the last century and the first decades of the present century undoubtedly forced many Fulani to settle in West Africa; in East Africa their impact on tribal movements played a major role in excision of the White Highlands from lands which appeared to be largely empty but which had previously been used regularly, primarily by the Masai. Elsewhere increased occupance by agriculturalists has restricted grazing lands, while population-growth of pastoral tribes has also forced the abandonment of traditional migration.

There are relatively few examples in Africa of the kind of transhumance one finds in the Alps, namely a vertical movement reflecting the availability of summer pastures in areas that are too cold or, in some cases, too dry in winter. It does exist, however, as for example in the Atlas Mountains of the Maghreb, which are snow-capped in winter, and in parts of the highveld of Swaziland, Lesotho, and South Africa. In recent years, however, substantial areas of the highveld in Swaziland and the Republic that were formerly used for trek sheep have been planted to exotic forests.

Population Movements in Africa   *169*

Migration in Response to Insect and Disease Infestation. This second type of physically motivated movement is less likely to be regular than those of the nomadic graziers, though the example of seasonal pastoral movement into and out of tsetse-infested regions has already been cited. An important example of this class of migration is the evacuation of large areas as the result of particularly virulent attacks of human trypanosomiasis, which has been necessary on occasion in Nigeria, Congo (Kinshasa), and from the islands and low-lying shores of Lake Victoria in Uganda and Kenya. Three major epidemics of sleeping sickness were recorded in the Lake Basin of Kenya in the early part of the century; the shore belt has not fully recovered demographically since and tsetse "continues to be a major obstacle to a fuller utilisation of the land."[40]

Additional examples of this cause of migration are the population withdrawal from parts of the Volta River basin because of the heavy incidence of "river blindness," the shift in concentration of cocoa farmers in Ghana due to losses of producing trees from "swollen shoot disease," and the retraction of tribes in East Africa near the turn of the century due to a very serious rinderpest epidemic which decimated the herds of pastoral tribes.

Migrations Resulting from Soil Exhaustion. Many tropical soils deteriorate rapidly in structure and plant nutrients once they are opened to cultivation, the poverty of such soils being the main explanation for shifting agriculture or bush fallowing. The desirability of having a long fallow period means that a relatively large area is needed per family unit. It is only natural that there would be a migrational drift as new lands are opened up under the system of shifting agriculture. This may help to explain the gradual shifting of various tribal groups in the past; it is less likely to occur today since tribal boundaries have been more fixed and population growth leaves less vacant land to which to move.

Another example of migration caused by exhaustion of soils is that of Senegal where fragile soils used monoculturally for growing peanuts have had to be abandoned, resulting in a gradual landward shift in the production of this major crop.

In some extreme cases overuse of the land has led to severe soil erosion, which has forced at least a short-distance migration of peasants

[40] Simeon H. Ominde, *Land and Population Movements in Kenya* (Evanston, Northwestern University Press, 1968), p. 18.

inhabiting such regions. Examples can be found in Iboland, Rwanda and Burundi, Ethiopia, Lesotho, and Madagascar.

Migrations Resulting from Drought or Flooding. Large parts of Africa are characterized by marked variability in the amount and seasonal distribution of precipitation. It is not surprising, therefore, that periodic droughts, sometimes lasting for years, occur irregularly but certainly in many areas, and that these should influence migration to a greater or lesser degree. Excessive precipitation, which not infrequently succeeds a lengthy drought, may also lead to the forced evacuation of an area due to flooding.

Brooke, cites an example of drought-induced migration in the Biharamulo District south of Bukoba in Tanzania where "many people—not infrequently entire families—move to more favored parts of the Bukoba and Geita Districts during such times. . . . Nearly all return to their homes before the next season's planting. Temporary movement of this kind (in Swahili, *hemea*) is widespread in East Africa."[41]

As a second example, recruiters of labor throughout southern Africa have long recognized a direct correlation between the numbers seeking work and drought conditions in their home areas. In good rainfall years, on the other hand, a more intensive effort was required to secure an adequate supply.

Not all flooding is irregular or catastrophic in nature. There is a regular seasonal ebb and flow on the broad flood plains, in the swamps, and on vast flattish areas which are inundated each year in the rainy season. This requires regular seasonal movements on the part of grazier and farmer alike, which may be seen along parts of the Senegal, Niger, Nile, Zambezi, and lesser streams. In the case of the Barotse there is an annual migration from the flood plain to the borderlands and back in response to flooding, the move being made in a colorful procession of large canoes headed by the paramount chief.

Migration to Relieve Excessive Pressure on Existing Lands. This motivation, which may or may not be associated with other types of physically motivated migration, is sometimes called the demographic factor in migration. It is often involved in population movements whether or not it is the dominant force. Sorré, while accepting that the

[41] Clarke Brooke, "Types of Food Shortages in Tanzania," *The Geographical Review*, LIII, No. 3 (July, 1967), 340.

greatest cause of migration is economic, believes that the demographic factor almost invariably lies behind the economic reasons and suggests that migration studies are more and more becoming a branch of population theory.[42]

In the case of graziers, an increasing population pressure may lead to a wider seasonal migration, a search for new lands, or a switch to tillage agriculture. The demographic factor is undoubtedly of great significance in reducing the extent of nomadic grazing in Africa, though it is not possible to quantify the impact. There are, however, numerous examples of previously nomadic tribes which have moved partly or entirely to a settled existence. Their continued growth and that of tribes which were traditionally tillage farmers leads to a gradual reduction in the lands available to pastoralists. Nonetheless there are enormous areas which cannot be used for cropping and which will probably continue to be used by roving graziers. The carrying capacity of some such lands can be increased by providing more water points, but there are serious handicaps involved in other betterment programs such as control of cattle disease and division of land into ranching units. In Masai country in Kenya, for example, the advent of the KAG vaccine led to a large-scale increase in numbers of cattle which in turn brought "deterioration of the land from good grass over most of the area . . . to the almost desert conditions existing over the whole area in 1961."[43]

The fact of the matter is that most extensive grazing lands in Africa are already overworked. With many tribes now experiencing more rapid rates of population growth it must be concluded that there will have to be an increasing number of people moving to other activities.

Among settled farmers increasing pressure on the land may lead to the gradual expansion of tilled areas, to a hiving off of communities to settle elsewhere, to a gradual drift of the whole group, or to a greater participation in the more familiar labor migration. Each of these results involves some population movements; pressure on the land may also, of course, stimulate more intensive land-use practices which reduce the necessity for migration.

Vermeer gives an illustration of migratory drift due to population pressure in his account of the Tiv:

[42] See Sorré, *Les Migrations des Peuples: Essai sur la Mobilité Géographique.*
[43] J. H. B. Prole, "Pastoral Land Use," in W. T. W. Morgan, ed., *Nairobi: City and Region* (Nairobi, Oxford University Press, 1967), p. 96.

Population Movements in Africa

From a traditional homeland lying to the southeast, the Tiv have been progressively spreading northward primarily and have incorporated peripheral territory into their tribal area. Toward the northern, eastern, and western margins. . . , population densities are relatively low. But in the south, where the Tiv are bounded by areas already densely populated and where therefore penetration is impeded, population densities have increased markedly and are quite high.[44]

And according to Colson, the great attention given to studies of labor migration in recent decades has tended to overshadow

the very considerable movements which have continued to take place as people have moved to seek new land either for subsistence purposes or for the planting of cash crops. . . .[45]

Examples of migrations which are in considerable measure related to pressure on the land are numerous and include movement from many of the major source areas given in the regional discussion earlier in this chapter:

1. The drift from Saharan oases, many of which are grossly overpopulated, to the cities of North Africa.
2. Migration from the rural areas of the Maghreb.
3. Movement of Nubians from the south of Egypt where available land per capita has been only 44 percent of the already low level in the north.
4. The large-scale movement of Mossi from their densely populated and poorly endowed homeland in Upper Volta; involving about 20 percent of the working population of the country at any one time.
5. Movement from the smaller nodes of high density in West Africa such as the Korhogo Cercle in Ivory Coast, several areas in northern Ghana, and the Atakora Mountains in northern Dahomey.
6. The seasonal flow from parts of northern Nigeria where many migrants are called *masu cin rani* (Hausa), "men who eat away the dry season." By going elsewhere they help to conserve the

[44] Donald E. Vermeer, "Population Pressure and Crop Rotational Changes among the Tiv of Nigeria," paper presented at the annual meeting of the African Studies Association, Los Angeles, October, 1968, p. 22.
[45] Colson, "Migration in Africa: Trends and Possibilities," p. 61.

Rural scene in Burundi, showing evidence of overgrazing and overcropping. Rwanda and Burundi, two densely populated highland countries, have long been major source areas for migrants to adjacent countries.

limited supplies of food, alleviating some of the pressure on the land.[46]

7. The large numbers of Ibo who migrated to adjacent lands and to other regions of Nigeria prior to the recent Nigeria-Biafra conflict;
8. The out-migration of the Bamiléké in Cameroon.
9. The large-scale and well-known migration of Rwandans and Rundi to adjacent countries, especially Uganda.

[46] Prothero, *Migrants and Malaria*, p. 2.

10. The migration from some of the pressure areas of Uganda such as Kigezi or the West Nile Province where "the higher the density of population of a Lugbara county . . . the higher is the rate of labour migration."[47]

11. Movement of the Kikuyu, Luo, and other tribes in Kenya, primarily from high-density portions of the Highlands and of the Lake Victoria

[47] J. F. M. Middleton and D. J. Greenland, "Land and Population in West Nile District, Uganda," *The Geographical Journal*, CXX, No. 4 (December, 1954), 453.

Muslim chief on pilgrimage to Mecca.

Population Movements in Africa 175

basin, where some locations and sublocations have 40–50 percent of the adult males absent at any one time.

12. The expansionary drift of Chagga from their preferred zone on the slopes of Mt. Kilimanjaro to less desirable lands at lower elevations.

13. The large-scale international migrations of Malawians, Sotho, and Tswana to South Africa and Rhodesia.

14. The migration of Africans from the reserves of Rhodesia and South Africa which cannot support the population that is now resident in them.

15. The emigration of peoples from the densely populated Indian Ocean island groups.

SOCIO-CULTURAL FACTORS IN MIGRATION

Falling within this category are migrations motivated by religious considerations, tribal ties, and by modernizing influences.

Religious Considerations. Modern migrations have been related both to religious beliefs and to conflict among religious groups. Included in the former is the pilgrimage of Muslims to Mecca, which probably involves over 100,000 people from Africa annually. Numerous pilgrims can scarcely be classed as migrants, however, since their trips may be made by plane in a short period. The example of the "Westerners" who work in the Sudan on one or both directions of their pilgrimage may very justifiably be so characterized. Shorter pilgrimages are made by thousands of Muslims to local holy cities such as Moulay Idriss in Morocco or Kairouan in Tunisia, but again there is some question regarding the labeling of such movements as migrations.

An example of emigration caused in part by religious antipathy is that of the Jews from North Africa. Here some of the predominantly Muslim countries have adopted a somewhat ambivalent position, subjecting the indigenous Jew to considerable pressure and yet forbidding him to leave the country on the grounds that this would strengthen Israel. Many Jews from North Africa have in fact gone to Israel, larger numbers have gone to France, and some have moved to colonies in South America.

Religious considerations have also been involved in certain other modern population movements, though other elements in a complex of factors were probably of greater weight, namely the flight of refugees from southern Sudan, that of the Ibo from Northern Nigeria after 1966, and occasional flights of Muslim tribesmen from the Eritrea Province of Ethiopia into Sudan.

Tribal Considerations in Migrations. The tribal factor is involved in most types of migration to a greater or lesser extent. Its most important influence is to make migration circular or reciprocal (i.e., "one foot in the village, one in the modern economy"). The tribal system links together kinship relations, cultivation rights, religious beliefs, allegiance to the chief, and many other features in a single system so that each element is inextricably involved in every other. The persistence of the tribal system, then, accounts for the vast bulk of migrants returning regularly to their tribal areas and hence the persistent importance of migration and migrant labor in the African scene. It should be noted that some of the attachment to indigenous areas has a significant economic component, particularly the rights to land and the social-security content of kinship and tribal relations.

There is considerable difference of opinion as to the continuing strength of tribalism in sustaining reciprocal migration. Some observers claim that it is as strong as it has ever been. But many modernist Africans consider tribalism as stultifying, as something which should be abandoned in favor of a national allegiance, as something which is to be condemned along with colonialism and imperialism. An increasing number of urbanized Africans can no longer be thought of as tribally committed or oriented. Certainly many African residents in the cities of South Africa have been more or less completely detribalized. The same is true on a reduced scale for some of the urban dwellers and educated elite in tropical Africa.

Even when a periodic return to the tribal area remains as the ideal, however, it may not always be possible. In some cases no land is available for some tribesmen (e.g., parts of Kikuyuland). In other cases economic considerations may weigh heavily against migrating back to the reserve. Evidence from the Copperbelt, for example, indicates that the average length of service of mine employees is increasing very rapidly. At the same time, the number of available jobs is decreasing and it is probable that some Africans are constrained not to migrate back to their villages for fear of losing their jobs and not finding an opening if they return at a later date. Comparable situations very likely exist in an increasing number of employment areas in Africa. A politico-economic element is sometimes added, as in the case of Mozambique migrants to

Rhodesia. In 1958, Rhodesia placed restrictions on additional in-migration of foreign Africans, which meant that any Mozambiquis who returned home might not be able to reenter Rhodesia for employment. In these circumstances the pressure to remain would be very great, for employment in Rhodesia was generally preferable to working in Mozambique, or, for many, to working in the mines of South Africa.

An example cited by Plotnicov is pertinent to the discussion:

Tribal immigrants in Jos also manifested the ideal of returning to their ancestral homes upon retirement age, but the data suggest that social forces in their tribal areas and in the city prevent the fulfillment of this ideal. Further the data suggest that the children of these immigrants are developing into confirmed urbanites who do not manifest the tribal loyalties of their parents.[48]

Occasionally, tribal influences will work counter to the usual patterns. Respondents to queries on the reasons for their migrating sometimes give such answers as the desire to escape from taboos or witchcraft, the wish to escape from the authority of the tribe, or the desire not to live under an unpopular chief. Similarly an individual may wish to leave because he would like to achieve a position of leadership or authority which would not be open to him in the tribal setting under traditional age and status restrictions. Conversely, some individuals may not wish to return to their homes because they do not wish to fulfill their farming commitments or share their earnings, or because their education, success, and position have made the gap between them and their home groups too wide to be readily bridged.

Tribal conflicts, which were a major cause of migrations in the past but which were generally suppressed under colonialism, have again caused a number of migrations in post-independence years. Examples of intertribal conflicts which have led to migrations in recent years, sometimes on a large scale, are those of the Baluba and Lulua in Congo (Kinshasa), the Ibo and Hausa-Fulani in Nigeria which led to the displacement of an estimated 1.8 to 2 million people even before the outbreak of the Federation-Biafra war, the Tutsi and Hutu in Rwanda, and the north-south conflict in Sudan.

On the other hand, types of migration which started after the suppression of intertribal conflict, such as the "downhill" movements

[48] Leonard Plotnicov, in *African Urban Notes*, I, No. 1 (April 1966), 5-6.

Population Movements in Africa

already referred to, continue to take place and in most areas there has been a continuing if not increasing intermingling of tribal groups in the years following independence.

A tribal-cultural factor of interest in a study of migration is that of migration as a kind of *rîte de passage* or initiation to adult life—something which must be done to prove one's prowess or fearlessness and to make one acceptable in marriage. Thus emigration would replace the former necessity to kill a lion, steal some cattle, or display valor in warfare as a test of manhood.[49]

This factor has probably been greatly over-stressed as a motivation for migration. Indeed it may have been seized upon by some apologists as an acceptable rationalization for continuing practices which were otherwise less acceptable. The same may be said for the justification of bachelor wages on the grounds that the male worker wished to come alone and return to his reserve. Such rationalizations could also be used to justify continuance of the compound system and failure to provide certain social-security benefits and arrangements.

In his writing about Thonga migration some of Harris' comments have interest in this discussion: "Heavily in the balance on the side of the mines was the greater glamour and prestige associated with work on the Transvaal. . . . The belief that enlistment for mine labour is proof of manhood was certainly spread as much by the recruiters as by the recruits."[50]

Skinner says with respect to the Mossi of Upper Volta that "migration itself brings little lasting prestige. It is not seen as a *rîte de passage* or even an unusual feat. The men who go away are not considered brave but poor."[51]

Leistner's inquiry among urban Africans in South Africa had only a 1 percent response in reasons for seeking employment under the category "initiation to adult life."[52] On the other hand, Rouch writes that "from the middle of the last century it has become accepted in the savan-

[49] I. Schapera, *Migrant Labour and Tribal Life: A Study of Conditions in Bechuanaland Protectorate*, p. 116.

[50] Marvin Harris, "Labour Migration among the Moçambique Thonga: Cultural and Political Factors," *Africa*, XXIX, No. 1 (January, 1959), 59-60.

[51] Skinner, "Labour Migration and Its Relationship to Socio-Cultural Change in Mossi Society," p. 383.

[52] G. M. E. Leistner, "Patterns of Urban Bantu Labour," *South African Journal of Economics*, XXXII, No. 4 (December, 1964), 253-77.

nah that to be a real man you have to have been to the Coast at least once,"[53] and Schapera held that labor migration had come to be widely regarded as a form of initiation into manhood, that social prestige was gained by making such a trip.

Another important factor where youths are concerned is the marked preference shown by girls for those who have been abroad. On their return home, well-dressed and perhaps with money to spend on gifts, such men, with their airs and glamorous stories, exercise a far greater attraction than those who have never been away. . . . The desire to find favour in the eyes of the other sex instead of being treated with contempt may thus be a powerful stimulus to labor migration.[54]

Schapera concludes, however, that economic necessity is a more universal and the most important cause of migration. Similarly, Mitchell writes that

in many tribes a trip to the town has become a recognized symbol of a boy's becoming a man. He achieves adult status when he pays his first tax and goes off to earn his first wages: he shows by this that he is now an independent person in the tribe.[55]

One wonders, however, if another statement on the same page does not capture the position more accurately, possibly reducing somewhat the emphasis on the initiation aspect of migration; he writes that

labour migration has been *accepted* as part of the ordinary way of modern tribal life and is behaviour *expected* of young men when they reach adulthood.[56]

And elsewhere the same author says that migration has become accepted behavior, a part of the normative system of the society.[57]

Non-economic Motivations Associated with Modernizing Influences. It is appropriate to include under the category of sociocultural influences on migration certain motivations associated with modernization. For example, it has sometimes been claimed that migrants are as

[53] Jean Rouch, "Second Generation Migrants in Ghana and the Ivory Coast," in Southall, ed., *Social Change in Modern Africa*, p. 301.

[54] Schapera, *Migrant Labour and Tribal Life*, p. 117.

[55] J. Clyde Mitchell, "Labour and Population Movements in Central Africa," in Barbour and Prothero, eds., *Essays on African Population*, p. 237.

[56] *Ibid.*, my italics.

[57] See Mitchell, "The Causes of Labor Migration."

much attracted by the bright lights, gay life, and excitement of urban communities as by economic incentives. Again, this motivation has probably been exaggerated. Mitchell suggests that the bright-lights theory is patently inadequate,[58] and Skinner reported from his study of two Mossi villages that no one claimed to have migrated for the pleasures of Kumasi.[59] Even where the individual is attracted by the promised glamour of the urban center it is likely that other considerations overshadow this element in his decision to migrate.

Much more important, according to numerous surveys, is the desire to seek education for oneself or one's children which is often more readily available in the urban than in the rural setting. Plauvert notes that many children leave their families and sometimes their region to go to school, and calls attention to a study of Kabré and Ewe in Togo which showed that 40 percent of elementary and intermediate students lived outside the family group, and to a study in the Palimé region which concluded that 20 percent of all absent inhabitants were students.[60]

On the other hand, the receipt of education has tended to stimulate out-migration, some would say excessively, and there is no doubt that migrants to urban centers in most African countries have a better than average educational achievement level. Lewis believes that accelerated schooling in the countryside has produced a disequilibrium between expectation and reality which has become a major reason for the flow to the towns.[61]

Very few Africans appear to have the kind of love for the land attributed to many peasants in Europe; many of those who receive even a primary education feel that farm work (or sometimes any physical labor) is no longer an appropriate occupation. Indeed some youth see the life of the peasant as the most miserable that exists, one which subjects the individual to physical forces beyond his control, while town life appears to offer rapid social advance and liberation.[62]

[58] *Ibid.*, pp. 262-63.
[59] Skinner, "Labour Migration and Its Relationship to Socio-Cultural Change in Mossi Society," p. 380.
[60] See Plauvert, "Migrations et Éducation," pp. 467-75.
[61] W. Arthur Lewis, "Unemployment in Developing Countries," *World Today*, XXIII, No. 1 (January, 1967), 14.
[62] Robert Descloitres, "Problèmes d'Urbanisation en Afrique," *Industries et Travaux d'Outre-Mer*, No. 168 (November, 1967), pp. 1059-61. See also Robert Descloitres and J. C. Reverdy, "Recherche sur les attitudes du sous-prolétariat algérien à régard de la société urbaine," *Civilisations*, XIII, Nos. 1-2 (1963).

That this attitude is not entirely of recent origin is suggested by a conclusion of a survey of the Ashanti made in 1945 that "the Ashanti ideal is to be an absentee farm-owner, and hence any Ashanti who can afford to do so directs his children's upbringing in such a way as to qualify him for any other occupation than cocoa farming."[63]

The problem of "school leavers" who migrate to towns only to find that there is no employment considered acceptable has become a serious one in numerous countries. It does not necessarily follow, of course, that there should be a cutback on education in order to alleviate this problem.

An explanation of some migration to urban areas is sickness of some member of the family and the resultant desire to move to a place where hospital or other medical facilities are more readily available. Before more advanced medicinal treatment was developed there was also a partially enforced migration of persons suffering from leprosy to special, isolated leprosariums.

POLITICAL MOTIVATIONS IN MIGRATIONS

Political factors have been of increasing importance in causing migrations in the years immediately preceding and following independence in a considerable number of African areas. These factors are, however, often interwoven with tribal and racial considerations, religious animosities, nationalist sentiment, and economic motives, making an intricate complex which is not subject to ready classification.

Certain generalizations regarding politically motivated migrations may be hazarded, however. First, many such migrations are likely to have a one-way character, though in the case of non-Africans leaving the continent there is some justification in considering the movements as the return trip of a very long-range and even multi-generational circular migration, and in some cases of African refugee movements it may be assumed that there will be a return trip in the indefinite future. Second, the element of force or at least of legal prescription is often present. Third, political steps have acted both to cause migration by people who did not wish to migrate and to stop migrations of people

[63] Meran McCulloch, "Survey of Recent and Current Field Studies on the Social Effects of Economic Development in Inter-Tropical Africa," in UNESCO, *Social Implications of Industrialization and Urbanization in Africa South of the Sahara* (Paris, 1956), p. 94.

who did wish to. And fourth, the trends would appear to be toward increases both in the numbers involved in forced migrations or migrations motivated by fear of political repression and in the restrictions on the movement of African migrants across international borders. Some efforts have also been made to restrict internal migration, particularly to overcrowded urban centers, but these have generally been markedly unsuccessful.

Some political migrations have involved the exodus of non-Africans either voluntarily, as in the emigration of whites from Kenya, Tanzania, and other countries, or under fear of reprisals, as in the large-scale abandonment of Algeria, or because of increasing restrictions placed on them, as in the case of Asians in Kenya where laws passed in 1967 gave preference to Africans in job placement. The position of Asians in Kenya has been most difficult because Britain has refused to admit all of those who hold British passports, most do not wish to move to India or Pakistan, and even the status of those who have been granted or who have applied for citizenship in Kenya is distinctly uncertain. The Asian exodus from Kenya reached panic levels in early 1968 and again in 1969, but it will not be easy to cause the emigration from Kenya of all of the Asian population, which is estimated to have increased from 177,000 in 1962 to 192,000 in 1967. The most recent example of a European exodus is the flight of about 600 people from Equatorial Guinea in February–March, 1969 following disputes with the newly independent government.

Other political migrations have involved the flight or expulsion of foreign Africans from a number of countries. Cornevin notes that some of these have been caused by more or less spontaneous nationalist reflexes,[64] including the displacement in 1959 of about 17,000 Dahomeyans and 7,000 Togolese from the Ivory Coast after riots stemming from resentment over their holding positions in commerce and administration that might have been held by Ivoriens. Again in 1966 Houphouet Boigny, President of the Ivory Coast, was forced to retract the extension of dual citizenship to Upper Voltans and others in the Entente shortly after it was promulgated in December, 1965 due to the aggressive reaction of Ivoriens. Similar reflexes led to the expulsion of

[64] See Robert Cornevin, "Le grave problème des réfugiés en Afrique Noire," *Europe France Outremer*, 420 (January, 1965), pp. 32-35.

Dahomeyans from Niger and the mutual expulsion of peoples from Gabon and Congo (Brazzaville).

There are also examples of officially sponsored expulsion of foreign Africans, as in the case of the Congo (Kinshasa) decision of August, 1964 to expel all persons from Congo (Brazzaville) and Burundi; this decree might have affected some 100,000 people but led to an actual exodus of perhaps 30,000. And in late 1969 Ghana ordered the repatriation of all foreign Africans not possessing proper work permits, resulting in the exodus of perhaps 200,000 people in the ensuing two week period.

More and more common, too, are the restrictions placed on the immigration of foreign Africans by various destination countries, several of which, including Rhodesia and South Africa, have rather smugly noted that the ability to shut off foreign migrants gives them an enviable cushion against employment difficulties which might arise in an economic recession.

Another type of political migration is that related to tension between the government of an independent country and minority groups of indigenous people in that country. The flight from Ghana in 1959 of about 2,000 Ewe who had been accused of plotting with their brothers in Togo is an example. A second example is the large-scale flight of southerners resulting from the political-racial-religious difficulties between the Muslim northerners and the Christian-pagan southern tribes in the Sudan; their emigration has persisted for over a decade and has involved movement to Uganda, Congo (Kinshasa), the Central African Republic, Chad, and Ethiopia and from the last into Kenya after Ethiopia and Sudan concluded an extradition agreement. In mid-1966 it was estimated that 115,000 Sudanese refugees were present in six countries.

A third example of this type of migration is the flight of about 145,000 Tutsi from Rwanda up to December, 1963 following a decision by the Hutu-dominated government to sequester some of their traditional pastoral lands. The failure of a counterinvasion of Tutsi refugees led to much bloodshed and to further flights of the remaining Tutsi. Congo (Kinshasa) provides several additional examples, including the emigration under pressure of several thousand Katangese gendarmes to Angola and thousands of other refugees who fled to the surrounding countries. In mid-1966 an estimated 70,000 Congolese refugees were in Uganda, Burundi, Tanzania, and the Central African Republic.

Other examples of the emigration of indigenous peoples belonging to minority groups include: the exodus of Jews from Morocco and other northern African countries; the flight of Muslim Ethiopians to Sudan, estimated in March 1967 at 17,000 by the Eritrean Liberation Front and at 1,500 by the Ethiopian government; the flight of some 3,000 Rundi to Rwanda; and the emigration of Arabs from Zanzibar after they were deposed by the dominant Afro-Shirazis.

Still another type of political migration is associated with revolts against Portuguese authority in its African territories or with the desire of Africans to escape repression from any of the remaining white-dominated countries in Africa. The flight from Portuguese areas began on a large scale after the beginning of the revolt in Angola in 1961 and has been estimated to have involved 400,000–500,000 persons, almost all seeking refuge in Congo (Kinshasa). Some 61,000 refugees from Portuguese Guinea were reported to be in the Casamance area of Senegal in 1968. The number of refugees from Mozambique has been lower, having been estimated at 22,000 in 1968 in Tanzania and Zambia. These last two countries also now harbor a number of Rhodesian refugees. Africans who have left South Africa and South-West Africa differ from those who have emigrated from other countries in their generally higher level of educational achievement and in being scattered not only in other African countries but in Western Europe and the United States. Their total numbers are not particularly large. A rather special example of political migration was the French expulsion in March, 1967 of some 4,000–8,000 Somalis from what is now the French Territory of Afars and Issas prior to the election in that country, presumably to help assure a vote favorable to the French presence.

Udo reports an interesting type of political migration, which he calls census migration, whereby an estimated 15,000 Nigerians migrated in 1962 from their place of employment and residence to their villages to be counted there, in the expectation that the benefits promised in government promotional campaigns would accrue to those communities.[65]

The number of refugees in various parts of Africa has been increasing in recent years. In 1964, it was estimated to be 250,000, most of whom were from Congo (Kinshasa); in October, 1966 the U.N. High Commission for Refugees placed the number at 672,000; in mid-1967 it was put at 785,000; and in July, 1968 the High Commission listed

[65] Reuben K. Udo, "Census Migrations in Nigeria," paper delivered at the annual meeting of the African Studies Association, Los Angeles, October, 1968.

845,000 refugees. The actual number may be considerably larger, since many refugees can be absorbed in their host countries and go unrecorded in official statistics. Shamuyarira, for example, estimated the 1967 total to be some 215,000 higher than the U.N. figure,[66] while the U.S. Commission for Refugees estimated in late 1969 that the total number of refugees in Africa was no fewer than 5,206,213.

The countries which were recipient of the largest number of refugees in 1968 were Congo (Kinshasa), with 414,000 or about 49 percent of the official total, and Uganda, with 163,000 or about 19 percent of the total. Of the Congo total, 84.5 percent were from Angola, 9.7 percent from Sudan, and 5.8 percent from Rwanda. Other countries with significant percentages were Senegal (7.3 percent), Tanzania (4.7 percent), the Central African Republic (3.2 percent), and Zambia (1.3 percent). Sudan, which was the source of a large number of refugees, was also the receiving area of the largest number of new refugees in 1967–1968, with an estimated 28,600 Eritreans moving to Kassala. In September 1967 Sudan reported its intention to move about 123,000 Eritrean refugees from camps near the Ethiopian border in order to reduce tension on that border and to resettle them on agricultural projects near Khashm-el-Girba.

It should be recalled that much of the migration of Africans in South Africa is controlled by various political regulations, though economic needs have constrained the government from applying the laws as rigidly as it might like. Population movements other than the usual migration to and from the reserves are also associated with the apartheid policy in South Africa, including the removal of Africans from the so-called "black spots," or enclaves of African occupation within the European areas, and the obligatory shifting of numerous people in the North Transvaal as the government attempts to sort the Venda, Tsonga, Pedi, and Sotho into separate "Homelands."

ECONOMICALLY MOTIVATED MIGRATION

There is increasing agreement and evidence that the predominant motive behind most decisions to migrate is economic need and desire. Greater earnings in cash and/or in kind are desired for a variety of purposes, such as the payment of taxes or of the bride price, provi-

[66] Nathan Shamuyarira, "Political Refugees in Africa," *Institute of Race Relations Newsletter*, May, 1967.

sion of daily needs, the purchase of cattle or of more sophisticated material goods, and the satisfying of nonmaterial wants such as education. It should be reiterated that an economic component is also frequently present in migrations listed under other categories.

Several arguments have prevailed which have pertinence to a discussion of economic migration. Brief attention has already been given to one of these, namely the relative importance of economic and cultural forces, particularly that of the *rite de passage*. Gulliver summarizes the position when he writes that incentives in labor migration are "primarily and preeminently a desire for cash and material wealth which are not available at home," and that such migrations are *"not* explicable in terms of men seeking travel and adventure, new experiences, the wonders of the white man's world, wanderlust, the evasion of filial duties or political obligations, or others of the employers' and white man's stereotyped myths."[67] He further suggests that proof of this contention is seen in at least two ways: that many individuals who migrate to low-wage rural employment select it in order not to become involved in an unknown urban situation, and that many of those who migrate to high wage jobs state that they would prefer to stay home but cannot afford to.[68]

Another argument is concerned with whether Africans view migration as essentially repugnant or not. Gulliver's comments above also relate to this issue. Skinner's evidence (see page 171) would suggest that they do to some degree, although it is apparently undertaken in good spirit. Brooke's evidence is somewhat contradictory since, on the one hand, "movement out of the home area is a traumatic experience," but, on the other, "far more Sukuma migrate of their own volition than by compulsion."[69] Richard's findings relative to migration of Rwanda and Rundi to Uganda suggest that they would prefer not to migrate since "they usually figure in the reports as shy, timid and easily scared . . . [and] most spoke of sheer economic necessity or of particularly unpleasant political or kinship obligations they wanted to avoid."[70] Mitchell examines the work of numerous authors, including several who had been considered as weighting cultural over economic motivation, and con-

[67] Gulliver, "Incentives in Labour Migration," p. 161.
[68] *Ibid.*, pp. 161-62.
[69] Brooke, "Types of Food Shortages in Tanzania," p. 350.
[70] Audrey I. Richards, ed., *Economic Development and Tribal Change*, p. 52.

cludes that their evidence and analyses overwhelmingly support the contention that economic factors are predominant, but not the only ones.[71] One has reservations regarding the attitudes toward migration of persons moving from the South African reserves to employment in European areas. Certainly the rigid prescriptions which prevail, the mandatory separation of families due to the prohibition on wives and children moving to urban centers, and the harassment of the pass laws have been subjects of severe criticism and have created unnecessary distortions in the social fabric of African life in the Republic.

A third argument has to do with whether Africans respond to economic incentives in the way in which Westerners are supposed to, particularly with regard to the level of wages, the length of working period, and the satisfaction of wants. The claims have been that the African will work less at higher wage rates and more at lower ones primarily because he is a target worker who will work just so long as is required to amass the necessary funds to pay for the target item, be it a bicycle, a radio, three cattle, or the bride price. While there may have been some justification for these claims in the early years of contact or under a system of coercive migration, there is little evidence that they apply today. Berg found that most Africans do not change their period of stay in relation to changes in wage rates or village income. In the first place, many are contract laborers who are committed to work for a set period. Numerous other migrants plan their trips to accord with the agricultural calendar in the village. In any case, goals and wants undoubtedly change both in the short and long run. Finally, surveys and experience have shown quite clearly that, where possible, the African is selective with respect to the employment he seeks and responsive to incentive payments, wage rates, and provision of amenities. Berg even concludes that "it is most unlikely that for any given country . . . the aggregate supply of labor was ever negatively elastic with respect to wage rates."[72]

Types of Economic Migration

There are many different types of economic migration, not all of which are classified as "labor migration." Examples according to the type of work or profession involved include:

[71] Mitchell, "The Causes of Labour Migration," pp. 259-80.
[72] Berg, "Backward-Sloping Labor Supply Functions in Dual Economies," p. 492.

1. Traders. Several tribal groups are well known for their penchant as traders and market vendors. Some travel far afield from their home communities and may be kept in stock by a more or less elaborate supply system. Examples are: the itinerant "rug salesmen" from Morocco and other North African countries who are encountered in various West European cities; the Hausa, who may be seen selling handicrafts anywhere from Dakar to Lubumbashi; Yoruba "market mammies" found in the Northern Region and in Ghana, where they comprise 40 percent of the female vendors in the Kumasi market; the domination by Nigerians of the waterfront stores at Winneba in Ghana and the Fadama motor parts market outside of Accra; Gao yam vendors from Mali in the Kumasi market; woodcarvers from Machakos in Kenya who set up stalls or sidewalk displays in many East African cities; and itinerant artists and craftsmen encountered in almost any major African city outside of South Africa. A special type of trader is the cattle drover, who sometimes drives cattle and other livestock long distances to market, not infrequently into the tsetse-ridden areas where cattle cannot be raised and where there is likely to be a considerable demand for meat.

2. Fishermen. Several groups operate considerable distances from home waters, and not for the supply of their own areas. Dahomeyans and Togolese, for example, work along various stretches of the Guinea Gulf coast, fishing off the Ivory Coast being handled almost entirely by Togolese and others. On the other hand a considerable number of Ghanaians are reported as fishing regularly in Dahomey, while as many as 5,000 Fanti from Ghana fish from Liberian ports.

3. Military personnel. Service in the military forces of individual nations often involves migration outside one's home district, in part because it is frequent practice not to assign a unit to the place of origin of most of its personnel. During the last war numerous African troops were used outside the continent; Moroccan and Senegalese troops were dispatched after the war to help quell the revolt in Madagascar; and several countries sent troops to participate in the U.N. operations in Congo following the collapse of authority in that country after independence.

4. Prostitutes. A number of tribes specifically train girls to engage in prostitution, usually in towns somewhat removed from the home area. Examples include the Ouled Naïl of North Africa and the Kotokoli and Bassari of northern Togo. Froelich and others writing about the two

latter tribes state that many women go southward to such places as Accra to engage in prostitution. Often married in their native country they seek to become attached to a rich client, bear children, and one fine day disappear to return to Togo with their savings and their children.[73]

5. Laborers. Most of the remaining types may be classified under labor migration. Certain characteristics of interest may be noted with respect to specific types of employment:

a. Stevedoring. Several ethnic groups have developed specializations in this strenuous work, including Kru from Liberia and Sierra Leone who are picked up on the outward voyage by vessels which are proceeding to load logs and lumber in Gabon and Congo (Brazzaville), and dropped off on the return trip. In Madagascar, the French found it difficult to find enough men strong enough and willing to serve as stevedores and regularly brought Yemeni to the island on a contract basis.

b. Agricultural workers. A very large number of Africans have migrated to work both on the farms of Africans and on the estates and plantations of non-Africans, although it is not always possible to distinguish between numbers of workers secured from adjacent areas and those who have migrated from some distance. Sometimes the migrants go solely for the harvesting or other peak-period work, as is the case of the *navétanes* and "strange farmers" moving to the peanut harvest in Senegal and Gambia or of urban residents of Omdurman who go to the Gezira in the cotton-picking period. In some cases the migration can be done without interruption to the farming cycle at home; the example of the *masu cin rani* of Northern Nigeria has already been cited. In other cases the migration is of longer duration or may even become permanent as, for example, for some Voltans in Ghana, "Westerners" in the Sudan, or Rwandans in Uganda, some of whom have acquired their own land or tenancies. In some areas migrant farm workers are given space for a kitchen garden and allowed to keep a specified number of livestock.

c. Mine workers. Migration to mining areas occurs on a very large scale in Katanga, the Copperbelt, Rhodesia, and in South Africa, and provides the main opportunity for employment of numerous tribes, some of which develop something of a specialization in specific types of work involved in mining, such as the Sotho in the sinking of shafts on the South African gold fields. It should be noted that not all migrants go to

[73] Jean-Claude Froelich, Pierre Alexandre, and Robert Cornevin, *Les Populations du Nord-Togo* (Paris, Presses Universitaires de France, 1963), p. 25.

the mining areas for employment in the mines or by mining companies; substantial numbers are involved in service jobs or in catering in one way or another to the local population. Satellite communities often develop near to the company towns, and the total African population essentially dependent on the mining industry may be considerably greater than is represented by the mining employees and their families.

The patterns and characteristics of migrations to mining areas vary considerably. Alluvial diamond and gold fields in Sierra Leone and Ghana have attracted large numbers of bachelor migrants who "pothole" on their own account. In South Africa the vast bulk of mine workers come on set contracts on a bachelor basis and are housed in compounds, often reserved for specific tribal groups; most of those employed on the gold and coal mines are foreign Africans who are willing to work for the low prevailing wages. In Congo (Kinshasa), on the other hand, efforts were made to reduce the reliance on migrant workers by encouraging the permanent settlement of families. The two big Copperbelt companies had policies somewhere between those of Congo and South Africa with the trend more and more toward that of Congo, and there has, in fact, been a remarkable reduction in the turnover rate of African employees on the Copperbelt.

d. Urban workers. Very large and increasing numbers of migrants have been attracted to the urban centers of Africa, particularly to the primate cities and to those cities which have participated most fully in the process of modernization. Employment is sought in industry, government, commerce, transport, and services, but most large cities have received more migrants in recent years than the number of jobs available. Some of the characteristics and problems associated with migration to urban centers are treated in Chapter 4.

The Impact of Migration

Considerable attention has been given to assessing the faults and assets of migration, and there is a wide difference of opinion on various points. What is seen by one observer as a desirable economic phenomenon is interpreted by another as slavery. As is true for so many African subjects, the available data are often inadequate to permit a really thorough assessment or to justify rigid conclusions. In the following sections an attempt will be made to summarize the favorable and unfavorable

Workers on a tea estate in Malawi. African and expatriate farms and plantations have attracted very large numbers of migrants in several parts of the continent.

aspects of migration as they affect the source areas, the destination areas, and the migrant and his family, and as they relate to more pervasive social and economic conditions.

IMPACT ON THE SOURCE AREAS

It has frequently been claimed that the loss of able-bodied men from the rural areas may result in a reduction in the amount and variety of food produced and in a deterioration of the local agricultural system. Schapera concluded from a study in Botswana in 1943 that traditional animal husbandry and cropping had suffered because of inadequate attention resulting from the migration of men to South Africa, though he also observed that agriculture had benefitted from new methods learned abroad and from ploughs and other implements brought back by the migrants.[74] Skinner noted in 1965 that Mossi officials and the

[74] Schapera, *Migrant Labour and Tribal Life: A Study of Conditions in Bechuanaland Protectorate*, pp. 162-3.

Population Movements in Africa

French before them in Upper Volta believed that migration adversely affected the production of basic food crops but that the actual situation is not known. He agreed that migration has affected the cultivation of some crops, such as cotton, in part because their production did not jibe with the migration cycle, but he observed that it was more profitable to migrate than to grow cotton. He also indicated that returned migrants tend to give up farming since it provides no cash income as it does in Ghana and the Ivory Coast, and he noted that dry-season nonagricultural activities, such as weaving, well-cleaning, and hut-building are adversely affected, though the loss is offset by cloth brought back, government well-sinking programs, and imported construction materials.[75] Kay writes concerning the reserve areas of Rhodesia that

the dearth of adult males between the ages of 20 and 40 and the relatively heavy burden of young and aged dependents in the tribal areas are clearly shown. It is clear that the system of labour migration saps the vigour of the population in the tribal areas, and presents serious problems to those anxious to develop these areas.[76]

Margaret Read noted in her study of the Chewa in 1939 that migration had adversely affected the supply of labor. She wrote that

among the Chewa, the young bridegroom used to be expected to work for his wife's parents, hoeing their land as well as his own. . . . Today he gets out of this obligation whenever he possibly can by going off to work on the mines or farms.[77]

It is also suggested by several authors that where heavy work is required in agriculture, such as in the pollarding of trees under the *chitemene* system of the central African savanna woodlands, the absence of males will lead to a reduced opening of new plots, a simplified crop complex, and a general deterioration of indigenous farming.

On the other hand, numerous observers believe that as labor migration has become an accepted part of the tribal system, accommodations to it have minimized the unfavorable impact on local farming.

[75] Elliott Skinner, "Labor Migration among the Mossi of the Upper Volta," in Kuper, ed., *Urbanization and Migration in West Africa*, pp. 70-73.

[76] George Kay, "The Distribution of African Population in Southern Rhodesia: Some Preliminary Notes," *Rhodes-Livingstone Communication*, No. 28 (Lusaka, 1964), p. 16.

[77] Margaret Read, "Migrant Labour in Africa and Its Effects on Tribal Life," *International Labour Review*, XLVI, No. 6 (June, 1942), 627.

Brush ready for firing under the *chitemene* system of cultivation, northern Zambia. Involving a good deal of heavy labor in felling or pollarding the trees and assembling the cuttings, the system is likely to be adversely affected by large-scale out-migration of young males.

Colson suggests that official thinking was strongly influenced, indeed misled, by the contention that rural life would be disrupted if more than 5 or 10 percent of the able-bodied males were away; she concludes to the contrary that attempting to stabilize the population by stemming migration has frequently failed because it was working against more powerful forces which were good for the area.[78]

Watson, studying the Mambwe in Zambia, concluded that disruption of the farming system was offset because men in a kinship group worked out a system whereby work was shared while others were away, including the lopping of trees done by a few men working together once a year. He notes that many men traditionally did not work in agriculture anyway, since they were required to protect the village and its livestock. *Pax britannica* liberated these men, who could now be involved in migration without loss to agricultural production. Watson does sug-

[78] Colson, "Migration in Africa: Trends and Possibilities," p. 63.

gest that when the ratio of women to men exceeds 2 to 1 disruption of the economic and social life may begin.[79]

Van Velsen notes that the Tonga in northern Malawi have been able to adjust to an unusually heavy rate of migration (with 60–75 percent of the adult males away at any one time), because "their staple diet is cassava, the cultivation of which does not require much labor" and because "subsistence cultivation is almost entirely in the hands of the women and does not seem to be adversely affected by the exodus of male labor."[80]

Gulliver found that among the Ngoni, who had about a third of all males away at any one time and up to 50 percent in some areas, "the principal role of local agriculture—the provision of the family food supply—is . . . only slightly affected by the temporary absence of the average migrant."[81] Adjustments to the absence of migrants, who are usually away for from nine to eighteen months, include the assumption of work by women, the use of community working parties who usually work for beer supplied by the owner of the field, and by the preparation of any new fields required for the following season before the migrant leaves. Only when the migrant is away for two or more consecutive seasons may the wife experience serious difficulties, but she may then be aided by her husband's brothers or her own kinsmen.

In West Africa migrants from the savanna zones may migrate to the cocoa-coffee zones of the south in the slack season and thus minimize the effect on local farming. Skinner states that 80 percent of Mossi annual migrants return home for the agricultural season, and that they have set up an ideal pattern for migrations designed to minimize any upsetting impact on the sociocultural system of the tribe.[82]

Some of the concern expressed about the impact of migration on local farming seems to be based on the questionable assumption that "if, somehow, men would only dedicate themselves to transforming sub-

[79] See William Watson, "Migrant Labour and Detribalisation," in C.C.T.A., *Migrant Labour in Africa South of the Sahara*, pp. 281-97.
[80] J. Van Velsen, "Labor Migration as a Positive Factor in the Continuity of Tonga Tribal Society," *Economic Development and Cultural Change*, VIII, No. 3 (April, 1960), 267.
[81] P. H. Gulliver, *Labour Migration in a Rural Economy, East African Studies*, No. 6 (Kampala, East African Institute of Social Research, 1955), p. 33.
[82] Skinner, "Labour Migration and Its Relationship to Socio-Cultural Change in Mossi Society," p. 381.

Population Movements in Africa

sistence agriculture, the revolution would occur. . . . [But] the development of agriculture in the relatively poorly endowed areas from which most of the migrants originate entails a wide range of auxiliary investment, from roads to research,"[83] and to restrict migration would seriously damage many source areas.

This argument may be illustrated by the patterns in such countries as Gabon and Liberia where developments in plantations, forestry, and mining have appeared to result in lowered domestic production and heavier reliance on imports of food. The question is, is this necessarily bad? The activities noted have permitted a far more rapid rise in productivity, exports, and government revenues than would have been likely in modernizing and commercializing indigenous agriculture and hence the potentiality for a more rapid economic development. Furthermore, both countries have sparse populations and are not particularly favored for agriculture. The adjustment that has taken place is, therefore, both logical and the one to be expected. While it makes sense to be aware of the impact on domestic agriculture, the trends are not necessarily to be deprecated. What might be appropriate is to give attention to commercial forms of agriculture to help meet domestic needs rather than to worry too much about the effect on subsistence-farming areas.

Whatever the exact accounting may reveal there can be little doubt that migration does have the effect of draining away from the rural areas, either temporarily or permanently, some of the strongest, most able, most energetic young men and, as has already been noted, there is a tendency for those with a better education to leave their indigenous communities or to eschew assignment in government, education, and other services in rural areas.

On the other hand, migrants often provide income and material goods which their home villages would otherwise not have. These permit at least somewhat greater material comforts and, directly and indirectly, greater expenditures on education and other social needs. Read observed in her 1939 study that "it would be impossible to find in the Northern Province of Nyasaland today a village where a number of inhabitants had good houses, furniture and clothes, and other goods, and where none of them had been away at work."[84]

It is appropriate to distinguish between source areas on a local

[83] Berg, "The Economics of the Migrant Labor System," pp. 177-78.
[84] Read, "Migrant Labour in Africa and Its Effects on Tribal Life," p. 626.

scale and national source areas from which numerous migrants move to positions in other countries. In some cases it may be concluded that these laborers could have been used more effectively in work which would have helped to develop their own countries. Swaziland, for example, has sufficient potentialities to absorb the relatively small number of migrant workers who go to South Africa. The same is probably true of Mozambique, though development here would have required a greater willingness on the part of Portugal to accept investment by other nations.

For countries which are poorly endowed or densely populated or both, however, one must conclude that without the remittances and other earnings of migrants, these nations would be much worse off than they are. Upper Volta, Rwanda, Burundi, Malawi, Botswana, and Lesotho would be included. The foregone consumption of the large numbers of migrants is also a substantial benefit to these countries.

IMPACT ON DESTINATION AREAS

While it is obvious that much of African development has required the migration of workers to places where they were needed and hence that the effect on destination areas weighs heavily in favor of the system, it does not follow that there are not undesirable features associated with it.

First, the level of migration may be excessively high, resulting in unemployment, sometimes of serious proportions, in the destination areas. This has been an almost universal phenomenon in recent years as far as migration to major urban centers is concerned. The rise of a large class of unemployed workers in the cities places an unbearable load on the local and national governments and may contribute to political tensions inimical to stability and the most effective employment of limited available funds.

Migration may also be excessive to rural areas. In the Gezira, for example, the penchant of tenants to avoid physical labor to gain leisure time and prestige has led to the use of far more hired labor than would be necessary, resulting in a long-run decrease in agricultural productivity. This penchant has been related to the heritage of past use of slaves. Somewhat similarly, inhabitants of the fertile crescent of Uganda have not always taken advantage of the available opportunities for increasing earnings and productivity because they too have chosen to employ workers when they might have done the work themselves. An offsetting advantage of these seemingly undesirable situations is the spreading of wealth that prevents one region or group from drawing too far away from other regions and groups.

Another effect of migration on recipient areas which has been criticized is the segregation of migrants into communities or quarters, present even in small towns and villages where a foreigner is anyone from outside the local tribal group. Thus, while groups are brought together, they do not fuse, and divergence is maintained. This may result in resentment against those with superior skills, as in the tragic case of Ibo migrants in Northern Nigeria, or resentment against the influx of poor, unskilled workers if there is serious unemployment and local groups are willing to accept the available unskilled jobs.

The frequent reliance on *reciprocal* migration has additional disadvantages. It leads to lower labor productivity because rapid turnover rates do not permit adequate selection or full development of skills. This may not be a vital factor in occupations which require little skill, particularly in agriculture and, to a lesser degree, in mining, or when it is possible to hire several workers to do the work of one higher-waged, more skilled person. But it is undoubtedly a detriment in numerous jobs, particularly in manufacturing, which is not well suited to high turnover rates. This helps to explain why in such countries as Rhodesia and South Africa local labor tends to be used for more skilled jobs and extra-territorial or domestic migrants for work requiring less skill.

There are, however, advantages in circular migration for some employers and some receiving areas in that migrants are likely to be more docile, willing to accept lower wages, and to require less expenditure on social services, housing, and family support. It is also possible, by paying on a piecework basis, to shift the concern with efficiency to the worker or subcontractor.[85]

IMPACT ON THE MIGRANTS AND THEIR FAMILIES

One of the major objections to the migrant labor system is that it frequently involves the separation of families, which is socially disruptive at home, and which contributes to the rise of prostitution and homosexuality abroad. A scathing indictment of migration was given in

[85] Berg, "The Economics of the Migrant Labor System," p. 175.

the 1965 report of the Cape Synod of the Dutch Reformed Church which listed the disadvantages of the system as it exists in South Africa as: the neglect and moral disintegration of the workers' families in the homelands; attacks on African, Coloured, and White women and girls; the economic poverty of the families in the reserves; the frightening increase in homosexuality, especially in the compounds and single quarters; and the fighting and stabbing that go hand in hand with these social deviations and frustration. The report concludes that the disadvantages

can be summarized as the complete break-up of family life, a religious and social problem of the gravest moment. Under the conditions of migratory labour it is absolutely impossible to build a stable social life, and a peaceful and happy community.[86]

Houghton, using 193 employment histories, tells about a hypothetical typical migrant in South Africa who spent 64 percent of his working life from age 16 to 47 in employment away from home. In these 31 years he had 34 different jobs, remaining an average of 47 weeks in each job.[87] He concludes that

migratory labour cannot suddenly be abolished because the very survival of both White and Black depend upon it, but it should be recognized for what it is—an evil canker at the heart of our whole society, wasteful of labour, destructive of ambition, a wrecker of homes and a symptom of our fundamental failure to create a coherent and progressive economic society.[88]

The situation in South Africa is, of course, atypical, although conditions in Rhodesia, Angola, and Mozambique have some of the same earmarks and at least four other countries are affected to an important degree by the character of the South African migratory system because of the numbers of their citizens who work in the Republic. Major contrasts between the southern African system and that prevailing elsewhere are its rigid legalistic restrictions, the much wider sex ratios existing in receiving areas largely due to proscriptions on family move-

[86] As quoted in Peter Randall, "Migratory Labour in South Africa," South African Institute of Race Relations, *Topical Talks*, 1 (March, 1967), p. 6.

[87] D. Hobart Houghton, "Men of Two Worlds: Some Aspects of Migratory Labour in South Africa," *South African Journal of Economics*, XXVIII, No. 3 (September, 1960), 180-81.

[88] *Ibid.*, p. 189.

ment, the higher percentage of migrants living under controlled conditions in compounds, and the lack of freedom of choice and of representation on the part of the migrants. Many if not most of those individuals who are involved in circular migration in tropical Africa are doing so because they wish to keep one foot in the home area; in southern Africa there is usually no alternative.

Another criticism of the migratory system is that the migrant may be subjected to unsatisfactory employment conditions, particularly in those cases where he has no freedom of choice among employers and where labor legislation and unionization provide little or no protection. In many cities, however, various stranger groups have their own political associations and mutual assistance societies and the new migrant can move to such cities and still retain close ties with his tribe and culture.

Advantages accruing to the migrant worker to a greater or lesser degree include the acquisition of new skills and experience, increased earnings, and the possibility of achieving vertical mobility for himself and his children.

GENERAL CONSIDERATIONS REGARDING THE IMPACT
OF LABOR MIGRATION

There are a number of theoretical and actual ways in which migration may have an impact not necessarily associated with source and destination areas. Belgian and French colonial officials commonly complained, for example, that the separation of families resulted in undesirable national demographic trends which threatened the future well-being of certain countries. The Belgians were probably overly concerned in this matter because of early incorrect interpretations of birth and death rates. Both they and the French also appeared to feel that the population should be increasing everywhere regardless of supporting capacity. Colson reported that studies have failed to uncover any evidence that birthrates declined in areas subject to heavy migration. And today, there should certainly be more concern regarding excessively high rather than excessively low birth rates.

Migration has had both desirable and undesirable effects on health and the incidence and spread of disease. Moving from one region to another may subject the migrant to new diseases to which he may have no immunity. Prothero notes that "migrant labourers, when they return each year from working in and passing through malarious areas, bring fresh malaria infections and thus build up and maintain the reservoir of malaria transmission" and that nomadic movements "are a major hazard in eradication programmes."[89] Migrants may also be exposed to occupational diseases such as respiratory ailments, which caused a large number of deaths among miners before precautions were taken. And the high and sometimes increasing incidence of venereal diseases is also, at last in part, related to migration. On the other hand many large employers, such as miners and plantations, or urban centers, typically have superior medical and health facilities and records indicate that employees frequently leave work in better health than when they arrived. In some cases, supplementary food is provided to improve the health and stamina of workers.

It is sometimes objected that migration subjects the worker to economic conditions which are less stable than traditional life, such as the greatly reduced job opportunities which exist in times of recession or in a bad crop year or when a mine closes down upon exhaustion. While the assumption that rural life is stable is subject to question, there is little doubt that a modern economy is subject to greater dynamism and fluctuation than most subsistence situations. But to move from that point to suggest that migration to the modern economy is bad would be to suggest that it was better to sustain a subsistence economy to assure stability. This consideration might be taken as a reason to sustain the circular migratory pattern, to retain the security of the home farm while still participating in the modern economy, but even this reasoning is too simplistic.

It is difficult to see how modernization could have or can continue to take place without migration. From the social standpoint it has helped to break down some of the undesirable aspects of tribalism and, with some notable exceptions, to lower the barriers among ethnic groups. From the economic standpoint it has stimulated the growth of the money economy; even the reciprocal character of migration has advantages here in that it creates easier labor supply conditions and makes it easier for Africans to enter into wage employment without abandoning their interests at home. Furthermore, it tends to have a leveling influence, helping to counteract disparities which might lead to social and political problems later if not now. As Kay points out

[89] Prothero, *Migrants and Malaria*, pp. 4, 31.

the major gulf . . . is that between the money and indigenous economies. It is bridged by the system of labour migration. . . . The present system allows large numbers to participate from time to time and for varying periods in the money economy and to derive some benefits therefrom. It also provides much needed succor to the rural society in the form of remittances from absentees in paid employment. Such transfers of cash and goods constitute both "poor relief" and "social insurance payments."[90]

Berg, writing on the economics of migration in West Africa, says that

the migrant labor system represents an "efficient" adaptation to the economic environment in West Africa. Historically, it permitted West Africa to enjoy more rapid economic growth than would otherwise have been possible. It continues to benefit both the labor-exporting villages and the recipient areas. . . .

It is not, as is so frequently claimed, a cause of West Africa's economic ills; it is rather a symptom of a specific set of economic circumstances. . . . And until the economy does change, migration continues to make good economic sense, from the point of view both of the individuals concerned and the economy as a whole.[91]

Future Trends in the Level of Migration

The types of migration prevailing in Africa and the complex of factors bearing upon it make it extremely hazardous to predict whether it is tending to increase or decrease. The availability of data on past and present migrations is not adequate to develop more than very dubious projections. And in any case, certain elements are changing with such rapidity while others are so uncertain that prediction is impractical.

It is possible, however, to list a number of factors and forces which are tending to increase or decrease various types of migration. Factors which are likely to stimulate a continuing or increasing migration include the following:

1. Improving economies and modernization in some countries, creating more job opportunities. This would be true for such countries as Libya, the Ivory Coast, Ethiopia, Swaziland, and South Africa and to a lesser degree for another ten or twelve countries.

2. Increasing pressure in many source areas, either because of deteriorating capacity or increasing population or both.

[90] Kay, *A Social Geography of Zambia*, p. 148.
[91] Berg, "The Economics of the Migrant Labor System," pp. 160, 181.

3. Improving educational achievement levels, which lead to disenchantment with rural life and a desire for new types of employment.

4. The apparently growing attraction of urban communities.

5. An increasing interest in and desire for material and non-material purchases.

6. An increasing tendency in many areas for migration to shift from a bachelor to a family basis.

7. The likelihood that internal conflicts and repression will increase the flow of refugees. This is perhaps the most uncertain factor of all, since tensions can erupt rapidly and lead to some very large-scale movements.

Factors which are likely to depress migration levels include the following:

1. Tightened labor markets. The replacement of expatriates, which created a large number of jobs in the years following independence, is well advanced in most countries. More important, the economies of many countries have been stagnant if not retrogressive in recent years, witness the situation in such countries as Algeria, Senegal, Guinea, Dahomey, the Central African Republic, Rwanda, Burundi, Rhodesia, Botswana, Lesotho, the Seychelles, and the Mascarenes. In other countries the rate of growth has not been adequate to stimulate a satisfactory expansion in job opportunities, including Egypt, Sierra Leone, Togo, Mali, Upper Volta, Niger, Chad, and Somalia. A good many nations appear, in any case, to have very limited capacities to provide jobs outside the agricultural sector.

2. The desire by employers to stabilize labor to permit improved productivity.

3. The distinct trend in many receiving areas for migrants to remain for longer periods or even to remain permanently in order to assure retention of their jobs.

4. The rising population and labor pool in areas of employment and increasing unemployment in many urban centers.

5. In a few cases, improving economies in source areas, reducing the necessity and propensity to migrate.

6. Increased detribalization, modernization, and urbanization, leading to one-way rather than circular migration.

7. Nationalist reaction to the employment of foreign Africans, likely to reduce the opportunities for international migration.

8. Efforts by numerous countries to reduce the influx to cities, to be discussed in Chapter 5.

9. Ethnic conflicts which restrict the opportunities of some groups to migrate.

Several predictions may be hazarded. One, that circular migration will decline in significance except where it is retained by compulsion as in South Africa and except where it fits the needs of both migrant and employer as in the case of many agricultural jobs. Two, that migration to cities is likely to continue apace for some years despite the growing problems of absorption. Three, that there will be increasing efforts to control the movement of people but that these controls will continue to be largely ineffective. And four, that population movements will be such as to increase the disparity and contrasts among regions.

Bibliography

Adam, A. "Problèmes Sociaux du Maroc: l'Exode Rurale," *Sessions d'Études Juridiques, Politiques et Économiques* (Rabat, Faculté du Droit du Maroc), IV (1959), 67.

Apthorpe, R. J., ed. *Present Interrelations in Central African Rural and Urban Life.* Eleventh Conference, Rhodes-Livingstone Institute for Social Research, Lusaka, 1958.

Armor, Murray. "Migrant Labour in the Kalabo District of Barotseland (Northern Rhodesia)," *Bulletin, Inter-African Labour Institute*, IX, No. 1 (February, 1962), 5-41.

Barbour, Kenneth M. "Rural-Rural Migrations in Africa: A Geographical Introduction," *Cahiers de l'Institut de Science Économique Appliquée*, V, No. 9 (October, 1965), 47-68.

Battistini, R. "Note sur l'Agriculture Autochtone et les Déplacements Agricoles Saisonniers dans le Delta du Mangoky," *Mémoires Institut Scientifique Madagascar*, Sér. C, V (1959), 215-31.

Berg, Elliot J. "Backward-Sloping Labor Supply Functions in Dual Econo-mies," *The Quarterly Journal of Economics*, LXXV, No. 3 (August, 1961), 468-92.

——. *The Recruitment of a Labor Force in Subsaharan Africa.* Cambridge, Harvard University, 1960.

Berque, J., *et al.* "Nomads and Nomadism in the Arid Zone," *International Social Science Journal*, XI, No. 4 (1959), 481-585.

Bouvier, P. F. "Some Aspects of Labour Migration in the Belgian Congo," *Bulletin, Inter-African Labour Institute*, VI, No. 6 (November, 1959), 8-55.

Bohannon, Paul. "Migration and Expansion of the Tiv," *Africa*, XXIV, No. 1 (January, 1954), 2-16.

Brooke, Clarke. "Types of Food Shortages in Tanzania," *The Geographical Review*, LVII, No. 3 (July, 1967), 333-57.

Caldwell, John C. *African Rural-Urban Migration: The Movement to Ghana's Towns.* New York, Columbia University Press, 1969.

——. "Determinants of Rural-Urban Migration in Ghana," *Population Stu-*

dies, XXII, No. 3 (November, 1968), 361-77.

Carreira, A., and A.-M. de Meireles. "Quelques Notes sur les Mouvements Migratoires des Populations de la Province Portugaise de Guinée," *Bulletin de l'Institut Français d'Afrique Noire*, XXII, Sér. B, Nos. 3-4 (July-October, 1960), 379-92.

Clarke, John I. "Summer Nomadism in Tunisia," *Economic Geography*, XXXI, No. 2 (April, 1955), 157-67.

Collomb, Henri, and Henri Ayats. "Les Migrations au Sénégal: Étude Psychopathologique," *Cahiers d'Études Africaines*, II, No. 8 (1962), 570-97.

Colson, Elizabeth. "Migration in Africa: Trends and Possibilities," in Frank Lorimer and Mark Karp, eds. *Population in Africa*. Boston, Boston University Press, 1960, pp. 60-67.

Commission for Technical Cooperation in Africa South of the Sahara. *Migrant Labour in Africa South of the Sahara*. Abidjan, 1961.

Cunnison, Ian. *Baggara Arabs: Power and the Lineage in a Sudanese Nomad Tribe*. Oxford, Clarendon Press, 1966.

David, Philippe. "Fraternité d'Hivernage (Le Contrat de Navétanat): Théorie et Pratique," *Présence Africaine*, No. 31 (April-May, 1960), 45-57.

Davies, H. R. J. "Nomadism in the Sudan: Aspects of the Problem and Suggested Lines for Its Solution," *Tijdschrift voor Economische en Sociale Geografie*, LVII, No. 5 (September-October, 1966), 193-202.

——. "The West African in the Economic Geography of Sudan," *Geography*, XLIX, No. 3 (July, 1964), 222-35.

Deniel, Raymond. *De la Savane à la Ville: Essai sur la Migration des Mossi vers Abidjan et sa Région*. Aix-en-Provence, Centre Africain des Sciences Humaines Appliquées, 1967.

Descleres, R. "The Manpower Problems of the Ivory Coast and the Solutions Thereto," *Bulletin, Inter-African Labour Institute*, VII, No. 2 (March, 1960), 39-50.

Diop, Abdoulaye. "Enquête sur la Migration Toucouleur à Dakar," *Bulletin de l'Institut Français d'Afrique Noire*, XXII, Sér. B, Nos. 3-4 (July-October, 1960), 393-418.

Elkan, Walter. "Circular Migration and the Growth of Towns in East Africa," *International Labour Review*, XCVI, No. 6 (December, 1967), 581-89.

——. "Migrant Labor in Africa: An Economist's Approach," *The American Economic Review*, XLIX, No. 2 (May, 1959), 188-97.

——. *Migrants and Proletarians: Urban Labour in the Economic Development of Uganda*. London, Oxford University Press, 1960.

——. "The Persistence of Migrant Labour," *Bulletin, Inter-African Labour Institute*, VI, No. 5 (September, 1959), 36-43.

Engmann, E. V. T. "Population Movements in Ghana: A Study of Internal Migration and Its Implications for the Planner," *Bulletin of the Ghana Geographical Association*, X, No. 1 (January, 1965), 41-65.

Etienne, Pierre, and Mona Etienne. "L'Émigration Baoulé Actuelle," *Les Cahiers d'Outre-Mer*, XXI, No. 82 (April-June, 1968), 155-95.

Gleave, Michael B. "The Changing Frontiers of Settlement in the Uplands of Northern Nigeria," *The Nigerian Geographical Journal*, VIII, No. 2 (December, 1965), 127-41.

——. "Hill Settlements and Their Abandonment in Tropical Africa," *Institute of British Geographers, Transactions*, No. 40 (December, 1966), 39-49.

——. "Hill Settlements and Their Abandonment in Western Yorubaland," *Africa*, XXXIII, No. 4 (October, 1963), 343-52.

Grossman, David. "Migratory Tenant

Farming in Northern Iboland in Relation to Land Use." Unpublished Ph.D. dissertation, Columbia University, 1968.

Gulliver, P. H. "Incentives in Labour Migration," *Human Organization*, XIX, No. 3 (Fall, 1960), 159-63.

———. *Labour Migration in a Rural Economy: A Study of the Ngoni and Ndendeuli of Southern Tanganyika*. Kampala, East African Institute of Social Research, East African Studies, No. 6, 1955.

———. *Land and Social Change amongst the Nyakyusa*. Kampala, East African Institute of Social Research, East African Studies, No. 11, 1958.

———. "Nyakyusa Labour Migration," *The Rhodes-Livingstone Journal*, No. 21 (March, 1957), pp. 32-63.

Hailey, Lord. *An African Survey*. Rev. ed. London, Oxford University Press, 1957.

Hamrell, Sven, ed. *Refugee Problems in Africa*. Uppsala, The Scandinavian Institute of African Studies, 1967.

Harris, Marvin. "Labour Migration among the Moçambique Thonga: Cultural and Political Factors," *Africa*, XXIX, No. 1 (January, 1959), 50-64.

Harrison, Robert S. "Migrants in the City of Tripoli, Libya," *The Geographical Review*, LVII, No. 3 (July, 1967), 397-423.

Hassoun, I. A. " 'Western' Migration and Settlement in the Gezira," *Sudan Notes and Records*, XXXIII (1952), 60-112.

Henin, Roushdi A. "Economic Development and Internal Migration in the Sudan," *Sudan Notes and Records*, XLIV (1963), 100-19.

———. "A Re-estimation of the Nomadic Population of Six Northern Provinces [in Sudan]," *Sudan Notes and Records*, XLVII (1966), 145-47.

Hill, Polly. *The Migrant Cocoa-Farmers of Southern Ghana*. London, Cambridge University Press, 1963.

Houghton, D. Hobart. "Men of Two Worlds: Some Aspects of Migratory Labour," *The South African Journal of Economics*, XXVIII, No. 3 (September, 1960), 177-90.

———. *The South African Economy*. Rev. ed. Cape Town, Oxford University Press, 1967.

Hurst, H. R. G. "A Survey of the Development of Facilities for Migrant Labour in Tanganyika during the Period 1926-1959," *Bulletin, Inter-African Labour Institute*, VI, No. 4 (July, 1959), 50-91.

Jarrett, H. Reginald. "The Strange Farmers of the Gambia," *The Geographical Review*, XXXIX, No. 4 (October, 1949), 649-57.

Johnson, Marion. "Migrants' Progress," *Bulletin of the Ghana Geographical Association*, IX, No. 2 (July, 1964), 4-27.

Jürgens, Hans W., Kenneth A. Tracey, and Peter K. Mitchell. "Internal Migration in Liberia," *The Bulletin, The Journal of the Sierra Leone Geographical Association*, No. 10 (1966), pp. 39-59.

Khuri, Fuad I. "Kinship, Emigration, and Trade Partnership among the Lebanese of West Africa," *Africa*, XXXV, No. 4 (October, 1965), 385-95.

Kuper, Hilda, ed. *Urbanization and Migration in West Africa*. Berkeley and Los Angeles, University of California Press, 1965.

Lee, Everett S. "A Theory of Migration," *Demography*, III, No. 1 (1966), 47-57.

Leistner, G. M. E. "Foreign Bantu Workers in South Africa: Their Present Position in the Economy," *The South African Journal of Economics*, XXXV, No. 1 (March, 1967), 30-56.

Lombard, J. "Connaissance du Sénégal: Géographie Humaine," *Études Sénégalaises*, IX, No. 5 (1963), 1-183.

———. "Le Problème des Migrations 'Locales': Leur Rôle dans les Changements d'une Société en Transition

(Dahomey)," *Bulletin de l'Institut Français d'Afrique Noire*, XXII, Sér. B, Nos. 3-4 (July-October, 1960), 455-66.

McDonald, James R. "Labor Migration in France, 1946-1956," *Annals of the Association of American Geographers*, LIX, No. 1 (March, 1969), 116-34.

Manshard, W. "Land Use Patterns, Settlements and Agricultural Migration in Central Ghana," *Bulletin of the Ghana Geographical Association*, V, No. 2 (July, 1960), 23.

——. "Land Use Patterns and Agricultural Migration in Central Ghana (Western Gonja)," *Tijdschrift voor Economische en Sociale Geografie*, LII, No. 9 (September, 1961), 225-30.

Maquet, J.-J. "Motivations Culturelles des Migrations vers les Villes d'Afrique Centrale," *Folia Scientifica Africae Centralis*, II, No. 4 (1956), 6-8.

Mason, P. "Inter-Territorial Migrations of Africans South of the Sahara," *International Labour Review*, LXXVI, No. 3 (September, 1957), 292-310.

Mayer, P. "Migrancy and the Study of Africans in Towns," *American Anthropologist*, LXIV, No. 3, Pt. 1 (June, 1962), 576-92.

Mitchell, J. Clyde. "The Causes of Labour Migration," C.C.T.A. *Migrant Labour in Africa South of the Sahara*, Publication No. 79, Abidjan, 1961, 259-80.

Niddrie, David. "The Road to Work: A Survey of the Influence of Transport on Migrant Labour in Central Africa," *The Rhodes-Livingstone Journal*, No. 15 (1954), 31-42.

Noin, D. "Répartition de la Population et Mouvements Migratoires dans la Plaine du Tadla," *Revue de Géographie du Maroc*, VII (1965), 53-69.

Ominde, Simeon H. *Land and Population Movements in Kenya*. Evanston, Northwestern University Press, 1968.

——. "Population Movements to the Main Urban Areas of Kenya," *Cahiers d'Etudes Africaines*, V, No. 20 (1965), 593-617.

Oppong, C. "Local Migration in Northern Ghana," *Ghana Journal of Sociology*, III, No. 1 (February, 1967), 1-6.

Panofsky, Hans E. "Migratory Labour in Africa: A Bibliographical Note," *The Journal of Modern African Studies*, I, No. 4 (1963), 521-29.

——. "The Significance of Labour Migration for the Economic Welfare of Ghana and the Voltaic Republic," *Bulletin, Inter-African Labour Institute*, VII, No. 4 (July, 1960), 30-45.

Piault, M. P. "The Migration of Workers in West Africa," *Bulletin, Inter-African Labour Institute*, VIII, No. 1 (February, 1961), 98-110.

Plauvert, J.-Cl. "Migrations et Éducation," *Bulletin de l'Institut Français d'Afrique Noire*, XXII, Sér. B, Nos. 3-4 (1960), 467-75.

Prescott, J. R. V. "Migrant Labour in the Central African Federation," *The Geographical Review*, XLIX, No. 3 (July, 1959), 424-27.

Prothero, R. Mansell. "Labour Migration in British West Africa," *Corona*, IX, No. 5 (May, 1957), 169-72.

——. *Migrant Labour from Sokoto Province, Northern Nigeria*. Kaduna, Government Printer, 1959.

——. "Migrant Labour in West Africa," *Journal of Local Administration Overseas*, I, No. 3 (July, 1963), 149-55.

——. *Migrants and Malaria*. London, Longmans, Green and Co., 1965.

Randall, Peter. "Migratory Labour in South Africa," South African Institute of Race Relations, *Topical Talks*, No. 1, 1967.

Raulin, H. *Mission d'Étude des Groupements Immigrés en Côte d'Ivoire: Problèmes Fonciers dans les Régions de Gagnoa et Daloa*. Paris, ORSTOM, 1957.

Ravault, François. "Kanel: l'Exode Rural dans un Village de la Vallée du Sénégal," *Les Cahiers d'Outre-Mer*,

Population Movements in Africa

XVII, No. 66 (January-March, 1964), 58-80.

Ravenstein, E. G. "The Laws of Migration," *Journal of the Royal Statistical Society*, XLVIII, Pt. 2 (June, 1885), 167-235.

Read, Margaret. "Migrant Labour in Africa and Its Effects on Tribal Life," *International Labour Review*, XLV, No. 6 (June, 1942), 605-31.

Richards, Audrey I., ed. *Economic Development and Tribal Change: A Study of Immigrant Labour in Buganda*. Cambridge, W. Heffer, 1952.

Rouch, J. "Migrations au Ghana," *Journal de la Société des Africanistes*, XXVI, Nos. 1-2 (1956), 33-96.

Sautter, Gilles. *L'Exode vers les Grands Centres en Afrique Équatoriale Française*. Brazzaville, Institut d'Études Centrafricaines, 1950.

Schapera, Isaac. *Migrant Labour and Tribal Life: A Study of Conditions in the Bechuanaland Protectorate*. London, Oxford University Press, 1947.

Scott, Peter. "Migrant Labor in Southern Rhodesia," *The Geographical Review*, XLIV, No. 1 (January, 1954), 29-48.

——. "The Role of Northern Rhodesia in African Labor Migration," *The Geographical Review*, XLIV, No. 3 (July, 1954), 432-34.

Silberman, L. "Somali Nomads," *International Social Science Journal*, XI, No. 4 (1959), 559-71.

Skinner, Elliott. "Labour Migration and Its Relationship to Socio-Cultural Change in Mossi Society," *Africa*, XXX, No. 4 (October, 1960), 375-401.

——. "Strangers in West African Societies," *Africa*, XXXIII, No. 4 (October, 1963), 307-20.

Soper, Tom. "Labour Migration and Labour Productivity: Some Aspects of Experiences in East, Central, and Southern Africa," *Race Relations Journal*, XXV, Nos. 3-4 (July- December, 1958), 3-18.

——. "Labour Migration in Africa,"

Journal of African Administration, II, No. 2 (April, 1959), 93-99.

Sorré, Max. *Les Migrations des Peuples: Essai sur la Mobilité Géographique*. Paris, Flammarion, 1955.

Southall, Aidan, ed. *Social Change in Modern Africa*. London, Oxford University Press, 1961.

Spengler, J. J. "Population Movements and Problems in Sub-Saharan Africa," Chapter 10 in E. A. G. Robinson, ed. *Economic Development for Africa South of the Sahara*. New York, St. Martin's Press, 1964, pp. 281-311.

Stapleton, G. Brian. "Nigerians in Ghana," *West Africa*, XLIII (February 21, 1959), 175.

Stenning, Derrick J. *Savannah Nomads*. London, Oxford University Press, 1959.

——. "Transhumance, Migratory Drift, Migration: Patterns of Pastoral Fulani Nomadism," *Journal of the Royal Anthropological Institute*, LXXXVII, No. 1 (January-June, 1957), 57-73.

Thomas, Louis-Vincent. "Brève Typologie des Déplacements de Populations au Sénégal," *Cahiers de Sociologie Économique*, XX (May, 1964), 247-84.

Van Velsen, J. "Labor Migration as a Positive Factor in the Continuity of Tonga Tribal Society," *Economic Development and Cultural Change*, VIII, No. 3 (April, 1960), 265-78.

Vogel, William. "Is Labor Migration of Decreasing Significance in the Economies of East Africa?" Occasional Paper 34, Program of Eastern African Studies, Syracuse University, February, 1968.

Watson, William. "Migrant Labour and Detribalisation," *Bulletin, Inter-African Labour Institute*, VI, No. 2 (March, 1959), 8-32.

Wood, Eric W. "The Implications of Migrant Labour for Urban Social Systems in Africa," *Cahiers d'Études Africaines*, VIII, No. 29 (1968), 5-31.

CHAPTER 4

URBANIZATION IN AFRICA

AFRICAN CITIES, or at least the main cities in each country, are to an unusual degree the centers of modernization on the continent. They are the intellectual and social capitals, the seats of government, the main foci of political activity of all sorts, and the economic capitals of their respective countries. In the economic sphere they are the major transport centers, the main assembly and break-of-bulk points, the great markets, and the major financial nodes, and they contain the vast bulk of the newer market-oriented manufacturing establishments as well as a considerable share of the raw-material oriented plants. Indeed, one of the notable characteristics of many African countries is the rapid fading away of the signs of modernity as one leaves the urban centers.

The predominance of the primate cities in a variety of activities may be illustrated by a few examples. Dakar, with about 16 percent of the population of Senegal, accounts for 70 percent of the country's commercial workers, over 50 percent of employees in transportation, administration, and other services, and 80 percent of those in manufacturing; it consumes about 95 percent of total electricity consumed in the country. The "Three Towns" of Sudan—Khartoum, Omdurman, North Khartoum—employ nearly all of the educated Sudanese, contain half of the public utilities and most of the nation's industry, the main commercial and financial establishments, the only university, and the national government offices; they account for 90 percent of the vehicles registered in the country. Lagos, although it is only one of many urban centers in Nigeria and contains only about 1 percent of its population, accounts for about 46 percent of electricity consumed, 56 percent of the country's telephones, 20 percent of its newspapers, 90 percent of its periodicals, 37 percent of its hospitals, and 38 percent of vehicle registrations, while it is also the Federation capital, its leading port, rail, and airways center, and its single most important industrial node. Casablanca, in a country with a long tradition of urbanization and with ten cities of over 100,000, accounts for over half of the commercial and financial enterprises in Morocco, for over 70 percent of the tonnage handled in international trade, 57 percent of workers employed in industry and three-fifths of industrial output; the city consumes 59 percent of Morocco's high-tension electricity and has one-third of its telephones and two-fifths of the medical doctors in the country. The dominance of a single city is even more marked in some countries, including Conakry in Guinea, Abidjan in the Ivory Coast, Lomé in Togo, Bangui in the Central African Republic, Bujumbura in Burundi, Blantyre-Limbe in Malawi, and Maseru in Lesotho.

This chapter is devoted to an overview of urbanization in Africa. Following a brief historical survey, data will be presented regarding the level of urbanization, the size patterns of African towns, and the estimated growth rates in recent decades. The next major section examines the various typologies of African cities, which is succeeded by a discussion of a number of the problems besetting Africa's cities and towns. Finally, a brief comparison of present urbanization in Africa with that of the West at a roughly comparable stage of development is assayed. Chapter 5 presents thumbnail sketches of a number of African cities, illustrating the considerable variety and individuality that exists as well as the prevalence of certain problems in almost all cities.

Urbanization has not been an entirely modern development in Africa, some of the world's earliest cities having been formed in Egypt and along the Mediterranean shores. Most of the earliest cities are marked only by ruins if at all, but later cities have been built and rebuilt on or close to their sites. In sub-Saharan Africa some capitals of ancient kingdoms and emirates date to the tenth and eleventh centuries, and others were formed in the following centuries. Occasionally, modern centers have developed on or near them, but most of them have disappeared or remain as little more than villages. Dates of founding are frequently very hazy and vary by several centuries in different sources; estimates of early population levels are equally if not more uncertain.

ANCIENT CITIES

The earliest cities in Africa date to about 3,000 B.C., some centuries after the formation of urban centers in the Euphrates, believed to be the first in the world. It is not likely that any of these cities ever contained more than 5 or 6 percent of the population of their regions;[1] none of them exist today. Alexandria was a pharaonic city from about 1500 B.C. but it was founded with its present name in 332 B.C. by Alexander the Great who added a post to several existing villages; in the first century B.C., when cities began to dot the shores of the Mediterranean, it probably had about 20–30,000 residents. During the most advanced period of the Roman Empire, from the first to the third centuries A.D., a number of their cities in northern Africa attained great size and may have contained as many as 25 percent of the population of their regions. Roman centers were located at or near the sites of many present cities, including Rabat, Moulay Idriss, Tangier, Algiers, Bizerte, Tunis, which may have predated Carthage, and Tripoli. Almost all of these cities disappeared, however, before the present cities were built. The oldest cities in Africa which still exist would include Alexandria, Tripoli, Constantine, founded about 313 A.D. by the emperor of the same name on a Carthaginian town he had destroyed, and Aksum in the Eritrean Province

[1] See Amos H. Hawley, "World Urbanization: Trends and Prospects," in Freedman, ed., *Population: The Vital Revolution.*

of Ethiopia, which may date to the first century A.D. The last, however, is now only a relatively small and poverty-stricken town.

By the fifth century A.D. cities were in decline everywhere in the West and in northern Africa. The seventh century saw the rise of Gao in Mali, capital of the Songhai Empire, and Harar on the eastern massif of Ethiopia, while in the eighth century Muslim cities were developed along the Indian Ocean coasts, including Mogadishu in Somalia and Mombasa in Kenya, which remained a center of Arab trading and slaving into the sixteenth century.

In the eighth to tenth centuries Muslim cities developed in northern Africa and benefited from the revival of trade in the tenth century, including Fès (790), Kairouan (800), Oujda (944), Algiers (950), Meknès, and Cairo. Other cities with ancient origin, but which are difficult to assign to a specific date because of repeated sackings and transformations, include Ceuta, Melilla, Tlemcen, Oran, Sétif, Bejaia (Bougie), Annaba (Bône), Bizerte, Gabès, Suez, and Berbera.

In the tenth and eleventh centuries a number of towns appeared in West Africa, including Kano (999), Zaria, and Katsina in northern Nigeria and a number of the Yoruba towns in southwest Nigeria, including Ife, Iwo, Ilesha, Iseyin, Ede, Ilorin, and Ijebu-Ode. Benin, the ancient capital of the Bini and Tombouctou (1100) also date from this period. Many caravan centers developed from about the twelfth century, but most have since declined or disappeared and it is doubtful if any had as brilliant histories as are sometimes depicted.

THE PERIOD FROM THE SIXTEENTH TO THE LATE NINETEENTH CENTURY

This period witnessed the formation of a number of capitals of indigenous African kingdoms and ethnic groupings in sub-Saharan Africa, mainly in the west, though their exact founding dates usually remain shrouded in mystery. Among towns which still exist, some only as minor centers, are a number along the northern belt, including: Labé in Guinea, Sikasso and Ségou in Mali, the last the capital of the Bambara; Bobo-Dioulasso and Ouagadougou in Upper Volta, the latter having been the capital of the Mossi for centuries, Zinder in Niger, Abéché in Chad, and El Fasher, center of the Darfur Kingdom in Sudan, plus several early religious centers such as Louga in Senegal and Mopti in Mali.

Urbanization in Africa

Along the southern belt of West Africa, new indigenous centers included Kumasi (c. 1663) in Ghana, seat of the Asantahene of Ashanti, and Abomey and Porto Novo in Dahomey. Tananarive, capital of the Merina in Madagascar, is believed to date from about 1625. Ruins of stone buildings begun in the twelfth century but dating mainly from the mid-fifteenth century, are found in a belt running from Botswana across Rhodesia into Mozambique, indicating a considerably higher density of population than now exists in the area. Zimbabwe is the most spectacular example of these constructions, but neither it nor the other ruins necessarily indicate the presence of towns.

European influence in sub-Saharan Africa first became marked by settlements in the sixteenth century, usually small coastal forts, trading posts, or way stations. The Portuguese founded such settlements in this and the following century, sometimes adjacent to small African villages, at Rufisque (Rio Fresco) in Senegal, Kayes in Mali, Bissau in Portuguese Guinea, Porto Novo and Ouidah in Dahomey, Lagos in Nigeria, Luanda (1575), Benguela (1617), and São Salvador do Congo in Angola, Moçambique (1508) and Sena in Mozambique, Zanzibar City, and Mombasa in Kenya, while they also influenced construction at Gondar, the Ethiopian capital from 1650–1867.

Other Europeans, particularly the French, Danish, Dutch, and British, established forts along the West African coasts, including Saint-Louis (1659) and Gorée in Senegal, Conakry in Guinea, Sekondi, Cape Coast, and Accra in Ghana, and Calabar in Nigeria. The seventeenth century also saw the first European settlements on the Indian Ocean islands including Fort Dauphin, Madagascar (1643–1674), and Saint-Denis and Saint-Paul on Réunion; Port Louis on Mauritius dates from 1736.

Many of these early towns remained insignificant for centuries and some are still minor agglomerations. Saint-Louis had only 7,000 inhabitants in 1786, 10,000 in 1825, and 16,000 in 1878; Kumasi had a population estimated at 10,000 in 1874 which fell to only 3,000 in 1901; Gondar had only about 1,000 people in 1905; and Zanzibar City remained unimportant until 1832, when the Sultan of Oman transferred his capital there.

The eighteenth century saw relatively little town formation. Freetown, founded as a place to settle freed slaves plus some petty criminals from England, dates from 1787, but it had only 2,000 inhabitants in

1808 and 10,000 in 1833. The first half of the nineteenth century saw Alexandria and Tunis revive after centuries of neglect, and Cairo again became one of the great capitals of the world. French occupation of Algeria led to the renewal and modern expansion of Algiers, transformation of Oran and other cities into modern centers, and development of European towns at Sidi bel Abbès, Skikda (Philippeville), and elsewhere. In West Africa, Monrovia was founded in 1822 as another settlement for freed slaves, the British established a post at Bathurst, Gambia, in 1816, and most of the present large Yoruba towns were founded in this period, which was marked by attacks by Fulani from the north and fighting among the Yoruba tribes themselves. Important among the newer towns were Oshogbo (1825), Ibadan (1829), Abeokuta (1830), and Oyo (1839); many of the Yoruba towns visited by Clapperton in 1826 and Lander in 1830 have, however, disappeared. Some were destroyed in warfare or in slave raiding by other groups, some by epidemics, especially of smallpox. Once a settlement built of traditional materials has been abandoned, the elements and insects quickly obliterate its remains and the site reverts to the bush.

Elsewhere in Africa, El Obeid and Khartoum in Sudan were founded in 1821 and 1829, Mwanza and Tabora in Tanganyika as Arab-Swahili posts by 1840, Tamatave on Madagascar in 1811, Libreville in Gabon in 1843, and Lobito in Angola in the same year by residents of Benguela who sought a more healthful site and a better harbor. The period also saw the founding of a number of South Africa's major cities —Port Elizabeth in 1799, Durban (then Port Natal) in 1824, Pietermaritzburg in 1838, Bloemfontein in 1846, East London about 1847, and Pretoria in 1855.

THE MODERN PERIOD

It was not until the last of the nineteenth century and the present century that many of sub-Saharan Africa's major urban centers were developed, and most of these remained quite small for several decades. Before partition of the continent European interest had been confined largely to small coastal locations; in sub-Saharan Africa the only significant indigenous agglomerations were the capitals of the larger kingdoms and chiefdoms, best developed in West Africa and across the sudan belt; Arab towns in the east were comparatively small.

The early period of colonial expansion temporarily halted the growth of some towns, such as Abeokuta and Zanzibar and its mainland satellites. It brought new direction to a number of centers such as those in Egypt, Tunisia, and somewhat later, Morocco. A few centers which had long existed as small posts surged to the then large size of 20,000 and more. And numerous new centers arose, many of which have since become major urban agglomerations if not the primate cities of their respective countries. Some of these were located adjacent to indigenous villages but to all intents and purposes they were new creations, built under European stimulus and direction.

Important centers founded in the latter part of the nineteenth century or in the first two decades of the twentieth century include:

Northern Africa
Sudan: Omdurman, 1885; Atbara, 1899; Port Sudan, 1905.

West Africa
Mali: Bamako, 1883
Ivory Coast: Bouaké, 1896; Abidjan, 1903
Nigeria: Jos, 1903; Maiduguri, 1907; Enugu, 1909; Port Harcourt, 1912; Kaduna, 1913.

Middle Africa
Chad: Fort Archambault, 1899; Ft. Lamy, 1900.
Central African Republic: Bangui, 1899.
Congo (Brazzaville): Brazzaville, 1883.
Cameroon: Yaoundé, 1889.
Congo (Kinshasa): Kinshasa (Leopoldville), 1881; Matadi, 1887; Kisangani (Stanleyville), c. 1900; Albertville, 1915; Lubumbashi (Elisabethville), 1910; Likasi (Jadotville), 1917.
Angola: Nova Lisboa, 1912.

Eastern Africa
Ethiopia: Addis Ababa, 1886; Asmara, 1897; Dire Dawa, 1904.
French Territory of Afars and Issas: Djibouti, 1892.
Kenya: Nairobi, 1899.
Uganda: Kampala, 1890; Jinja, 1901.
Tanzania: Dar es Salaam, 1862.
Malawi: Blantyre, 1876; Zomba, 1890; Limbe, 1909.

Rhodesia: Fort Victoria, 1890; Salisbury, 1890; Umtali, 1890; Bulawayo, 1893; Gwelo, 1894.
Zambia: Kabwe (Broken Hill), 1902; Livingstone, 1903.
Mozambique: Lourenço Marques, 1887; Beira, 1891.
Madagascar: Diégo-Suarez, 1884; Majunga, 1897.
Réunion: Le Port, 1882.

Southern Africa
South Africa: Kimberley, 1871; Johannesburg, 1886.
Lesotho: Maseru, 1869.
Swaziland: Manzini, 1885; Mbabane, 1887.

Of these new towns only Omdurman and Addis Ababa could be described as indigenous creations; Kampala was founded next to one of the few important African agglomerations in Eastern Africa, Mengo, capital of the Kingdom of Buganda.

Despite the impressive number of now-important cities dating from the late nineteenth century to World War I the population growth in most of them was quite slow during this period. Only northern and southern Africa had cities of substantial size, with the exception of a number of indigenous agglomerations. In the north, Algiers reached 103,000 in 1886, Oran 106,000 in 1906, at which time there were eight other cities in the country over 20,000; Cairo had an estimated 590,000 in 1897, Alexandria had 316,000, and there were six other cities over 20,000 in Egypt. In South Africa, mushrooming Johannesburg reached 237,000 in 1911, when Cape Town had 162,000 and Durban 90,000, while eleven other towns had populations exceeding 25,000. Relatively large indigenous towns included: Kumasi (19,000 in 1911), those of Yorubaland (e.g., Ibadan, 200,000 in 1900), northern Nigeria (Kano, 30,000 in 1903), Addis Ababa (35,000 in 1908), Omdurman (59,000 in 1910), and Tananarive (65,000 in 1914). Almost all intertropical cities were under 20,000 and many had not yet reached 10,000. Cities which had fewer than 10,000 in the period 1900–1916 include: Bamako (7,000 in 1910), Conakry (6,000 in 1910), Cotonou (2,000 in 1910), Bangui (8,000 in 1916), and Kinshasa (5,000 in 1908). Cities with from 10–30,000 population totals include Dakar (19,000 in 1904), Accra (27,000 in 1901 and 30,000 in 1911), Lagos (18,000 in 1901, but mushrooming to 74,000 in 1910), Mombasa

(30,000 in 1906), Nairobi (12,000 in 1906), Dar es Salaam (20,000), and Lourenço Marques (10,000 in 1907).

The period between the two World Wars saw a steady if not spectacular increase in the large cities of northern Africa. Cairo became the first million city in Africa by the time of the 1927 census. South Africa saw continued strong growth, interrupted only briefly by the early depression years. Jo'burg had 519,000 people at the time of the 1936 census, which recorded four other cities over 100,000 and nine between 50,000 and 100,000. Tropical Africa witnessed considerable growth in a number of centers. By World War II, Dakar, Kaolack, Lagos, Addis Ababa and Tananarive exceeded 100,000; Freetown, Accra, Kinshasa, Luanda, Asmara, Mogadishu, Nairobi, Mombasa, Dar es Salaam, Salisbury, and Lourenço Marques had over 40,000 inhabitants. But many centers of substantial size today remained below 20,000. The size of the major towns of tropical Africa at the end of the period seems small in relation to their present development, but the rapidity of growth was not expected and the administrations had not planned adequately to accommodate the increasing numbers of migrants who displayed, for the first time, a substantial degree of spontaneity. New towns formed in the interwar years were largely secondary and tertiary centers; exceptions include Niamey, capital of Niger, Pointe Noire, port and terminus of the Congo-Ocean Railway in Congo (Brazzaville), and new mining centers in Zambia, Congo (Kinshasa), and South Africa.

Since the last war the pace of urbanization has accelerated markedly and continues to do so in most countries. Particularly in tropical Africa this has been a period of mushrooming growth. Only a few new towns have been created in the period. They include: Marsa Brega, one of the petroleum processing and exporting points on the Gulf of Sirte in Libya; Nouakchott, the new capital of Mauritania; a number of relatively small mining centers in Guinea, Liberia, Gabon, Tanganyika, and Zambia; the new port of Tema in Ghana which is actually part of the Accra agglomeration; Kariba in Rhodesia; and, in South Africa, Welkom and other mining towns on the Orange Free State goldfields, and Sasolburg, an important petrochemical center. In South Africa, urbanization has continued apace despite increased restrictions and regulations governing the free movement of Africans. Map 23 depicts the towns and cities of over 10,000 in Africa by various size ranges for the most recent year permitted by the available data.

Urbanization in Africa

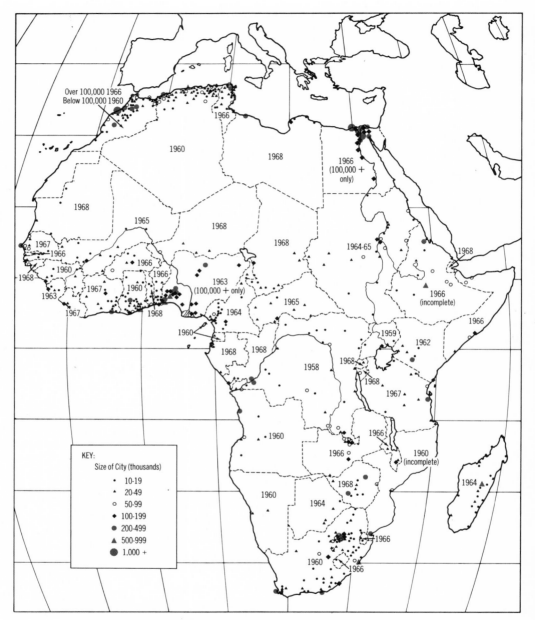

KEY:
Size of City (thousands)
- • 10-19
- ▲ 20-49
- ○ 50-99
- ◆ 100-199
- ● 200-499
- ▲ 500-999
- ● 1,000 +

MAP 23. The towns and cities of Africa by selected size ranges.

Urbanization in Africa

Africa is the least urbanized of the inhabited continents by whatever definition of urban is accepted and whatever cutoff point is used to distinguish villages, towns, and cities (Tables 11 and 12).[2] The major

Table 11

PERCENT OF URBAN POPULATION FOR MAJOR
WORLD REGIONS, 1950

	Percent in Cities of	
	20,000+	*100,000+*
World	21	13
AFRICA	9–10	5
Oceania	47	41
North America	42	29
Europe (except U.S.S.R.)	35	21
U.S.S.R.	31	18
South America	26	18
Middle America	21	12
Asia (except U.S.S.R.)	13	8

SOURCE: Kingsley Davis, "The Origin and Growth of Urbanization in the World," in H. M. Mayer and Clyde F. Kohn, eds., *Readings in Urban Geography* (Chicago, University of Chicago Press, 1959), p. 64.

explanation for the relative level of urban development in Africa is the lateness of its exposure to modern economic and technologic forces. This is reflected in the high percentage of people in many regions still in the subsistence sector and conversely the relatively poorly developed exchange economy over much of the continent, the early stage of industrial development in most countries, and the generally restricted representation of many services which are characteristically urban based. Additional explanations include the small size of numerous political entities, the fragmented market and population patterns of several countries, and the remote and land-locked character of several nations and many regions.

[2] It should be noted that use of statistics on African towns and cities is affected by a variety of difficulties including lack of census and survey data, the rapidity with which figures are outdated, wide variations in estimates given for the same year, failure to distinguish between city and metropolitan area, and changing town boundaries over a period of years.

Continental averages for levels of urbanization, however, conceal wide differences among the major regions. As would be expected, South Africa has the highest percentage of its total population urbanized. Northern Africa, with a long heritage of urban living in several coun-

Table 12

ESTIMATED PERCENT OF URBAN POPULATIONS
FOR AFRICA AND THE WORLD, 1920-1967

| | 5,000+ | 20,000+ | | 100,000+ | |
	Africa	*World*	*Africa*	*World*	*Africa*
1920		14	5		
1930		16	6		
1940		19	7		
1950		21	9–10	13	5
1960		25	13		9.7
1967	19.8		15.8		11.5

SOURCE: U.N. Department of Economic and Social Affairs, *International Social Development Review*, No. 1, *Urbanization: Development Policies and Planning* (New York, 1968), p. 11. Estimates for 1967 from city data compiled by author.

tries, also ranks relatively high, while tropical Africa has the lowest urban representation (Table 13). Within tropical Africa, West Africa has the highest urban percentage, reflecting both the significance of numerous indigenous communities, some of very substantial size, and the presence of more modern cities in a region which has had considerably longer contacts with the Western world than other parts of tropical Africa. Middle Africa, with only a third of the total population of West Africa, also has a relatively high percentage of urban residents, reflecting in part the unusually high percentages in such countries as Congo (Brazzaville), the Central African Republic, and Gabon. Almost all towns and cities in this subregion were European creations. Eastern Africa has a markedly lower urban percentage, explained in part by the relatively low development in its most populous country, Ethiopia, and unusually low percentages of urbanization in the five East African countries—Kenya, Uganda, Tanganyika, Rwanda, and Burundi—which had 40 percent of the total population of the subregion.

Each of the major subregions of Africa has marked internal differences from country to country, as is illustrated in Table 14 and

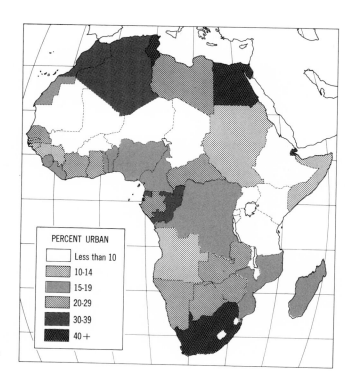

PERCENT URBAN

	Less than 10
	10-14
	15-19
	20-29
	30-39
	40+

MAP 24. Estimated percentage of urban population by country, 1967.

Map 24. In northern Africa all of the major countries except Sudan have relatively high urban percentages; Sudan follows the sub-Saharan pattern more closely, as is to be expected from the location of most of its productive regions and the large Three Towns conurbation.

In West Africa urban percentages range from around 6 to over 29. Several patterns may be delineated. One, the landlocked countries—Mali, Upper Volta, Niger—have among the lowest levels of urban development. Two, the three countries with large portions of their territory in the Sahara, including two of the landlocked countries plus Mauritania, are at the low end of the scale. Three, Senegal and the Ivory Coast, which contain the two largest cities in francophone West Africa and which have received far more attention from France than other former colonies, rank close to the high end. Dakar in Senegal owes a portion of its size to its former position as capital of French West Africa, which brought it the largest expatriate population of any city in West Africa.

Table 13

ESTIMATED PERCENT OF URBAN POPULATIONS
IN MAJOR AFRICAN REGIONS, 1960 AND 1967

Size in thousands	1960		1967		
	20+	100+	5+	20+	100+
Northern Africa	23.8	19.0	32.2	24.9	19.7
West Africa	10.7	7.3	16.6	11.7	8.8
Middle Africa	8.4	4.7	15.6	11.4	7.5
Eastern Africa	5.6	3.1	9.5	7.6	4.9
Southern Africa	34.5	23.6	41.3	40.7	29.6
AFRICA	13.7	9.7	19.8	15.8	11.5

SOURCE: Calculated from data on individual cities and from national censuses.

It now suffers from a reduced hinterland and one which is potentially far less productive than that of Abidjan in the Ivory Coast; nonetheless it continues to grow with no close relation to expansion of its national economy. The phenomenal growth of Abidjan dates from about 1948 when it had a population of 54,000; ten years later it was about three times as large and probably reached ten times that figure in 1968 or 1969. Four, Ghana, which has had the highest per capita income in West Africa, has the highest urban percentage. Nigeria, despite the large number of indigenous cities and the nearly unique urbanization of the Yoruba in the southwest, has a total percent close to the average for West Africa. This may be explained in part by the relatively slower rate of growth of most of these cities plus the relative unimportance of colonial centers comparable in dominance to those of other countries.

Middle Africa shows some of the same patterns as West Africa. Chad, landlocked and having a large bite of the Sahara, has the lowest level. Congo (Brazzaville) compares to Senegal with Brazzaville having been the capital of French Equatorial Africa and the main focus of the expatriate population; the urban population of that country is further increased by the necessity of having ocean and river ports at each end of the Congo-Ocean axis. Nonetheless it appears to have an urban ratio more out of proportion to the level of economic development than any African country. The Central African Republic also seems to have experienced an inexorable pull to its capital city, Bangui, whose population bears little relation either to its economy or that of the country.

Urbanization in Africa

Table 14

ESTIMATED PERCENT OF URBAN POPULATIONS
IN AFRICAN COUNTRIES, 1967

Country	Percent urban	Country	Percent urban	Country	Percent urban
NORTH AFRICA		Ghana	26.0	Tanganyika	5.5
Span. Sahara	16	Togo	19.5	Zanzibar	32.8
Morocco	30.7	Dahomey	16.1	Rwanda	1.2
Algeria	32.5	Nigeria	17.0	Burundi	3.4
Tunisia	40.0	Subtotal	16.6	Malawi	5.0
Span. N. Africa	100	MIDDLE AFRICA		Rhodesia	21.0
Libya	26.0	Chad	9.0	Zambia	21.0
U.A.R. (Egypt)	41.0	C.A.R.	27.0	Mozambique	17.0
Sudan	12.0	Congo (B)	38.5	Madagascar	15
Subtotal	32.2	Gabon	23.0	Comoro Is.	8
WEST AFRICA		Cameroon	17.0	Seychelles	27
Mauritania	6.1	Equat. Guinea	30.0	Réunion	35
Senegal	29.2	São Tomé, Príncipe	15.0	Mauritius	50
Mali	9.6	Congo (K)	15.0	Subtotal	9.5
Upper-Volta	6.6	Angola	12.0	SOUTHERN AFRICA	
Niger	6.1	Subtotal	15.6	South Africa	45
Gambia	13.0	EASTERN AFRICA		South West Africa	20
Port. Guinea	12.5	Ethiopia	5	Botswana	27
Guinea	10.0	F.T.A.I.	90	Lesotho	3
Sierra Leone	13.0	Somalia	13	Swaziland	7
Liberia	15.0	Kenya	9	Subtotal	41.3
Ivory Coast	22.0	Uganda	6	CONTINENT	19.8

SOURCE: National censuses; data on individual cities.
   N.B. Some extrapolations are based on partial data and are subject to considerable error.

Eastern Africa is an extremely diverse region and it is not surprising that urban levels should vary considerably. Striking features of the pattern are: the low urban levels of Ethiopia and the East African countries, plus Malawi; the markedly higher percentages for Rhodesia and Zambia; and the generally high ranking of the small, densely populated islands such as Zanzibar, Mauritius, and Réunion. The low levels in the three main East African countries is explained in part by restrictions on movement to the cities under the colonial regimes. Low urbanization in Uganda, Rwanda, and Burundi reflects the preferred dispersed settlement patterns of the major ethnic groups. The relatively high percentage for Rhodesia is explained by the high proportion of Europeans who reside in the cities, which are all European initiated, plus the unusually well developed industrial and service sectors. Zambia's ranking mirrors the large-scale character of mining on the Copperbelt which has given rise to one of the most urbanized zones in Africa; the post-independence boom in Lusaka must also be mentioned.

In southern Africa, the Republic has the highest urban percentage of any major country on the continent. This reflects not only the significance of manufacturing in Africa's leading industrial nation, but the massive and concentrated character of the gold and coal-mining regions, and the great port activity, particularly at Durban and Cape Town. Botswana's percentage is misleadingly high; most of its centers are large agrotowns, while the three European-created towns had a total of only 22,000 inhabitants at the time of the 1964 census. Gaborone (Gaberones), the new capital, has since grown rapidly but was still under 20,000 in 1968.

It is clear that correlations may be made between the level of urbanization and such factors as location, political significance, development of mining, manufacturing, and services, and expatriate population for individual countries, but the patterns break down on a continental scale. Rank correlations between urban level and country population, population density, value of exports, and GNP per capita are not significant.

Size of Communities

Most Africans probably live in agglomerations of some kind rather than in dispersed, individual family huts and houses, though every gradation of "rural" settlement may be found from the large agrotowns of Yorubaland through the *ksour* (fortified villages) of Saharan Morocco, encampments of nomadic herdsmen, and groupings of huts in hamlets and villages, to the remarkably dispersed shambas of Rwanda and Burundi. Traditional forms may show a clear relation to physical and politico-cultural conditions. Defensive considerations influenced the *ksour*, the hilltop villages of the Maghreb, the cliff dwellings of the Dogon, and the large Yoruba agglomerations. Difficulty of clearing high forests may have led to settlement in hamlets and compounds of the Guinean forests to facilitate joint farming efforts on a small scale.[3] In drier areas con-

[3] See W. B. Morgan, "Settlement Patterns of the Eastern Region of Nigeria," *The Nigerian Geographical Journal*, I, No. 2 (December, 1957), 23-30.

Urbanization in Africa

Table 15

VILLAGES AND SMALL TOWNS IN SELECTED AFRICAN COUNTRIES

Country	Year	Size range	Number	Population (thousands)
NORTHERN AFRICA				
Morocco	1960	1–5,000	33	87
U.A.R. (Egypt)	1960	"villages"	4,021	
WEST AFRICA				
Mauritania	1966	2–5,000	4	10.5
Senegal	1966	"villages"	13,000	
Upper Volta	1960–1961	"villages," av. -600	7,000	4,200
Niger	1968	3–5,000	48	
	1962–1963	"villages," av. -278	8,533	2,372
Gambia	1966	1–2,000	28	40
		2–5,000	6	20
Sierra Leone	1963	1–5,000	129	
Togo	1959	"villages," 370-1180	1,979	1,158
	1968	2–5,000	6	24.3
Dahomey	1961	"villages," av. -685	2,806	1,922
Nigeria	1952	nucleated villages		dominant, except in Iboland, Tivland
MIDDLE AFRICA				
Chad[a]	1962–1963	"villages," av. -260	9,048	2,352
C.A.R.[b]	1961–1963	"villages," av. -186	2,910	542
Congo (B)	1966	"villages"	c.4,200	
Gabon	1966	"villages," av. -122	4,503	549
Angola	1960	2–5,000	11	42
EASTERN AFRICA				
Kenya	1962	2–5,000		11
Tanganyika	1967	2–5,000	13	42.9
Malawi	1966	2–4,000	3	8.7
Zambia	1963	100–2,500	46	49
		2.5–5,000	4	17
Madagascar	1960	2–5,000	56	156
South Africa	1960	2–500	76	27
		500–2,000	246	274
		2–5,000	202	611
South West Africa	1960	500–2,000	11	9
		2–5,000	11	37
Swaziland	1966	1–5,000	9	17.7

SOURCE: National censuses and surveys.
[a] In area surveyed, estimated to include 72.1 percent of total population.
[b] 76 communes in 4 Prefectures.

Urbanization in Africa

Some African settlements. Most Africans live in agglomerations of some kind rather than in dispersed huts and houses.

Banani IV, a Dogon village in Mali.

Aït ben Haddou, Morocco, a ksour or fortified village in the eastern flanks of the High Atlas near Ouarzazate.

Urbanization in Africa

centrations may be explained by limited availability of water supplies. Clustering is also sometimes explained by the greater ease of clearing a vegetation belt to reduce incidence of disease. Elsewhere well-nucleated villages were most appropriate for such activities as fishing, production of craft articles, and local or regional trade. Colonial governments also frequently required or encouraged movement from dispersed settlements to villages or to roadside locations where street villages were formed, and at least one independent country, Tanzania, attempted to agglomerate rural residents in a villagization program.

Table 15 presents fragmentary data on villages and small towns in a number of African countries, suggesting the significance of these settlements and agglomerations in most of Africa. The term "village" as used in francophone countries is somewhat misleading, however, since it refers to a small political subdivision; some villages in these countries are, in fact, fractionated into smaller hamlets, while the populations of some villages become itinerant in the dry season. The "village" may, however, consist basically of groups of kinsmen linked to a nodal village which houses the local headman.

Ganvié, Dahomey, a stilt village devoted primarily to lagoon fishing.

Part of Wau, a village in the southern Sudan.

Table 16

PERCENT OF URBAN POPULATION OF SELECTED AFRICAN COUNTRIES (BY SIZE RANGE IN THOUSANDS)

	Year	Urban total	Urban population as percent of total urban population					Urban population as percent of total country population				
			5–9	10–19	20–49	50–99	100+	5–9	10–19	20–49	50–99	100+
NORTHERN AFRICA												
Morocco	1960	28.3	4.9	9.8	11.1	5.1	69.0	1.4	2.8	3.1	1.4	19.5
Algeria	1960	31.1	3.6	14.1	15.9	19.3	47.0	1.1	4.4	5.0	6.0	14.6
Tunisia	1966	37.1	19.7	20.3	21.1	11.0	28.0	7.3	7.5	7.8	4.1	10.4
WEST AFRICA												
Senegal	1960	24.0	7.9	2.1	14.5	25.3	50.1	1.9	0.5	3.5	6.1	12.0
Mali	1965	9.4	15.8	18.6	27.2	21.1	38.4	1.5	1.8	2.6	1.4	3.6
Upper Volta	1966	6.5	29.8	8.1	6.8		34.2	1.9	0.5	0.4		2.2
Niger	1968	5.4	21.3	11.2	31.5		36.0	1.2	0.6	1.7		1.9
Sierra Leone	1963	12.7	21.9	13.0	9.2		55.8	2.8	1.7	1.2		7.1
Ivory Coast	1967	22.0	10.4	12.2	19.5		57.8	2.3	2.7	4.3		12.7
Ghana	1960	23.8	24.4	18.8	16.8		40.1	5.8	4.5	4.0		9.5
Togo	1966	19.5	6.7	47.7	6.1		39.4	1.3	9.3	1.2		7.7
Dahomey	1965	15.8	12.0	16.8	18.4	20.6	32.1	1.9	2.7	2.9	3.3	5.1
MIDDLE AFRICA												
Chad	1962–1963	8.3	6.5	19.8	35.7		38.0	0.5	1.6	3.0		3.2
Cameroon	1963–1964	15.5	18.2	21.6	23.8	12.4	23.9	2.8	3.4	3.7	1.9	3.7
Angola	1960	9.8	10.6	17.6	13.2	10.6	47.8	1.0	1.7	1.3	1.0	4.7
EASTERN AFRICA												
Kenya	1962	7.8	10.3	3.6	12.4		73.9	0.8	0.3	1.0		5.8
Uganda	1959	3.3	12.5	16.7	13.9		56.9	0.4	0.5	0.5		1.9
Tanganyika	1967	5.5	7.1	6.2	35.1	9.4	42.2	0.4	0.3	1.9	0.5	2.3
Malawi	1966	3.8	5.2	25.2			69.7	0.2	1.0			2.6
Rhodesia	1961–1962	19.1	7.0	8.1	13.8		71.1	1.3	1.5	2.6		13.6
Zambia	1966	20.5	4.3	2.5	9.3	36.6	47.4	0.9	0.5	1.9	7.5	9.7
Madagascar	1964	13.0	12.0	15.7	30.3		42.0	1.6	2.0	3.9		5.5
SOUTHERN AFRICA												
South Africa	1960	41.0	6.8	7.0	11.7	9.8	64.6	2.8	2.9	4.8	4.0	26.5
South West Africa	1960	13.4	31.4	17.1	51.4			4.2	2.3	6.9		
Botswana	1964	25.4	15.9	13.0	71.0			4.0	3.3	18.0		

SOURCE: National censuses and other estimates.

There usually is no clear-cut distinction between the village and the small town; indeed several agglomerations of city size are essentially large villages housing more farmers than those employed in other occupations. At the same time there are examples of "villages" of quite small size which have urban functions, particularly in East, Central, and South Africa.

THE RANGE OF URBAN AGGLOMERATIONS

The selection of any cutoff point in the definition of urban centers is bound to be arbitrary, especially in Africa. It is seldom possible to utilize sophisticated criteria or data on functions to define categories of agglomerations, though this has been done for a number of countries. The 1966 Algerian census, for example, classified towns of over 5,000 into six groups from most to least urbanized on the basis of the following criteria: number of inhabitants, minimum relative density index, a minimum proportion of nonagricultural activities to total activities, an absolute minimum number of nonagricultural activities, and a maximum proportion of undifferentiated activities as compared to total nonagricultural activities. As other examples, Grove and Huszar used seventeen kinds of services to define five clusters of Ghanaian towns,[4] and Josephine Abiodun classed southwest Nigerian cities into five groups on the basis of presence or absence of selected central functions and discrete population levels.[5]

Table 16 gives, within selected size ranges, the percent of urban populations of a number of African countries to the total population of agglomerations above the 5,000 level and to the total country populations. The patterns shown vary greatly from country to country, but several generalizations may be made. First, a relatively high percent of urban populations is found in cities of 100,000 and over in the countries shown; their share of total national populations varies much more widely and tends to be highest in northern Africa, South Africa, and in tropical countries which have been more strongly influenced by European presence and investment, including Rhodesia, the Ivory Coast, and

[4] David Grove and Laszlo Huszar, *The Towns of Ghana: The Role of Service Centres in Regional Planning* (Accra, Ghana University Press, 1964).

[5] Josephine Olu Abiodun, "Urban Heirarchy in a Developing Country," *Economic Geography*, XLIII, No. 4 (October, 1967), 347-67.

Senegal. Second, in many tropical African countries, a gap appears between cities below 50,000 and those above 100,000, suggesting the powerful attraction of the largest cities.

The rank-size rule[6] applies rather poorly to African countries, especially in the tropical portion. Here it is common for the largest city to be from four to eight times the size of the second-ranking city, rather than twice as large. In those tropical countries with two large cities the drop in size to the third city is frequently quite sharp. This may reflect immaturity in the development of urban hierarchies.

LARGE URBAN COMMUNITIES AND PRIMATE CITIES

The U.N. classifies urban centers as agglomerations of over 20,000, with those from 20-99,000 ranked as towns, those from 100-499,000 as cities, and those above 500,000 as big cities. Focusing upon cities of 100,000 and over it may first be noted that about 1960 Africa had 84 such agglomerations or 6.4 percent of the world total; the number exceeded those of South America (71), Central America (35), and Australia (11), but was well below the numbers of North America (217), Europe (327), and Asia (442).[7]

The estimated number of 100,000 and 1,000,000 cities in Africa during recent decades is given in Table 17, and Table 18 breaks down the 100,000 cities for tropical Africa by decade and level. As may be

Table 17

NUMBER OF CITIES OVER 100,000, 1930s TO 1970 (ESTIMATED)
(of which, cities over 1 million in parentheses)

Period	Africa	Northern Africa	Tropical Africa	South Africa
1930s	21 (1)	13 (1)	4	4
1940s	39 (1)	19 (1)	13	7
1950s	72 (2)	28 (2)	34	10
Year				
1960	85 (3)	31 (2)	44	10 (1)
1967	111 (3)	37 (2)	62	12 (1)
1970 est.	114 (5)	37 (4)	64	13 (1)

SOURCE: National censuses, surveys, and estimates.

[6] See, for example, Chauncy D. Harris, "City and Region in the Soviet Union," in Beckinsale and Houston, eds., *Urbanisation and Its Problems*, pp. 277-80.

[7] Beaujeu-Garnier and Chabot, *Urban Geography*, p. 18.

Urbanization in Africa

Table 18

NUMBER OF CITIES OVER 100,000 IN TROPICAL AFRICA,
BY LEVEL, 1930s TO 1970 (ESTIMATED) (*population in thousands*)

Period	100	200	300	400	500	600	700	800	900
1930s	2	1	1						
1940s	10	1	1	1					
1950s	25	4	4	1					
Year									
1960	31	5	3	4		1			
1967	39	10	5	3	2	2	1		
1970 est.	33	14	6	3	1	2	4	1	

SOURCE: National censuses, surveys, and estimates.

seen, tropical Africa has made a notable gain in number of cities relative
to northern Africa and South Africa, though it is still not credited with
any million cities. Some observers believe that Kinshasa may now have
1.1–1.3 million inhabitants, but this has not yet been confirmed by official estimates. It is quite possible that Dakar, Abidjan, Accra-Tema, and
Lagos will reach the million mark in the 1970s. Notable concentrations of
100,000+ cities are found in Morocco (10), Egypt (16, of which Cairo
has over 4 million and Alexandria close to 2 million), southwest Nigeria
(Lagos and the large Yoruba towns, many of which are cities only by
size), and the Katanga-Copperbelt region. Aside from these areas the
100,000+ cities are rather sparsely scattered over the face of the
continent.

Table 19

PRIMATE CITIES OF AFRICAN COUNTRIES; THEIR PERCENT OF TOTAL
URBAN POPULATION AND OF RESPECTIVE NATIONAL POPULATION

Country	City	Year	Population (thou.) A=agglom.	% of urban pop.[a]	% of national pop.
NORTH AFRICA					
Morocco	Casablanca	1960	965	29.3	8.3
Algeria	Algiers	1960	884A	26.3	8.2
Tunisia	Tunis	1966	463	28.0	10.4
Libya	Tripoli	1964	214	55.7	13.7
U.A.R. (Egypt)	Cairo	1967	4,500A	35.5	14.6
Sudan	Khartoum	1956	114	17.5[b]	1.1
WEST AFRICA					
Mauritania	Nouakchott	1966	22	35.5	2.1
Senegal	Dakar	1960	374A	50.1	12.0

TABLE 19 (*Continued*)

Country	City	Year	Population (thou.) A=agglom.	% of urban pop.[a]	% of national pop.
Mali	Bamako	1965	165	38.4	3.6
Upper Volta	Ougadougou	1966	110	34.2	2.2
Niger	Niamey	1968	71	36.0	1.9
Gambia	Bathurst	1966	43	100.0	12.8
Guinea	Conakry	1967	197A		2.4
Sierra Leone	Freetown	1963	163A	55.8	7.1
Liberia	Monrovia	1962	81	64.8[b]	8.0
Ivory Coast	Abidjan	1967	400	45.4	10.8
Ghana	Accra	1960	338	21.1	5.0
Togo	Lomé	1966	129	39.4	7.7
Dahomey	Cotonou	1965	120	32.1	5.1
Nigeria	Lagos	1963	665A	19.4	1.2
MIDDLE AFRICA					
Chad	Ft. Lamy	1962–1963	100	38.0	3.2
C.A.R.	Bangui	1967	150	38.1	10.3
Congo (B)	Brazzaville	1968	200	59.5[b]	22.9
Gabon	Libreville	1968	62	67.4[b]	13.0
Cameroon	Douala	1963–1964	187	23.9	3.7
Congo (K)	Kinshasa	1966	508		3.2
Angola	Luanda	1960	225A	47.8	4.7
EASTERN AFRICA					
Ethiopia	Addis Ababa	1966	600A		2.6
Somalia	Mogadishu	1966	170		6.6
Kenya	Nairobi	1962	315A	47.7	3.7
Uganda	Kampala	1959	123A	56.9	1.9
Tanganyika	Dar es Salaam	1967	273A	42.2	2.3
Zanzibar	Zanzibar City	1967	95A	81.9	26.9
Rwanda	Kigali	1967	25	73.5	0.8
Burundi	Bujumbura	1967	100		3.0
Malawi	Blantyre-Limbe	1966	108	69.7	2.6
Rhodesia	Salisbury	1961–1962	310A	42.4	8.1
Zambia	Lusaka	1966	152	19.6	4.0
Mozambique	Lourenço Marques	1964	300		4.4
Madagascar	Tananarive	1964	322	42.0	5.5
SOUTHERN AFRICA					
South Africa	Johannesburg	1960	1,153A	17.5	7.1
South West Africa	Windhoek	1960	35	51.4	6.9
Botswana	Gaborone	1968	19		3.1
Lesotho	Maseru	1966	18	100.0	2.1
Swaziland	Mbabane	1966	14	58.0	3.7

SOURCE: National censuses, surveys, and estimates.
[a] 5,000 and above. [b] 10,000 and above.

Much attention has been focused upon the primate cities of African countries, which are sometimes thought to have an excessive share of a nation's total urban population. Table 19 gives data regarding the size and relative importance of the primate cities; for those countries for which data are available the primate cities most frequently account for from 30 to 60 percent of the total urban population (over 5,000) but for under 10 percent of the total population of their countries. In fact, they account for under 5 percent in half of the countries and under 15 percent in all but two.

Table 20 checklists the largest urban community in 33 main countries by their significance as cities and as transport, manufacturing, and administrative centers, and thus provides a crude measure of their primacy. As may be seen, twelve of the cities are the only major cities in their respective countries and twelve are one of only two major centers. Twenty-seven are considered to be the dominant centers in their countries. Sixteen are the main general-cargo ports of their countries, about half are the leading rail centers, and all have the most important airfield. All but two rank as the leading industrial city, and all but four are the political-administrative capitals of their countries.

The concentrations noted are probably a sign of the early stages of urbanization and of development in general in many countries. The primacy developed in most cases by the focusing of transport routes; it was reinforced by the availability of power and other services, by formation of the country's largest single market for consumers' goods, and in some countries by budding advantages to manufacturers of various externalities. It is a case of strength building on strength, of nothing succeeding like success. And judging from available growth rates it appears that this stage of urbanization has by no means run its course, at least in many countries of tropical Africa.

Some observers see the concentration in the primate cities as excessive in relation to other urban centers. Spengler, for example, writes that

economic and social welfare in African states will be promoted more effectively if the urban population is kept distributed among smaller cities rather than concentrated in a few very large cities.[8]

[8] Joseph J. Spengler, "Africa and the Theory of Optimum City Size," in Miner, ed., *The City in Modern Africa*, p. 77.

Table 20

MEASURES OF IMPORTANCE OF PRIMATE CITIES IN 33 COUNTRIES

City	Only major city	One of 2 major cities	Dominant city	First port	River port	Main airport	Main rail center	Main road focus	Main manufacturing center	Main administrative center	Total checked
NORTHERN AFRICA											
Casablanca			x	x		x	x	x	x		6
Algiers			x	x		x	x	x	x	x	7
Tunis	x		x	x		x	x	x	x	x	8
Tripoli		x	x	x		x		x	x	x	7
Cairo			x		x	x		x	x	x	6
Three Towns	x		x		x	x		x	x	x	7
WEST AFRICA											
Dakar	x		x	x		x	x	x	x	x	8
Bamako	x		x		x	x		x	x	x	7
Ouagadougou		x	x			x		x		x	5
Niamey	x		x		x	x			x	x	6
Freetown	x		x	x		x	x	x	x	x	8
Monrovia	x		x	x		x	x	x	x	x	8
Abidjan		x	x	x		x	x	x	x	x	8
Accra			x	x		x			x	x	5
Lomé	x		x	x		x	x	x	x	x	8
Cotonou		x	x	x		x	x	x	x	x	8
Lagos				x		x	x		x	x	5
MIDDLE AFRICA											
Ft. Lamy	x		x		x	x			x	x	6
Bangui	x		x		x	x		x	x	x	7
Brazzaville		x	x		x	x			x	x	6
Libreville		x		x		x			x	x	5
Douala		x		x		x	x		x		5
Kinshasa					x	x			x	x	4
Luanda			x	x		x			x	x	5
EASTERN AFRICA											
Addis Ababa		x	x			x	x	x	x	x	7
Nairobi		x	x			x	x	x	x	x	7
Kampala-Entebbe		x	x		x	x		x		x	6
Blantyre-Limbe	x		x			x	x	x	x	x	7
Salisbury		x	x			x		x	x	x	6
Lusaka						x		x	x	x	4
Lourenço Marques		x	x	x		x	x		x	x	7
Tananarive	x		x			x	x	x	x	x	7
SOUTHERN AFRICA											
Johannesburg						x		x	x		3
Totals (33 cities)	12	12	27	16	9	33	16	23	31	30	

A U.N. publication takes a somewhat ambivalent view. Noting that urbanization is crucial to the development process, it suggests that

in those areas which still lack a major urban center to serve as the building ground for economic growth, the highest priority for public development might become the creation of a viable urban centre with its attendant institutions and infra-structure in as short a time as possible. . . . But in countries where the problem of concentrated migration and economic growth is severe, there must be a commensurate effort on the part of the national Government to select new loci for concentration away from the main centres.[9]

There are several difficulties associated with these prescriptions. First, it is extremely difficult to determine objectively what an optimum size primate city should be. Second, it is equally difficult to turn off the migration to such centers as the selected size is approached, much less to limit the natural growth of the resident population. Third, there are very few African countries which can afford the luxury of subsidizing the development of new towns, new poles of growth. Fourth, it is not at all clear that priority should be given to urban development as opposed to investment in agriculture and other productive enterprises. And fifth, there are advantages to a considerable degree of concentration of modern activities through savings in scale and the development of externalities. It has been shown for South Africa, where concern has been expressed for some years regarding the marked concentrations on the Witwatersrand and at the major ports, that a period in which centripetal development was dominant has been succeeded by a period in which centrifugal forces are benefiting the less-developed parts of the country.

In any case the situation with respect to primate cities should not be exaggerated. Many countries have a number of actual or potential growth poles tending to reduce the dominance of the largest city. Examples include:

Morocco: Casablanca, Rabat-Salé
Algiers: Algiers, Oran, Annaba
Libya: Tripoli, Benghazi
Egypt: Cairo, Alexandria
Upper Volta: Ouagadougou, Bobo-Dioulasso
Ghana: Accra-Tema, Kumasi, Sekondi-Takoradi

[9] "Urbanization Development Policies and Planning," *International Social Development Review*, No. 1 (1968), p. 85.

Nigeria: Lagos, Ibadan, Enugu, Port Harcourt, Kano, Kaduna
Congo (Brazzaville): Brazzaville, Pointe Noire
Cameroon: Douala, Yaoundé
Congo (Kinshasa): Kinshasa, Lubumbashi, Kisangani
Angola: Luanda, Lobito, Nova Lisboa
Ethiopia: Addis Ababa, Asmara
Kenya: Nairobi, Mombasa
Uganda: Kampala-Entebbe, Jinja
Tanganyika: Dar es Salaam, Arusha (as headquarters of the East
    African Community)
Malawi: Blantyre-Limbe, Lilongwe (selected as new capital city)
Rhodesia: Salisbury, Bulawayo
Zambia: Lusaka, Copperbelt cities
Mozambique: Lourenço Marques, Beira, Nacala
Swaziland: Mbabane, Manzini

The Growth of Urban Centers

To understand the reasons for urban growth in Africa one must focus
upon factors such as economic change, growth of political and administrative bureaucracies, migratory trends and traditions, and conditions in
the rural regions. Psychological attitudes, which are very imperfectly
known or understood, are also of great importance.

The postwar years to about 1958 did see very satisfactory growth
in the economies of most African countries, a good deal of which was
focused in the cities or at least contributed to their employment opportunities. In the period 1938–1958 external commerce grew much more
rapidly in tropical Africa than in the world as a whole, increasing 620
percent at current prices as compared to 340 percent for the world.
Colonial powers devoted much more attention to their territories than
they had and France, in particular, took belated interest in the previously neglected colonies.

Since a main function of these powers was to assure connections
between the local economies and the metropoles, a substantial part of
investment reflected the determining role of external stimulants, preponderantly in administrative services, commerce, and transport. This
was particularly noticeable in the port cities, and for the 26 main countries fronting on the sea the main city is the first-ranking port in 19

cases, all but two of which are also the political capitals of their countries. Starting about the middle of the 1950s the economic advance became much more spotty, however, and many African countries have shown inadequate growth rates over the last decade. Some have, of course, progressed very satisfactorily by any of the usual measures and only a few have shown absolute regression.

Government staffs increased rapidly in the first decade after the war and again in the years following independence. Since the capitals were more often than not the main commercial and transport nodes, the primate cities naturally tended to grow more rapidly than lesser centers. A relatively new factor was the rise of market-oriented industries, which again contributed to the growth of the first-ranking cities which frequently provided a substantial share of their markets.

There were, then, incentives for rapid urban growth, but the growth has continued or even accelerated with the dissipation or reduction of these incentives. And, as was seen in the chapter on migration, the move to the cities appears to have become more or less inexorable.

RATES OF GROWTH

Measuring the growth of African cities is difficult because of the questionable character of much of the data, the lack of comparability of successive censuses, and the absolute lack of data for certain countries and cities. The data which follow must be accepted with these cautions. The following estimates of the annual rates of growth of cities of over 20,000 indicate that Africa now has one of the highest rates in the world:

	1920–1930	1930–1940	1940–1950	1950–1960
World	1.4	1.6	1.3	1.8
Developed countries	1.2	1.4	0.8	1.4
Developing countries	2.4	2.5	2.8	2.8
Africa	2.1	2	3	3.1

Other estimates give rates for Africa as follows: 1940–1952—3.9 percent; 1940–1950—4.5 percent and 1950–1960—5.4 percent; 1960—5.0 percent; 1959–1960—3.7 percent compared to 1.2–1.7 percent for the world.[10]

[10] See, for example, Breese, *Urbanization in Newly Developing Countries*, p. 33; Marc Nerfin, "Towards a Housing Policy," *Journal of Modern African Studies*, III, No. 4 (December, 1965), 543-65; U.N., *Housing in Africa*.

Table 21

ESTIMATED PERCENT OF URBAN POPULATION IN SELECTED AFRICAN
COUNTRIES AT SUCCESSIVE DATES

NORTHERN AFRICA		Mali		EASTERN AFRICA	
Morocco		1956	5.1	Kenya	
1900	5-10	1967	9.6	1948	5.2
1936	16	Upper Volta		1962	7.8
1952	22	1956	4.0	Tanganyika	
1960	28.3	1967	6.6	1931	1.2
Algeria		Niger		1957	4.1
1886	13.9	1956	2.7	1967	5.4
1906	16.6	1967	6.1	Zanzibar	
1926	20.2	Liberia		1958	26.5
1936	22.2	1950	1.2	1967	32.8
1948	23.6	1962	12.3	Malawi	
1960	31.1	1967	15.0	1950	1.1
Tunisia		Ghana		1966	4.0
1946	29.9	1921	8	Rhodesia	
1956	35.6	1948	14.3	1950	12.8
1960	37.0	1960	21.6	1961–1962	19.2
1967	40.0	Togo		Madagascar	
U.A.R. (Egypt)		1959	14.0	1913	5
1879	19.1	1966	18.2	1952	12.9
1917	21	Dahomey		1960	13.9
1937	24.3	1955	7.1	Mauritius	
1947	30.1	1965	15.8	1881	25.7
1960	35.4	MIDDLE AFRICA		1911	29.5
1965	40.0	Cameroon		1931	35.4
WEST AFRICA		1950	5.8	1952	46.9
Mauritania		1963	15.6	SOUTHERN AFRICA	
1956	4.5	Angola		South Africa	
1967	6.1	1940	5.4	1951	43
Senegal		1950	6.1	1960	47
1956	22.9	1955	7.4	South West Africa	
1967	29.2	1960	10.6	1946	8.8
				1951	12.5
				1960	23.4

N.B. Estimates may vary from those in Table 15 due to use of different
cut-off points.
SOURCE: National censuses, surveys, and estimates.

Turning to individual countries, we may illustrate the growth of
their urban populations by the estimated percentages of urban to total
population at successive periods (Table 21). My estimates of recent
rates of growth, using various periods from the late 1950s and the early

Urbanization in Africa

1960s to 1967, indicate that the rate of urban increase in tropical African countries is now about 3.8 times the rate of rural increase, and occasionally 6 to 8 times that rate. In north and south Africa the urban population appears to be increasing about 1.6 times as fast as the total population; in tropical Africa the ratio varies from 1.5 to 5.9 to 1 and averages about 3 to 1.

There are marked differences in the rates of growth among individual cities. The first-ranking city of most countries is characteristically growing at a substantially more rapid rate than the rest of the urban communities. Those primate cities which were relatively small at the time of independence appear to be growing at particularly high rates.

Table 22 gives the population of selected African cities over the past decades. As may be seen, many of these leading cities were very small indeed in the early decades of the century, and many were still quite small before World War II. The most rapid growth has come in the last two decades and a number of cities have seen a remarkable expansion in the last decade, associated in part with the increased influx of migrants following independence.

Taking 52 major cities in 33 countries, including the main city in each, cities for which reasonably good estimates were available for the 1930s and the 1960s, one finds that the cities grew from 1.4 to 50 fold over this period. The lowest increases were recorded for traditional cities such as Marrakech, Fès, Ibadan and Addis Ababa, all of which were rather large at the beginning of the period, plus Freetown. The cities of northern Africa and of South Africa generally grew about 2–3 fold, while most of those of tropical Africa grew by factors of 8 or more.

Cities with the highest increase were Abidjan (50.0×), Niamey (35.5×), Conakry (28.1×), Ft. Lamy (18.8×), and Lomé (16.9×). These and other francophone cities were generally much smaller than cities in anglophone Africa before World World II; their more rapid increase reflects in part the somewhat belated application of massive attention accorded by France to many of its African possessions. Other cities which increased many fold (10–15×) over the period include: Dakar, Ougadougou, Monrovia, Bouaké, Cotonou, Kinshasa, Asmara, Bujumbura, and Salisbury.

Figures 1 and 2 illustrate the growth of selected African cities on graph and semilog scales. The lines for Cape Town and Algiers are characteristic of cities which were already large early in the century. Ibadan,

Table 22

ESTIMATED POPULATION OF SELECTED AFRICAN CITIES, 1900-1969 (*population in thousands*)

City	1900s		1910s		1920s		1930s		1940s		1950s		1960s			
	Yr.	Pop.	Yr.	Pop.	Yr.	Pop.	Yr.	Pop.	Yr.	Pop.	Yr.	Pop.	Yr.	Pop.	Yr.	Pop.
NORTHERN AFRICA																
Casablanca	'07	25							'47	551	'52	682	'60	965	'66	1120
Rabat-Salé									'47	218	'52	203	'60	303	'66	370
Marrakech									'47	238	'52	215	'60	243	'66	275
Fès									'47	201	'52	179	'60	216	'66	255
Algiers	'06	174			'26	266	'31	319	'48	473	'54	570	'60	884	'66	943
Oran	'06	106			'26	153	'31	166	'48	265	'54	313	'60	370	'66	328
Constantine	'06	54			'26	89	'31	100	'48	114	'54	143	'60	238	'66	254
Annaba (Bône)	'06	41			'26	49	'31	66	'48	101	'54	112	'60	135	'66	169
Tunis									'47	365	'56	410			'66	463
Tripoli									'46	100	'54	130	'60	184	'68	330
Benghazi									'46	60	'54	70	'60	120	'68	130
Cairo	'07	678	'17	791	'27	1065	'37	1312	'47	2091			'60	3349	'66	4220
Alexandria	'07	354	'17	445	'27	573	'37	686	'47	919			'60	1516	'66	1801
Khartoum	'05	77	'16	59	'23	79	'32	104	'40	116	'50	125	'64	185	'66	198
Omdurman			'16	23	'23	31	'32	50	'40	45	'50	71	'64	175	'66	185
WEST AFRICA																
Dakar	'04	19	'10	25	'21	30	'31	54	'43	125	'55	300	'60	374	'68	600
Bamako	'00	6	'10	7	'21	14	'30	19	'43	32	'50	86	'60	130	'66	170
Ouagadougou					'26	12	'30	10	'44	18	'50	37	'61	58	'66	110
Bobo-Dioulasso			'10	8	'26	7	'30	12	'44	14	'50	38	'61	53	'66	68
Niamey					'26	3	'30	2	'43	6	'50	9	'60	30	'68	71
Conakry			'10	6	'21	9	'31	7	'44	25	'50	39	'60	113	'67	197
Freetown			'14	33	'21	44	'30	55	'47	65	'54	90	'63	128	'66	148
Monrovia					'26	5	'34	10			'53	27	'62	81	'67	100
Abidjan			'10	1	'21	5	'30	10	'42	36	'50	69	'60	180	'68	500
Bouaké			'16	4	'21	4	'30	10	'42	26	'50	32	'60	45	'65	105
Accra	'01	27	'11	30	'21	43	'31	61	'48	136			'60	492	'66	600
Kumasi	'01	3	'11	19	'21	24	'31	36	'48	78			'60	218	'66	301
Lomé					'26	6	'31	8	'46	29	'56	39	'60	91	'68	135
Cotonou	'02	1	'10	2	'21	4	'30	9	'43	15	'51	25	'61	78	'65	120
Lagos	'01	42	'10	74	'21	100	'31	127			'52	267	'63	665		

Note: This is a statistical appendix table (printed sideways). Each city is followed by a series of observations, given as a year (abbreviated, e.g. '06 = 1906) and an estimated population in thousands.

Region / City	Observations (year : population in thousands)
MIDDLE AFRICA	
Ft. Lamy	'06 — ; '36 8 ; '45 18 ; '51 21 ; '61 88 ; '67 150
Bangui	'16 5 ; '29 17 ; '31 20 ; '40 24 ; '50 63 ; '60 83 ; '66 150
Brazzaville	'16 8 ; '36 24 ; '45 43 ; '50 84 ; '60 120 ; '68 200
Libreville	'13 11 ; '22 3 ; '31 7 ; '45 11 ; '50 13 ; '61 30 ; '68 62
Douala	'13 7 ; '31 28 ; '40 31 ; '50 86 ; '61 150 ; '65 200
Yaoundé	'26 20 ; '31 30 ; '50 40 ; '62 93 ; '65 101
Kinshasa	'08 5 ; '22 18 ; '30 33 ; '40 49 ; '50 191 ; '61 420 ; '66 508
Lubumbashi	'23 16 ; '35 28 ; '40 30 ; '50 98 ; '61 190 ; '66 233
Luanda	'23 23 ; '30 51 ; '40 61 ; '50 142 ; '60 225 ; '66 250
Lobito	'40 16 ; '50 28 ; '60 39 ; '64 109
EASTERN AFRICA	
Addis Ababa	'08 35 ; '20 50 ; '38 300 ; '48 402 ; '58 400 ; '61 449 ; '68 620
Asmara	'05 9 ; '31 19 ; '47 95 ; '58 120 ; '66 200
Mogadishu	'06 12 ; '10 20 ; '20 15 ; '31 20 ; '53 61 ; '63 121 ; '66 170
Nairobi	'28 30 ; '36 50 ; '48 119 ; '57 222 ; '62 315 ; '69 478
Mombasa	'06 30 ; '26 40 ; '48 85 ; '59 152 ; '62 180 ; '69 246
Kampala	'11 32 ; '48 24 ; '59 123
Dar es Salaam	'00 20 ; '13 23 ; '21 25 ; '31 34 ; '48 69 ; '52 99 ; '62 150 ; '67 273
Zanzibar City	'48 45 ; '57 58 ; '67 95
Bujumbura	'45 10 ; '53 29 ; '60 47 ; '66 90
Blantyre-Limbe	'56 55 ; '66 110
Salisbury	'40 67 ; '58 233 ; '61–62 310 ; '68 380
Bulawayo	'46 43 ; '58 190 ; '61–62 210 ; '68 271
Lusaka	'56 58 ; '63 119 ; '69 343
Kitwe	'56 70 ; '63 116 ; '69 254
Ndola	'56 69 ; '63 90 ; '68 132
Lourenço Marques	'07 10 ; '27 37 ; '35 47 ; '50 94 ; '60 178
Tananarive	'00 50 ; '14 65 ; '21 64 ; '30 93 ; '41 123 ; '50 180 ; '60 248 ; '66 392
SOUTHERN AFRICA	
Johannesburg	'11 237 ; '21 288 ; '36 519 ; '51 919 ; '60 1153
Cape Town	'11 162 ; '21 207 ; '36 344 ; '51 632 ; '60 807
Durban	'11 90 ; '21 146 ; '36 260 ; '51 498 ; '60 681
Pretoria	'11 58 ; '21 74 ; '36 129 ; '51 285 ; '60 423

SOURCE: National censuses, surveys, estimates by local authorities, etc.

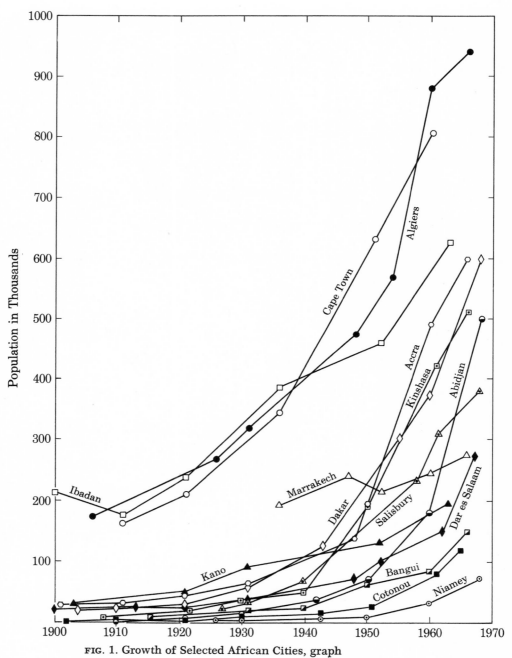

FIG. 1. Growth of Selected African Cities, graph

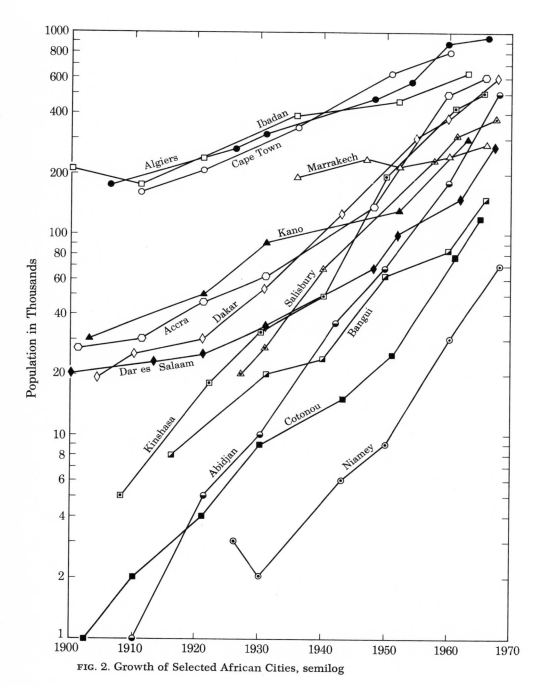

FIG. 2. Growth of Selected African Cities, semilog

Urbanization in Africa   243

Marrakech, and Kano are examples of indigenous cities. Both categories have had slower rates of growth than the newer cities. The remaining lines cover several cities in each major intertropical region; they reveal a characteristic sharp increase in the rate of growth, usually in the decade of the forties or fifties. Niamey illustrates a common pattern among primate cities which had remained small as late as the 1950s.

Classification of African Cities

Numerous systems have been used to classify African cities; each has some value, however limited, but much more data will be required before a satisfactory delineation can be achieved. Some of the present typologies are entirely too broad, others, by using several criteria, wind up with overlaps and contradictions or categories which are not mutually exclusive. Discussions of the various systems of classification follow.

CLASSIFICATION BY SIZE

Size is obviously an important characteristic and it has the advantage of simplicity. It is statistically the most convenient and permits dividing urban centers by internationally accepted rank-sizes, thus facilitating national and continental comparisons. Size is also important in central place hierarchical studies.[11] But taken alone it is not very helpful in analysis. As has been seen there are numerous centers of over 100,000 which are really agrotowns; their urban functions would compare unfavorably with many centers of less than 20,000.

CLASSIFICATION BY AGE

Mercier, Oram, Hauser, Little, Wallis,[12] and others have employed classes which use historical factors as a main consideration, though cultural elements are also often included. The divisions vary slightly but run about as follows: precolonial, preindustrial, traditional, or old towns; colonial or new towns; and more recent or mushroom towns.

[11] See Brian J. L. Berry and W. L. Garrison, "The Functional Bases of the Central Place Heirarchy," *Economic Geography*, XXXIV, No. 2 (April, 1958), 145-54.
[12] Mercier, *Sociological Problems of Urbanization*; Oram, *Towns in Africa*; Hauser and Schnore, eds., *The Study of Urbanization*; Little, *West African Urbanisation, a Study of Voluntary Associations in Social Change*; G. A. G. Wallis, "Factors Leading to Migration to Towns and Town Growth in Africa," Inter-African Conference on Housing and Urbaniaztion (Nairobi, CCTA/CSA, August 20, 1958).

The major difficulty in using the time factor is that a city could fall successively into each category. The cultural component of these systems may, however, be more valuable in drawing distinctions among cities, as will be seen below.

CLASSIFICATION BY CULTURAL CHARACTERISTICS

Several attempts have been made to develop a culturally oriented typology. Redfield and Singer classified cities according to their cultural role as either orthogenetic, in which an old culture is carried forward into systematic and reflective dimensions, and heterogenetic, in which the creating of original modes of thought might have authority beyond or in conflict with old cultures and civilizations.[13]

Cultural-genetic Classes. Holzner and others have proposed the development of a cultural-genetic city classification.[14] Their thesis is that every culture influences all aspects of its cities in a way different from all other cultures, which permits grouping cities by their cultural genesis rather than by functional or economic criteria. Using thirty-one "easily quantifiable criteria" divided into seven processes of city-cultural relationship, Holzner divides the continents into distinctive regions of which Africa has five: the north; a belt across the sudan and into Somalia; Ethiopia; Rhodesia and South Africa; and the huge remaining region including Madagascar. These divisions, while admittedly only preliminary and suggestive, are not particularly helpful in classifying African cities. The claim that the cultural-genetic classification will "reveal a pattern of homogeneous world regions of city similarity," preferable to the "haphazard" and "irregular" distribution of city types grouped according to functional characteristics,[15] is subject to question both as to the homogeneity of cultural regions and the haphazard distribution of cities defined functionally. For Africa, the indigenous and expatriate initiated cities represent very distinctive cultures but are found in the

[13] As described in Breese, *Urbanization in Newly Developing Countries,* pp. 49 f.

[14] See Lutz Holzner, E. J. Dommisse, and J. E. Mueller, "Toward a Theory of Cultural-Genetic City Classification," *Annals of the Association of American Geographers,* LVII, No. 2 (June, 1967), 367-81; Holzner, "World Regions in Urban Geography," *Annals of the Association of American Geograpers,* LVII, No. 4, (December, 1967) 704-12.

[15] Holzner *et al.,* "Toward a Theory of Cultural-Genetic City Classification," p. 379.

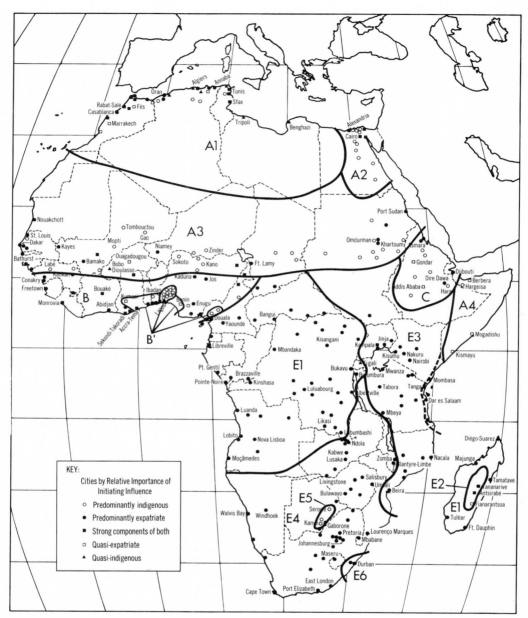

MAP 25. Selected towns and cities of Africa mapped according to the relative importance of indigenous and expatriate influence; tentative urban cultural regions. See additional key, opposite page.

Urbanization in Africa

same regions; furthermore, it is often very difficult to classify African cities which have both indigenous and expatriate components. There are also marked differences in city types defined culturally within individual countries of North Africa and West Africa, while indigenous Malagasy towns are distinctly different from those on the continent.

Map 25, whose preparation was stimulated by the Holzner proposition but which does not use the criteria he suggests, attempts to classify the cities of Africa according to the significance of the indigenous and expatriate influence in their present form, and outlines a number of regions which can justifiably be delineated as culturally homogeneous. It does not include indigenous towns which were once important but which have disappeared or lost their former glory.

Indigenous vs Expatriate Origin. The simplest classification based on cultural distinctions divides cities between those which were indigenous and those which were expatriate-initiated. The system has

---

Regions as delineated on map 25

A. Areas of Strong Muslim Influence
   A1. North Africa. Indigenous influence Arab-Berber; modern influence predominantly French in the Maghreb, Italian in Libya, Spanish in northern Morocco, Ceuta, and Melilla.
   A2. Egypt. Considerable European influence in modern cities.
   A3. Sudan belt. Emirate cities, old caravan centers; Arab and some Egyptian influence in east; modern influence French and British.
   A4. Somalia and East Coast. Arab influence strong.
B. Western Africa
   B. Modern cities dominate; French and British influence most important.
   B'. Subregions where indigenous cities are best represented.
C. Ethiopia. Mainly indigenous; Muslim and Italian influence in Eritrea; Italian influence in Addis Ababa, Gondar.
E. Middle, Eastern, and Southern Africa. European influence predominant in most areas.
   E1. Middle Africa. French, Belgian, and Portuguese influence dominant. On Madagascar French influence predominant in newer towns, some Arab influence at Majunga.
   E2. Merina and Betsileo towns of Malagasy Highlands; important French influence in new sections.
   E3. Eastern Africa. Modern influence primarily British; important Indian component in most towns; Arab influence on coast and islands.
   E4. Central and South Africa. Influence predominantly British; Afrikaner initiative in some South African towns.
   E5. Botswana. Agrotowns of Tswana.
   E6. Natal. Mainly British towns, strong Indian component.

disadvantages, including the difficulty of defining some cities because of the multiple influences affecting them, the fact that expatriate-initiated cities take on characteristics of indigenous cities and may in some quarters be indistinguishable from them, and the restricted number of classes unless a further breakdown is made among indigenous and expatriate groups. But it is of value because there are notable contrasts discernible depending on what peoples founded and developed the city.

Numerous observers have called attention to the relatively minor role of Africans in initiating the cities of sub-Saharan Africa, with major exceptions in the sudan belt from the Atlantic to Sudan and in parts of southern West Africa, and lesser exceptions in Ethiopia, Madagascar, and Botswana. Elsewhere in this portion of the continent practically all of the cities owe their origin in large part to non-Africans. These include not only Europeans but Arabs, who played a major role in such East African cities as Mombasa and Zanzibar, Indians, whose *dukas* at trading posts in East Africa later became the gazetted townships of the area and who are present in substantial numbers in the larger towns (also in Durban, South Africa), and Indonesians, who were the forebears of the more evolved portions of the Malagasy population and who founded such cities as Tananarive and Fianarantsoa, now classed as indigenous creations.

In West Africa there are numerous cities of indigenous origin, including the Yoruba towns, Benin, Ouidah and others in Dahomey, Kumasi, and many of the emirate cities of the sudan belt. Most of the commercial centers at the edge of the desert have declined greatly from their former glories and scarcely count among present cities; Tombouctou, for example, which is purported to have had as many as 45,000 residents in the sixteenth century is now a town of about 6,000. But even in West Africa many of the major cities are essentially European creations, including Dakar, Thiès, Rufisque, Bamako, Bobo-Dioulasso, Niamey, Conakry, Abidjan, Bouaké, Accra, Lomé, and Cotonou. In Nigeria the considerable number of existing towns reduced the necessity for founding colonial centers but, in addition to such pre-colonial creations as Lagos and Calabar, Europeans were responsible for initiating such centers as Port Harcourt, Sapele, Enugu, Jos, Minna, and Kaduna. In Equatorial, East, and South Africa it is hard to find a really indigenous city. Kampala could be considered an exception, or at least

Urbanization in Africa

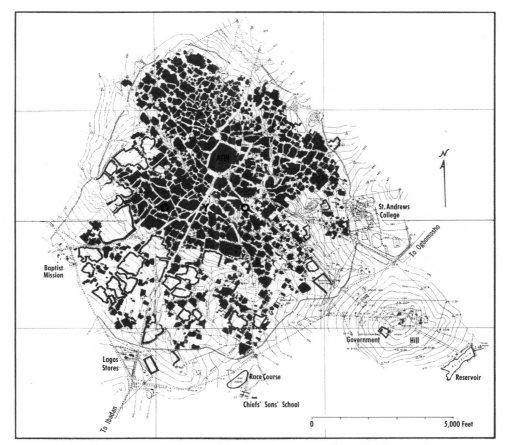

MAP 26. Oyo, Nigeria, a Yoruba town characteristic of large indigenous communities in Africa.

the Mengo quarter of that city, which was the tribal capital of the Kabaka of Buganda, as could the indigenous towns of Botswana which are really overgrown villages.

Indigenous towns. The distinguishing features of the indigenous towns are morphological, architectural, functional, demographic, and cultural.[16] Usually the main foci of the town are a central market, which

[16] See Mercier, *Sociological Problems of Urbanization*; Gideon Sjoberg, "The Pre-industrial City," *The American Journal of Sociology*, LX, No. 5 (March, 1955), 438-55.

Sofara, Mali.

Views of some indigenous towns.

may be quite large and usually without buildings, and the chief's quarters; in Muslim areas the main mosque is also centrally located (Map 26). Quarters of irregular size and shape surround the core, each likely to be grouped around subchiefs or clan heads. In some cities, as for example Addis Ababa, these take on the character of rural villages, each having its own small market. The quarters are likely to be congested and lacking in adequate sanitation and other facilities. The architecture of these towns is generally more uniform, of lower silhouette, and more in accord with indigenous tradition than that of expatriate towns. Surrounding many of the indigenous towns is, or was, a ditch, a crude wall, or in the case of the sudan-belt towns and Arab-Berber cities of northern Africa, more elaborate crenellated walls with large ornamental gates and towers.

A lane in Abeokuta, Western Nigeria.   A market in Bangangté, Cameroon.

Two or more main roads normally converge on the market place. The remaining routeways are often winding and irregular, suitable only as passageways for people and animals, and entirely inadequate for modern transport. Even the few main arteries that have been cut through the old towns are now likely to be constricted by drainage-ways, shops, and movable stalls, not to mention parked cars and numerous people walking on the roads.

Where expatriate quarters have been added to these towns they usually have been placed on the periphery. In Morocco, Lyautey decreed that the Muslim cities were to be as little disturbed as possible and modern French quarters were typically built outside the walls, as at Fès, Marrakech, or Rabat. In Algeria and Tunisia the old cities were less protected; some have been totally surrounded by newer structures while

Urbanization in Africa

*New expatriate-
initiated
additions to
old indigenous
cities.*

Algiers, where the old casbah has
been completely surrounded by the
modern, largely French-created
city.

others have disappeared. In West Africa the expatriate additions were
often confined to the offices and residences of the colonial administrators, a police station, and possibly some health facilities. Usually
an effort was made to select a high point, partially for protection, but
also to take advantage of whatever breezes might exist and to permit
separation from the unsanitary conditions prevailing in the indigenous
town. When the expatriate presence was more diversified and important

Urbanization in Africa

Part of the new town outside the walls of Marrakech, Morocco.

The commercial section of Tananarive, Madagascar, constructed under the French; part of the indigenous city is seen on the rocky eminence in the background.

the European quarter would frequently include a modern business and commercial district, often attracted to the rail line, if such existed, or to a main road bypassing the old town. Additions of schools, hospitals, churches, and playing fields were perforce placed on the periphery of the indigenous community because of the lack of available space within it.

Most of the indigenous towns have multiple functions, and the traditional economy remains important. Many such towns housed substantial numbers of farmers, a particular feature of Yoruba towns. Kano and other sudan-belt towns constructed their protective walls to include a large area where crops could be produced in the event of siege. Production of handicrafts is often well developed, many items being produced in individual homes or in tiny shops or stalls in the market, which also serve as the sales places for the crafted product. Inanimate power is rarely used, and crude techniques have frequently persisted to a surprising degree, though modern crafts (mechanical repairing, plumbing, electrical work, photographic studios, etc.) have often supplanted the traditional crafts in those towns which have been most exposed to modern influences. In northern Africa, Ethiopia, and to some extent in West Africa, craftsmen may be organized in guilds and there is frequently a correlation between specific crafts and ethnic groups, while distinct quarters are often occupied by different specialists.

The population of indigenous towns in sub-Saharan Africa is likely to be much more homogeneous than in expatriate towns. This is less true for the old towns of northern Africa, where wards may be rigidly segregated on social and ethnic grounds, sometimes separated from each other by internal walls. Both north and south of the Sahara indigenous people are likely to have a greater participation in commercial and other activities than in towns of expatriate origin. Growth rates of indigenous towns and cities are usually lower, and age structures and sex ratios are nearer to the national average.

Expatriate-initiated towns. Characteristics of the expatriate-initiated towns are the opposite of most of those mentioned for the indigenous towns, although many of them readily take on local character, particularly in the African residential areas. The central business district (CBD) is usually in a dominant position and contains types of stores and facilities rarely found in indigenous towns. African quarters were or are separate and segregated, sometimes by law, although separation would probably have developed for economic reasons without legal restrictions, and higher rents and costs tended to perpetuate the pattern even after independence. African quarters in expatriate towns were sections in a larger European-patterned city, wheras in the indigenous city it was the European quarters which were separate and distinct, covering a much smaller proportion of the total area. Much higher densities prevail in the African sections; indeed the terms "European" and "African" have frequently been replaced by "low density" and "high density" in post-independence Africa. Bidonvilles or shanty towns are more prevalent in the expatriate-originated towns, in part because of a much greater influx of migrants. Except for the CBD, separate facilities were likely to be provided for the various ethnic groups—in education, health, recreation, and other activities. Cities in East Africa may have three distinct types of marketing facilities, usually fairly separate from each other; these are the usually more modern and specialized European shops, Indian stores, and African markets.

The layout of expatriate towns is more often than not on a grid plan and in the newer towns the main roads are much wider. Rhodesian cities were fortunate in that an early decree stipulated that the main streets should be wide enough to permit turning a span of 16 oxen. Nonetheless, traffic problems are sometimes troublesome. Space allotted to the transport function, which was basic to colonial objectives, is much greater than in indigenous towns. In some of the older ports, however, not enough space was allotted to provide for vastly increased needs and considerable congestion results. Elsewhere, new port facilities constructed away from the earlier quays have greatly relieved the situation. This is true for: Accra, whose port functions have been moved to Tema; Lagos, with the enlarged quays on the mainland at Apapa; Cape Town, where filling-in of the foreshore has created new land; Dar es Salaam, and Mombasa with the mile-long Kilindini wharves and the newer Kipevu quay on the mainland. Provision of rail facilities has also frequently required much space and planning was not always far-sighted, resulting in undesirable interruptions of street patterns, lost space near city centers, and numerous grade-crossings.

Economically, expatriate-initiated towns contain few agriculturalists, their functions are more specialized though not necessarily very complex. The early main functions were commerce, administration, and transport and these have left indelible marks not readily altered as newer functions are added. Modern manufacturing has generally been a rather recent introduction which has frequently permitted its placement in clearly defined and planned industrial estates or districts.

The populations of the expatriate towns are usually diverse. Many tribes may be represented; there is often a greater representation of Indians, Pakistanis, Syrians, Lebanese, and other minority groups; and European populations were characteristically much more significant

Some European-initiated towns.

Cairo Road, Lusaka, Zambia, about 1950, when it was lined mainly with Indian-owned shops. Today, Lusaka is one of the most rapidly growing cities in Africa and many high-rise buildings have been constructed along this road.

Welkom, South Africa. Founded only in postwar years when the Orange Free State goldfields were developed, it is an entirely planned town.

Urbanization in Africa

Part of the CBD of Durban, Natal, South Africa. Nineteenth-century structures are interspersed with modern high-rise buildings.

Salisbury, Rhodesia.

than in the indigenous towns. Following independence the European component frequently changed rapidly, though departing colonial officers have often been replaced by expatriate advisers and diplomatic representatives. These towns and cities are likely to be growing more rapidly, to be receiving a substantial influx of migrants, to have a younger population, and to have a wider sex ratio among their inhabitants.

Distinctions can also be made among expatriate-initiated towns according to the metropolitan power which was involved. Each of the major powers tended quite naturally to bring in elements from their homelands which have left their mark in such things as traffic systems, architectural styles, and preferred types of recreation. Francophone cities are likely to have more prominent churches or cathedrals, formal parks with trimmed trees, open air cafés and bars, shuttered windows, red tiled roofs, multi-storied apartment houses, and large stadia.

Many French cities in North Africa have a characteristic Mediterranean style and Dakar is not dissimilar. British influenced cities in West Africa are often older than the French creations, which have a distinctly more modern appearance as a consequence.

British-initiated towns are likely to have somewhat more sedate architectural styles. The ideal residential pattern appears to be a single-family residence set in lawns and gardens as nearly reminiscent of England as possible. Among the early provisions were a race course, a golf course, and the inevitable private club. There is a commendable representation of museums and libraries.

Portuguese cities in Africa appear to emulate those in Portugal and the new settlement schemes replicate metropolitan styles precisely. The Italians in Libya, Eritrea, and to a lesser extent during their brief stay in Ethiopia favored one or more wide plazas, stately squares, and arcaded buildings on the major streets. South Africa has its Dutch- or Cape-colonial houses, much Victorian architecture, and a greater representation of new high-rise buildings than in other parts of Africa; the British influence is much more pervasive in the cities than that of the Afrikaner.

The question may be asked, just how African are the cities of Africa? The answer probably is that the Western influence has been outwardly stronger and will continue gradually to diminish the strictly African appearance of even the indigenous cities. On the other hand, the African genius and preferences for architectural and life styles will, of course, be of growing significance. Anyone who questions the predominance of the African influence in some expatriate-initiated cities should perhaps listen to the sounds and savor the odors of a city such as Lagos.

Some years ago I was a participant in a large panel discussing African cities when a member of the audience asked somewhat accusingly, "What are you people doing to preserve the soul of the African city?" Another member pusillanimously but perhaps wisely stated that he "would like to associate himself with that question." It is subject to some doubt whether any city may be credited with having a soul, though many observers would claim that each has a distinctive "personality." It was, of course, not the function of a group of American panelists to preserve the soul of African cities and they could scarcely have done so if charged with the task. Only the Africans can develop the collective personality or preserve such souls as their cities may have. One wonders, however, if African features in the architecture of the cities may not be more visible in detail work and ornamentation rather than in building design itself, for neither the baked mud house nor the grass hut is readily translated to modern structures. The personality would, therefore, emanate more from the bustling crowds on the streets and in the markets and from the gaiety, pride, dignity, and other qualities of the individual, rather than from the physical qualities of the city.

CLASSIFICATION BY FUNCTION

City functions are obviously important criteria for classifying cities, particularly for those concerned with economic development. A major difficulty arises, however, because available data make it very difficult to use any of the quantitative systems that have been applied to cities in the Western world.[17] Many towns are, of course, multifunctional and it would be desirable to employ ratios for different types of functions in developing a classificatory system.[18] Even the relatively simple distinction between basic functions (those activities producing

[17] For a brief description of some of these systems see Harold M. Mayer, "Making a Living in Cities: The Urban Economic Base." *Journal of Geography*, LXVIII, No. 2 (February, 1969), 70-87.

[18] See Chauncy D. Harris, "A Functional Classification of Cities in the United States," *The Geographical Review*, XXXIII, No. 1 (January, 1943), 86-99.

goods or services for export out of the urban area) and non-basic functions (those activities whose resultant goods and services are consumed within the city or urban agglomeration) is not readily applicable in the African context.[19]

Functions which may be used in a classification system include: commerce, or trade (retail, wholesale, diversified), transportation, manufacturing, agriculture, fishing, mining, finance, education, recreation (resort, retirement), military, religious, administration. Sometimes activities are combined into "services" and this function is employed.

Studies have been made in Morocco, Algeria, Ghana, Kenya, South Africa, and other countries which are related to the functional classification of urban communities. In Ghana, for example, Grove and Huszar used six functions to define cities into five "grades."[20] According to the 1960 Ghana census the localities under 5,000 had over half of their labor force employed in agriculture. Above that level the percentage of employment by function was a follows:

Town size	Agriculture (percent)	Commerce and transportation (percent)	Manufacturing (percent)	Services (percent)
5–10,000	41	30	13	8
10–20,000	22	33	15	12
20–50,000	12	42	15	17
100,000+	4	44	16	24

Manufacturing consisted mainly of one-man or family operations; only 1 percent of all manufacturing establishments employed six or more paid workers, but this 1 percent accounted for 81 percent of paid employment in manufacturing. A study by McNulty revealed the presence in Ghana of a relatively large number of functionally similar centers, which he took to imply that the country had not yet achieved a sufficiently high level of development to attain effective regional or urban specialization.[21]

Taking Zambia as a second example, Kay found that all of the

[19] See John W. Alexander, "The Basic-Nonbasic Concept of Urban Economic Functions," in Mayer and Kohn, Readings in Urban Geography, pp. 87-100.

[20] See Grove and Huszar, The Towns of Ghana: the Role of Service Centres in Regional Planning.

[21] Michael L. McNulty, Urban Centers and the Spatial Patterns of Development in Ghana (Evanston, Northwestern University, Ph.D. dissertation, 1966).

Urbanization in Africa

small towns were "service centers." In five of the "mining towns," 62 percent of the African employees, excluding domestic servants, and 56 percent of European employees worked in the mining industry while only 17 percent of Africans and 29 percent of Europeans were engaged in commerce, finance, and services.[22]

It is likely to be many years before sufficient data are available to permit a comparative assessment of African cities on the basis of quantified functional employment. Nonetheless the relative simplicity of most African cities, particularly in the tropical portion, permits a subjective assignment of cities over 20,000 to various functional types with some confidence. This is so in part because many of the possible functional classifications are not represented or are only rarely present. There are no fishing communities large enough to rate as towns, no educational or university towns, very few recreational towns, though some may well develop in North Africa and in East Africa in the next decade, no military towns, and only two or three religious towns, including Moulay Idriss in Morocco and Kairouan in Tunisia. There are also relatively few mining towns and these are quite readily distinguishable, and only a few manufacturing towns.

There are a number of towns and cities which have relatively high percentages of their inhabitants engaged in agriculture, especially in Yorubaland and, on a lesser scale, in Botswana. A survey of seven towns in Togo indicated that in two, Tsévié and Bassari, employment in agriculture was found to exceed nonagricultural employment, though for the seven towns combined agriculture accounted for only 14.9 percent of the total actively employed. The 1952 census of the Western Region of Nigeria indicated that ten of the twelve largest cities had percentages in agriculture and fishing varying from 14.3 to 46.1 percent; [23] Abeokuta had 6.9 percent and Lagos had only 2.5 percent. In three of the towns with populations exceeding 100,000, agriculture employed 60–70 percent of working males at the time of the sample count.

Craft production is another function which may have considerable

[22] See George Kay, "The Towns of Zambia," in Robert W. Steel and Richard Lawton, eds., *Liverpool Essays in Geography* (London, Longmans Green, 1967), pp. 347-61.

[23] William Bascom, "Some Aspects of Yoruba Urbanism," *American Anthropologist*, LXIV, No. 4 (August, 1962), 699-709.

significance in African towns, but it is rarely sufficiently important to justify classifying such towns as manufacturing centers. A feature of the employment pattern in many towns is the large number of petty traders; in northern Africa and the sudan belt these are predominantly males, while the southern belt in West Africa has a preponderance of women traders. Surveys in Ghana and Togo made about 1960 showed that 80 and 91 percent of those engaged in trade were women. The daily attendance in Accra markets, was given as over 25,000, of whom over 85 percent were women.

Some difficulty arises in measuring the administrative function and delineating administrative towns. Such small capitals as Nouakchott, Entebbe, Zomba, Gaborone, and Mbabane can probably be so labeled, but it is doubtful that the administrative function outweighs the commercial one in the larger and more complex capital cities, despite the importance of government employment.

The vast bulk of tropical African towns may, then, be classed as commercial centers, particularly if one uses the term commercial in its broadest form to include trade, finance, and transportation.

CLASSIFICATION BY LOCATION OR FUNCTION/LOCATION

It is also possible to classify most towns according to major locational characteristic, which is often intimately related to their major function. Possible classifications in a locational system include: ports (ocean, river, and lake); crossroad towns (including rail and road-rail junctions); resource-exploiting towns (mining centers, towns supporting large hydroelectric installations); and, for manufacturing towns, those whose industries are oriented either to the raw material, to low-cost energy, or to the market. The system has value in revealing the basic reason for the importance of many towns. It has shortcomings, however, in that administrative towns have no clear locational orientation (though, in fact, many of Africa's capitals have been located at or been moved to the country's major commercial center), because towns may fall into several classes, and because information about a town's actual functions is still needed to provide a fully satisfactory classification.

A more acceptable system may be achieved by combining function and location, with functions being used for the major classes and location providing a basis for subdivision of these classes. Applying this functional/locational system to towns over 20,000 in tropical Africa in the mid-1960s but excluding Nigeria and Mozambique because of the lack of current data, one finds that there were about 162 such towns with a total population of about 13.2 million, which could be broken down roughly as shown in Table 23. At the present stage of development it is not surprising to find that trading-commercial cities are by all odds the most important class, and that sea and inland ports account

Table 23

PERCENT OF POPULATION AND NUMBER OF TROPICAL AFRICAN[a] TOWNS BY FUNCTION/LOCATION CLASSIFICATION, MID-1960s

Function Location type	Percent of urban population in countries covered[a]	Percent of number of towns in countries covered[a]
Trading-Commercial		
Seaports	36.4	26.5
Inland ports	13.8	11.7
Cross-roads towns	38.3	44.4
Subtotal	88.5	82.6
Industrial		
Low cost energy oriented	0.2	0.6
Raw material oriented	0.2	0.6
Market oriented	1.0	1.9
Subtotal	1.4	3.1
Agricultural	1.5	3.7
Mining	8.3	9.3
Administrative	0.3	1.2
Total	100.0	100.0
Estimated actual	13.2 mil.	162

[a] Excluding Nigeria, Mozambique; including Madagascar.

for almost half of the urban population in the countries included. The significance of seaports may be further appreciated when it is noted that of the 24 tropical African countries fronting on the sea all but five have a seaport as their largest city. These frequently developed from early trading posts or forts; they were strengthened by becoming the coastal termini of rail lines and road nets. Usually being the largest communities they were logically selected as the political capitals,[24] and their size later

[24] All continental African countries facing the sea have their capitals on the coast except Cameroon, the two Congos, South West Africa, South Africa (the legislative capital is, however, at Cape Town), Kenya, Ethiopia, Sudan, and Egypt.

attracted the newer, market-oriented industrial establishments.[25] Their eminence also reflects the degree to which the modern economies are oriented to international trade. While the ports obviously have multiple functions, it is probable that in all cases it is the commercial function which is of prime importance.

Inland ports on rivers and lakes are less important than on other continents and only rarely is a seaport also a river port. The Congo River system, however, includes three important capital cities—Kinshasa, Brazzaville, and Bangui—while other inland ports are found on the Niger, the Senegal, the Nile, and Lakes Victoria and Tanganyika. In some cases the river or lake transport function is augmented by an important crossing point and other converging roads.

Inland crossroads towns are increasing in number as is to be expected on a massive continent with gradual improvements in overland transport routes. Examples of important crossroads towns include Kumasi, Kano and the other emirate cities, Kampala, Bulawayo, and Lusaka.

There are only a few towns in tropical Africa whose main function is manufacturing, and the classification of some of these is problematical. They include Rufisque, Jinja, Gwelo, and Que Que. Inclusion of northern Africa would add a few cities such as Mohammedia, Arzew, and Melhalla el Kubra; South Africa has a number of primarily industrial towns such as Vanderbijl Park, Vereeniging, Germiston, and Sasolburg.

Surprisingly few mining towns with populations exceeding 20,000 are found in tropical Africa. Most of them, and all those over 100,000, are associated with the Katanga-Copperbelt complexes. Lesser mining towns are scattered and include Obuasi, Jos, and Wankie. South Africa, of course, has a considerable number of mining communities, though some have evolved to a much more complex pattern than existed when mining was their major *raison d'être*.

Only Nouakchott, whose population is said to have reached 22,000 in 1968, and Zomba, the titular capital of Malawi, were classified as administrative towns, though it is possible that Yaoundé should have been so included. In all other cases it was considered that other functions superseded the political in the capital cities. Hamdan has

[25] The major port of 23 African countries is also the main manufacturing center in these countries.

Urbanization in Africa

called attention to the frequency of a marginal or eccentric location for African capitals, explained by the importance of their commercial functions and the maritime orientation of many.[26] The large number of capital cities which have disappeared or declined over the years and the substantial number of countries which have switched their capitals two or more times, more often than not because of the desire to be located at the principal commercial node, suggests the greater significance of the transport-commercial function than the political function in the development of existing capital cities.

Problems of African Cities

It is now appropriate to turn to a direct examination of some of the problems besetting Africa's urban communities. These include problems which may be related to the physical environment and those which fall into the social, political, or economic realms, but the bounds are often so fuzzy as to make classification somewhat arbitrary.

PROBLEMS ASSOCIATED WITH THE PHYSICAL ENVIRONMENT

Physical site problems affecting African cities are, fortunately, not too numerous. There are many examples of abandonment of early sites because of health problems, but modern engineering and medical technology now usually make it possible to provide for adequate drainage or to spray against disease-carrying insects.

A number of cities are handicapped by constricted sites. Bathurst, for example, is situated on two sand bars separated by a water-level depression and its redevelopment is made more difficult by shortage of space. The mountain behind Freetown limits its easy expansion as does Table Mountain at Cape Town. The entrances to the latter city are confined in a relatively narrow band, leading to problems of traffic congestion and its suburbs have extended in a somewhat inconvenient manner. Additional space for the port facilities and the CBD have, however, been created by filling in the foreshore behind Duncan Dock. A few cities are adversely affected by steep slopes and hilly sites, including Bukavu, parts of Addis Ababa and Nairobi, Mbabane, and, on Madagascar,

[26] See G. Hamdan, "Capitals of the New Africa," *Economic Geography*, XL, No. 3 (July, 1964), 239-53.

Cities handicapped by physical site.

Saint-Louis, Senegal. Silting at the mouth of the Senegal River precluded this old city from serving as the main port of the country. An islandic location was favored in early years for protective reasons, but later restricted growth.

Cape Town, South Africa, whose CBD (upper right) is situated in a constricted site below Table Mountain, leading to transport congestion.

Tananarive, Madagascar. The old town is huddled on the steep slopes of a rocky massif rising above the Betsimitatatra Plain.

Fianarantsoa, and Tananarive, which Robequain described as "terriblement incommode sur un site indomptable."[27]

In the period of early contact, sites for trading nodes were frequently selected with a prime consideration being easy defensibility. Small islands were favored, witness the installations at Saint-Louis, Ile de Gorée, Grand Bassam, Lagos, Moçambique, and Mombasa in tropical Africa. Peninsulas were also selected: Ceuta and Melilla are excellent examples of defensible peninsular sites which have been occupied for centuries, if not a millennium; in sub-Saharan Africa, Conakry, Monrovia, Luanda, and Durban are on peninsulas. The island stations usually declined as commerce increased since they required additional transshipments and were often too small in any case. Only when they could be more or less readily tied to the mainland, as at Lagos and Mombasa, have they survived and grown into large and important cities.

[27] Charles Robequain, *Madagascar et les Bases Dispersées de l'Union Française* (Paris, Presses Universitaires de France, 1958), p. 317.

Drainage continues to be a physical problem in some cities and provision for runoff from torrential showers requires unusually large storm drains in many areas. North Khartoum has a particular problem of drainage due to its extremely flat character, making disposal of noxious effluents in the industrial quarter very difficult. Many cities have overcome drainage problems, often at considerable expense, increased by early failure to provide facilities before much construction had been completed. Examples include Freetown, Lagos, Nairobi, Kampala, and Lusaka. A few towns have also been situated where flooding may occur, such as Bathurst and Saint-Louis.

Provision of potable water has proved to be a serious problem for some cities, including those of the Witwatersrand and other parts of the southern Transvaal. Dams on the Vaal and Hartz rivers have helped to meet the problem, while the Republic has agreed in principle to assist in financing the Oxbow Scheme in Lesotho in part to increase the flow of water to the deficit areas of the Republic. Mauritania's new capital, Nouakchott, was plagued with water shortages until recent completion of a desalinization plant.

Other physical problems include the risk of earthquakes (e.g., Agadir, Accra), sea erosion (Lagos, Luanda), and construction on unstable land requiring underpinning of some structures (Beira). Of course inadequate sanitation, drainage, and water supply exist in a great many towns without necessarily being related to inherent physical conditions of the site.

SOCIAL PROBLEMS

A good many of the social characteristics of urban communities are subsumed under the term instability. Instability is associated, among other factors, with the high percentage of newcomers in many cities, the sometimes large numbers of temporary migrants, the numerical preponderance of men and of young people, and ethnic differentiation, which tends to fraction the urban structure.[28]

The very recent and very rapid growth of most major cities means that many inhabitants do not possess an urban tradition and may integrate poorly with the new environment. This point should not

[28] See Little, *West African Urbanisation*; Eric W. Wood, "The Implications of Migrant Labour for Urban Social Systems in Africa," *Cahiers d'Études Africaines*, VIII, No. 29 (1968), 5-31.

be overstressed, however, because it is not applicable to most traditional towns and much less pertinent to smaller towns and cities whose major increase comes from reproduction of their residents. It certainly applies to those mushroom cities which are receiving immigrants at a rate two, three, and more times the rate of natural increase. In Abidjan in 1965, for example, only 29 percent of the total population and only 7.1 percent of the 112,000 persons over 20 years of age had been born in the city. In fifteen West and Middle African cities the percentage of the population which had been born there varied from a high of 62 percent for Bathurst to a low of 26 percent for Douala and Libreville and averaged about 42 percent in the early 1960s.

Lerner writes regarding the newcomers to urban centers that

the point that must be stressed in referring to this suffering mass of humanity displaced from the rural areas to the filthy peripheries of the great cities, is that few of them experience the "transition" from agricultural to urban-industrial labour called for by the mechanism of development and the model of modernisation. They are neither housed, nor trained, nor employed, nor serviced. They languish on the urban periphery without entering into any productive relationship with its industrial operations. These are the "displaced persons," the DPs of the developmental process . . ., a human flotsam and jetsam that has been displaced from traditional agricultural life without being incorporated into modern industrial life.[29]

The disproportion of males to females referred to above varies from relatively small ratio differences, about 100 males to 90–95 females in many West African cities, through 100 males to 80–90 females in middle Africa, to 100 males to 55–75 females in eastern Africa, and even lower ratios in some large cities of South Africa, the last a reflection of the influx policies which frequently prohibit wives and children from accompanying their migrant husbands and fathers. A few cities, such as Lomé in Togo, have more women than men; explanations given for this atypical situation include a higher infant-mortality rate for boys, a longer life span for women, and, probably more important, the fact that Lomé has itself been a source of emigrating young males. In any case, the disproportions do create abnormal family situations which doubtless contribute to crime and prostitution.

The effect of urbanization on women is not entirely to be depre-

[29] Daniel Lerner, "Comparative Analysis of Processes of Modernisation," in Miner, ed., *The City in Modern Africa*, p. 24.

cated, however. The cities have, to a degree, liberated African women, who play a substantial role in them, particularly in West Africa. An exception to this generalization is noted by Powdermaker for the Roan Antelope-Luanshya mine area of Zambia, where women in the townships no longer have the communal role characteristic of tribal life. Here they became more dependent on their husbands than before, which increased insecurity, led to disputes over money, and was a factor in the high number of divorces recorded.[30]

The patterns reported for African townships in South Africa suggest a substantially different situation from that reported by Powdermaker. In Soweto, the huge township associated with Johannesburg with a 1967 population of 377,249, it is common for both wives and hubands to be employed and to be away from home most of the day. Here the problem becomes one of unsupervised and unoccupied children. The South African Institute of Race Relations notes that, despite the adoption of the two-platoon system in the available schools and hiring by the residents themselves of extra teachers, there are not enough places for many, the result being "that hundreds of children roam the streets with no occupational interest and it is suggested that it is out of this group that delinquents and criminals are recruited."[31]

Ethnic differentiation is reflected in the characteristic grouping of tribes into distinct quarters and in the separation of Asian, European, and other non-African residents. Balandier writes regarding the contrasts between the districts of a town that they

serve to illustrate the social gulf that exists between the groups of which it is composed. This inequality in living conditions masks a numerical inequality which militates in the opposite direction; though the town is a zone of contact (and therefore of more diversified relationships) it is also . . . a setting in which social tensions are particularly liable to arise.[32]

Segregation often started either spontaneously or as a result of the desire to protect health conditions. Creation of the medina in Dakar,

[30] See Hortense Powdermaker, *Copper Town: Changing Africa* (New York, Harper and Row, 1962).

[31] South African Institute of Race Relations, "Crime in the Townships," mimeographed paper, 9 May 1967, p. 2.

[32] G. Balandier, "Urbanism in West and Central Africa: The Scope and Aims of Research," in UNESCO, *Social Implications of Industrialization and Urbanization in Africa*, p. 498.

for example, was initiated in 1914 in response to a severe epidemic of bubonic plague and a decision by the health authorities that its control required displacement of a sizable portion of the indigenous population. Betts writes regarding the establishment of this medina that

the arguments . . . were conventional: the inability . . . to decontaminate the poorly constructed native housing, the squalor which the native tolerated—particularly the cohabitation of people and domestic animals—and, on a cultural level, the fundamental difference of the African way of life from the French. Thus the basis for segregation was not racial in theory, even if it did later become so in substance.[33]

Somewhat comparably, Brelsford writes regarding the formation of African locations in Rhodesia that

the whole system rested on the assumption, well justified at the time, that there was a wide gap between European and African modes of life, and that the African would not wish to live permanently in the town. All he needed therefore was a place of temporary residence, where his doings might be supervised and where some minimum standards of health might be enforced.[34]

Once separation began, however, it almost invariably led to segregation, enforced more and more by various legal restrictions. Such restrictions have, of course, been removed since independence, except in Rhodesia and South Africa. Some African government officials have moved into the former European areas, though others have preferred lower-priced quarters. In some cities political leaders have been criticized for moving into sections where they are more likely to lose contact with the mass of the population. Despite changes prior to and at the time of independence, segregation by ethnic group continues common, including that between individual tribes.

In South Africa, where urban segregation has been most advanced, it is government policy that the several racial groups must live in separate areas, each of which should be large, have room for expansion, be separated from other areas by buffer zones, and have access to

[33] Raymond F. Betts, "The Establishment of the Medina in Dakar, Senegal, 1914," paper presented at the Annual Meeting of the African Studies Association, Los Angeles, October, 1968 (mimeographed), p. 1.

[34] W. V. Brelsford, ed., *Handbook to the Federation of Rhodesia and Nyasaland* (Salisbury, Government Printer, 1960), p. 72.

the places of work without passing through other racial areas.[35] This has involved the movement of thousands of people, sometimes from districts which they and their forebears had occupied for decades if not centuries.

The results of the social conditions found in urban communities include a higher rate of crime, separation of families, increased instability in marriage and family life, conflict among ethnic groups, and poorer nutritional standards.

With few exceptions, the evidence suggests that the towns and cities have distinctly higher rates of crime, particularly petty thieving, juvenile delinquency, and prostitution, but also crimes of more serious dimensions, all of which appear to be increasing. The rates of crime vary from town to town, in considerable part reflecting the degree of tribal authority maintained. Busia in his survey of Sekondi-Takoradi found much evidence of a maladjusted community, including a breakdown in the extended family system, increased crime, delinquency, prostitution, and frequent fighting, all of which were more marked in Takoradi than in Sekondi and which were attributed to the more mixed population of the former and the absence there of a "core of indigenous tribesmen with a tradition of tribal discipline and authority."[36]

The East Africa Royal Commission reported in 1955 that

where conditions of poverty and social instability exist, the incidence of crime is likely to be high. In the multi-tribal societies of the towns the restraints imposed by tribal codes do not operate. In many of the larger towns robbing with violence is carried out by organized gangs, and is of frequent occurrence. Africans suffer equally with members of other races, but they usually fail actively to support the police, whom they regard with distrust.[37]

Jenkins, in a brief survey of urban violence in Africa, writes that violence "seems to be strikingly urban in several ways. . . . Since World War II, there have been several dozen cases in which during a single day up to 120 persons were killed."[38] He sees this violence as a product

[35] T. J. D. Fair, "The Effect of Apartheid on Morphology of Cities in South Africa," paper presented to the Annual Meeting of the African Studies Association, Bloomington, Indiana, October 1966.

[36] As reported in UNESCO, *Social Implications of Industrialization and Urbanization in Africa South of the Sahara*, pp. 82-83, 86.

[37] *East Africa Royal Commission 1953-1955 Report*, p. 208.

[38] George Jenkins, "Urban Violence in Africa," *African Urban Notes*, II, No. 5 (December, 1967), 37-38.

of and responsive to the specialized, integrated urban structure and not as "a product of the recent rural-urban shift of population nor the response of up-rooted tradition-oriented individuals to the supposedly different nature of urban life."[39]

Crime appears to be particularly serious in some of the South African townships where the sex ratios and family life are subject to unusual distortions. In a one-year period, 1966–1967, the Johannesburg Municipal Area reported 891 cases of murder, 1,156 cases of rape, 7,747 assaults with intent to do grievous bodily harm, 8,075 common assaults, and 33,489 cases of theft. And in the four-year period 1962–1965 police in Soweto received reports of 1,192 cases of death by violence; of the 247 persons convicted in these cases 52.6 percent were under 21 years of age. A 1967 survey of Soweto indicated that its residents considered crime and lower moral standards the most serious problems in the township.[40]

References have been made in previous sections and chapters to the separation of families, increased family instability, and ethnic conflicts, and these elements of urban social conditions will not be further elaborated in this section. The impact of urbanization on dietal standards deserves brief mention. Hendrickse states that it is "becoming increasingly apparent that the nutritional problems of Africans in urban and periurban areas are as bad as, if not worse than, those encountered in rural areas,"[41] in part due to poverty and in part because migrant males do a poor job of feeding themselves. Higher incomes in urban centers are often more apparent than real and consumers are inclined to stint on food in favor of other purchases. Food expenditures are likely to absorb 50 to 80 percent of total domestic expenditures of many urban residents, which creates a strong incentive to reduce them if at all possible. Poleman reported that one of the most surprising findings of household budget surveys in the cities of Ghana was that the wealthier urban consumers apparently ate essentially the same type meal as their less fortunate neighbors.[42]

[39] *Ibid.*, p. 41.

[40] Market Research Africa (Pty.) Ltd., *An African Day: A Second Study of Life in the Townships* (Johannesburg, 1968).

[41] Hendrickse, "Some Observations on the Social Background to Malnutrition in Tropical Africa," p. 346.

[42] See Thomas T. Poleman, "The Food Economies of Urban Middle Africa: The Case of Ghana," *Food Research Institute Studies*, II, No. 2 (May, 1961), 121-74.

It is all too easy to focus on the problems and shortcomings evident in African urban communities and to lose track of the adjustments and achievements that ameliorate the undesirable features. Plotnicov reminds us that "urban life is complex but not disorganized as some . . . have suggested" and that what may appear inconsistent and irrational to an outsider is logical and appropriate from the point of view of the individual concerned,[43] while a U.N. study notes that

there are many old districts characterized not only by their slums but also by the social solidarity . . . of different social levels. . . . Despite their shabbiness and physical squalor, these slums enjoy the most integrated social life in the city.[44]

In many cities adjustment of the newcomer is eased by the assistance rendered by members of his extended family and by the existence of numerous voluntary associations. Aldous writes that in many cities

the extended family is indeed functional. Besides filling recreational, religious, legal or economic needs . . . kinsmen provide for the elderly and support the sick, the jobless and the destitute. They give the new arrival from the country shelter and food and help him to get work or an education and to adjust to the bustling city.[45]

Little indicates that voluntary associations play a significant role in adopting traditional institutions and in integrating the new social system in urban communities. He lists various types of such associations including specific tribal associations, syncretist cults, mutual aid societies, craft unions, recreation groups, and modern associations, all of which are bridges for the newly arrived migrant, providing fraternity, sense of identity, practical training, financial protection, religious guidance, and traditional social control.[46]

Other studies have also indicated satisfactory adjustments to urban conditions. Knoop reported from a 1965 demographic sample survey in Kisenso, a suburban squatting community of Kinshasa, results which he considered to deviate from the generally accepted picture,

[43] Leonard Plotnicov, *Strangers to the City: Urban Man in Jos* (Pittsburgh, University of Pittsburgh Press), pp. 290-1.

[44] "Urbanization: Development Policies and Planning," p. 42.

[45] Joan Aldous, "Urbanization, the Extended Family, and Kinship Ties in West Africa," in Pierre Van den Berghe, *Africa: Social Problems of Change and Conflict* (San Francisco, Chandler Publishing Company, 1965), p. 115.

[46] See Little, *West African Urbanisation*.

including a remarkably well-balanced sex ratio, a high employment rate, a very high rate of natural increase, and rapidly decreasing rural-to-urban migration.[47] Schaff found that urbanization in Uganda has not yet produced conditions of human degradation and disorder, but had rather, by bringing together peoples of different backgrounds and culture, been a positive factor in producing a modern nation and promoting economic development.[48] The Soweto survey previously noted also indicated, despite the listing of numerous problems, that 74 percent of those interviewed preferred life in Soweto (as compared to 63 percent in a 1962 survey), their main reasons being the better employment opportunities and better pay available and its better, more civilized life.[49]

Additional studies have noted increased socializing among races,[50] "change towards a more intimate and affectionate father-child relationship oriented towards the raising of fewer, more self-directed children,"[51] and development of a positive attitude to cooperation with different ethnic groups.[52]

The social problems of African cities are, in conclusion, serious and undeniable, but one should guard against exaggeration, particularly in drawing conclusions based on direct comparisons with non-African cities or from imputed attitudes based on imagined or expected reactions.

POLITICAL PROBLEMS

The major political problem related to African urbanization stems from demands of its residents for improvements which it is not always possible to provide. The jobless and the underemployed present a particular problem, leading to various coercive measures which have seldom proved successful (see page 277). The idle and semi-idle young men

[47] Henri Knoop, "Some Demographic Characteristics of a Suburban Squatting Community of Léopoldville: A Preliminary Analysis," *Cahiers Économiques et Sociaux*, IV, No. 2 (June, 1966), 119-49.

[48] Alvin H. Schaff, "Urbanization and Development in Uganda: Growth, Structure, and Change," *The Sociological Quarterly*, VIII, No. 1 (Winter, 1967), 111-21.

[49] Market Research Africa (Pty.) Ltd. *An African Day: A Second Study of Life in the Townships.*

[50] Plotnicov, *Strangers to the City: Urban Man in Jos.*

[51] Robert A. Levine, Nancy H. Klein and Constance R. Owen, "Father-Child Relationships and Changing Life-Styles in Ibadan, Nigeria," in Miner, ed., *The City in Modern Africa*, p. 252.

[52] H. Dieter Seibel, "Some Aspects of Inter-Ethnic Relations in Nigeria," *The Nigerian Journal of Economic and Social Studies*, IX, No. 2 (July, 1967), 217-28.

provide a fertile field for dissidents and have supported more than one of the many coups or attempted coups that have occurred since independence. The capital cities of Africa are the major sites of political action and intrigue; African leaders understand this clearly and may, therefore, feel obligated to adopt measures which certainly may not represent the economically most sensible allocation of limited available funds.

Dumont sees the gross inequalities between the standards of the African elite in the cities as compared with the mass of urban and rural dwellers as one of the most important hindrances facing African economic advance.[53] He suggests that Africans in rural areas have little incentive to work when they see the luxuries enjoyed by the special few in the cities, and recommends that the government elite should cut their salaries and extra emoluments. While this prescription may be a sound one, I doubt that it would have the predicted impact on increased effort by both the rural and urban masses.

ECONOMIC PROBLEMS

It is in the economic sphere, or so it seems to me, that the most intransigent dilemmas are revealed.

Unemployment and Underemployment. Foremost is the high incidence of un- and underemployment in most of Africa's major cities, caused more by an excess in-migration than by a lack of economic growth; indeed urban unemployment seems to be increasing more rapidly in some of the countries which have the most satisfactory rates of economic growth.

Data on unemployment are entirely inadequate to derive more than a rough estimate of its extent. The following fragmentary information will, however, provide some concept of its dimensions. Morocco had about 20 percent of its male urban labor force unemployed in 1964, while at least 50 percent of its male rural population was said to be either unemployed or under-employed. Unemployment was placed at over 50 percent in 1968. The Three Towns of Sudan had an eighth of their adult males without jobs in 1963. In 1965 an estimated 34 percent of the laboring force of Lagos was out of work. Frank, studying the

[53] René Dumont, *False Start in Africa,* translated by Phyllis N. Ott (New York, Praeger, 1966).

Urbanization in Africa

employment-creating aspects of Nigerian development plans, concluded that urban unemployment would become increasingly severe.[54]

Kinshasa was thought to have at least 150,000 people in 1968 without visible means of support. For Kenya, Segal estimated in 1965 that 51 percent of the country's population over 16 years of age was unemployed and he calculated that during the 1965–1970 plan 47 percent of the 720,000 young men reaching age 17 would be added to the ranks of the unemployed.[55] Nairobi had an estimated 40,000 unemployed in 1963, when Dar es Salaam was estimated to have one-fifth of its adult male population out of work. At least 27,500 Zambians were actively seeking work in 1963, a number equal to 18 percent of the actual labor force. That the problem is not entirely of recent origin is suggested by a 1951 report from Tanganyika which stated that "there is a perceptible drift of Africans from the country districts to the town, in many cases the numbers involved bearing no relation to the industrial or commercial opportunities offered by the towns."[56]

Lewis opines that current investment is generating small employment because much of it is wasteful and because it is more capital-intensive than it should be.[57] And Barber suggests that

cities in the underdeveloped world may not possess the "generative" properties [attributed by economic theory]. They may, on the contrary, be "parasitic," i.e., with effects that retard growth in the economy as a whole. . . . In the extreme case of "parasitic" urbanisation, the growth of cities might even provide a better index of economic deterioration than of economic progress."[58]

A variety of steps have been taken to offset the problem of unemployment in the cities. South Africa avoids it by influx control and endorsing out of the urban areas persons who have not found employment in a brief period. Despite the objections to such control of personal

[54] Charles R. Frank, Jr., "Industrialization and Employment Generation in Nigeria," *The Nigerian Journal of Economic and Social Studies*, IX, No. 3 (November, 1967), 277-97.

[55] Aaron Segal, "The Problem of Urban Unemployment," *Africa Report*, X, No. 4 (April, 1965), 17-21.

[56] *Report of Her Majesty's Government to the General Assembly of the United Nations on the Administration of Tanganyika, 1951.* Colonial No. 287. (London, H.M.S.O., 1952), paragraph 678.

[57] Lewis, "Unemployment in Developing Countries," p. 18.

[58] William J. Barber, "Urbanisation and Economic Growth: the Cases of Two White Settler Territories," in Miner, ed., *The City in Modern Africa*, p. 108.

movements many African states have, in fact, applied or attempted to apply comparable restrictions. In Niger, for example, a law passed in 1962 required that all unemployed urban youths either perform some service or return to their native villages. This was apparently ineffectual, the police were reported in April, 1963 as having rounded up all such youths in Niamey and shipped them back to the country; "again in June, 1964, the government ordered all unemployed men to take up agricultural work on pain of being prosecuted as vagabonds."[59] In January, 1959, the authorities in Bangui "more or less forcibly dispatched to a 2000-hectare area adjoining the road between Bangui and M'Baiki an initial contingent of 100 unemployed youths recruited in the capital city,"[60] while in 1964 a decree directed that all unsalaried citizens "living on their wits" in urban centers be rounded up and put to work rehabilitating old plantations. It was declared that this was not a violation of international work conventions, but rather an economic necessity. In the early 1950s Moyen Congo, now Congo (Brazzaville), offered inducements to urban idlers and vagabonds to return to their village in the form of free transport, exemption from taxes for a limited period, and loans of seed and equipment, but the program was not successful; in 1954 the government started a paysannat in the Niari Valley for unemployed urban youths. Riots in Brazzaville both before and after independence were attributed in part to that city's parasitic idlers.

In January, 1966, the Interior Minister of Congo (Kinshasa) ordered unemployed persons in the cities to return home by the end of the month; a similar order was announced in March, 1968, which also called for the deportation of all foreigners without sufficient means of support.[61] In 1964, Kenyatta announced that the Kenya government was taking immediate steps to return to the land all able-bodied people living in towns but without jobs, and Tanzania applied restrictions on travel to Dar es Salaam in the same year and ordered the police to pick up people who were unemployed and give them warrants to return home. As a final example, Ghana announced in 1969 that it planned to develop controls over internal migration to prevent the excessive movement of persons

[59] Virginia Thompson, "Niger," in Gwendolen Carter, ed., *National Unity and Regionalism in Eight African States* (Ithaca, Cornell University Press, 1966), p. 178.
[60] Thompson and Adloff, *The Emerging States of French Equatorial Africa*, p. 425.
[61] *The New York Times*, January 18, 1966, March 3, 1968.

to the cities. All of these efforts have objections on moral grounds, but they have also been expensive, unpopular, and ineffective.

Other attempted solutions to the problems of unemployment include the use of doles and labor projects in Tunisia and Morocco, both aided by partial payment of workers with American surplus food allotments; youth corps and workers brigades, as in Ghana, whose post-Nkrumahist government did not feel able to disband the program despite objections to it; compulsory employment schemes, as in Kenya, where a 1963 decree stipulated an increase in employment of 10 percent in the private sectors and 15 percent in the public sector in exchange for a one-year moratorium on wage increases and strikes; and raising taxes within the city far above the level prevailing in the nearby countryside, a system which was used in Ft. Lamy in 1954 and earlier in other towns of what was then French Equatorial Africa to encourage a return to the land.

In 1967, the Ivory Coast government announced a multi-pronged effort designed to increase the attractiveness of rural dwelling and thus reduce what was considered an excessive influx to Abidjan, including the formation of youth clubs, provision of open-air cinemas, a program to replace all mud-walled, thatched-roofed huts with cement-block, corrugated-metal-roofed houses, and propaganda campaigns. It will be interesting to see whether provision of additional amenities in the villages will, in fact, help to "keep 'em down on the farm" or whether it will merely whet the appetite for the bright lights of the capital city.

Instability. In addition to the social and cultural elements making for instability in African urban centers, a number of economic conditions contribute to make instability a socio-economic problem of great concern and complexity. Economically oriented explanations include the one-product character of African money economies which makes them subject to fluctuations largely beyond their control, the high percentage of unskilled among the new migrants and in the unemployed populace, the impact of modern industry and imports of manufactured products on components of the craft industrial sector, the existence of some unscrupulous bosses who see in a high turnover one way of avoiding the payment of higher wages or as a way of having workers who will cause less trouble, and the improving productivity of African workers which is resulting in keeping the rate of employment increase below the rate of economic growth. Indeed, Lewis cites an example from Kenya where, between 1954 and 1964 private real output in the money economy increased by 4 percent per annum while employment declined by 1 percent per annum.[62]

Instability is reflected in fluctuating employment levels in several activities, in the sometimes rapid demise of certain craft industries, and in high rates of turnover among employees in various establishments, which is, of course, also related to the continued importance of circulatory migration. Studies in the 1950s showed the high instability in job holding in a number of African cities. In Nairobi in 1953, for example, 48 percent of the African workers had less than one year service in their enterprises. In Salisbury in 1956 the average length of employment of African workers in any job was 12 weeks.

More recent studies suggest that turnover in jobs is declining, sometimes very rapidly, while it probably has long been lower in West African cities than in those of East and Central Africa. These studies reveal that turnover and absenteeism tend to decrease with age, family responsibility, level of skill, length of service, quality of lodgings and food, transport facilities, and health improvement.

The solution to unemployment and instability is somewhat euphemistically called the "normalization" of urban employment. The major questions become—how much can and how much should be done? Some suggest that the requirements are to slow urban movement, to stabilize workers in cities and to stimulate the return of a number of migrants to their villages. This implies that a major effort would have to be made to stimulate development in rural areas. But programs to get the unemployed to return to their source areas have not been successful and efforts to restrict the influx of new migrants also have not worked, both such efforts requiring an excessive degree of force which is politically dangerous to apply. Nor has anyone yet found the answers to rural development on a sufficient scale to reduce the lure of the cities.

The alternative, of course, is to create more jobs in the cities—in industry, construction, administration, and other services. But this, too, is not necessarily very easy—indeed, it is just the reverse, very difficult. One of the many dilemmas is that industry needs bigger markets and largely subsistence-farming hinterlands do not provide them.

[62] W. A. Lewis, "Unemployment in Developing Countries," pp. 17-8.

Provision of Physical Needs. The next problem is the provision of the necessary supplies, services, and housing for the urban communities. Generally speaking, the bigger the city the greater is the need for public services, though the cost per capita may go down. But the per capita consumption of water and electricity increases, the need for buses grows with the lengthening journey to work, and kitchen gardens and wood-fuel plots become more difficult to utilize, their output having to be replaced from more distant sources.

FOOD. Provision of food has become a serious problem in some cities and, as has been suggested, evidence indicates that many urban residents in Africa have poorer nutritional levels than rural residents. Domestic entrepreneurs have sometimes been slow to respond to rising demands and the cities find themselves increasingly dependent on foreign supplies, which also reflects a growing preference for western foods on the part of those who can afford them.

Perhaps the worst situations are in such cities as Algiers, Conakry, or Kinshasa whose residents have been reliant on foreign gifts for as much as a quarter to a half of total food supplies. The problem is revealed on a national scale in such countries as Sierra Leone, whose imports of food exceed the value of total agricultural exports, or Upper Volta, whose imports of food are greater than the value of total exports. It is ironic that many African countries, the vast bulk of which must look to agriculture as the main basis for development, are net importers of foodstuffs, animal products, and forestry products.

PUBLIC SERVICES. Provision of water, electricity, piped gas, and public transport, installation of roads, drains, and modern sewage systems are needs which often present difficult problems to African cities, not because of technological considerations but because of the lack of available funds and, to an extent, of the necessary specialists.

A rough idea of the position may be indicated by a few examples: in one *gourbi* in Tunis in 1960 only 6 percent of households had running water and 42 percent had neither electricity nor sewage; in the medinas of Casablanca half of the houses had no kitchen or toilet and two-thirds had no water; in Dakar in 1955 there was neither electricity nor water in 64 percent of the households; in Abidjan in 1965 only 29 percent of housing units had water-borne sewage systems; in 1955 only 1.5 percent of houses in Ft. Lamy had electricity; and in Tananarive in 1960 only one-third of households had electricity, while six-sevenths were without water or toilets.

A not atypical situation with respect to sanitation is described by Okediji and Aboyade for the core sector of Ibadan, which is "unhealthy, filthy, crowded and highly susceptible to any epidemics":

Most of the houses are built of mud and sticks, and cow dung is used to wipe the surface of the floors which are anything but smooth. Most of the houses have no kitchens and cooking is done in the corridor. An observer can see beads of carbon on the walls of the corridor. Most of the clay pots used for cooking are unwashed, and contain dirty water on the surface of which one sees dead flies and cockroaches. Cobwebs are common features of the various corners of the dwelling units. In places where there are separate kitchens, they are usually unswept and full of obnoxious odour. Aggravating this condition is the location of an uncovered salga directly behind the kitchen. There are usually bits of dried excreta all over the place. In some houses, the salga is used by all members of the compound and responsibility for cleaning it is not assumed by anybody. Standing water all over the place affords breeding grounds for mosquitoes and flies. Most of the gutters are uncemented and full of foul smelling water. Some houses have no salga and the inmates are not financially able to employ nightsoil men to serve them. In such cases, the members of the household go to the nearby bush or pit to excrete. To walk near the walls of any building is to experience the terrible odour of urine disposed there by the inmates or passersby. Most of the houses do not have pail latrines.[63]

While progress is being made in most cities in the extension of services, there remain numerous towns of considerable size which do not have electricity, a modern system of water supply, or water-borne sewage. Millions of houses and huts are dependent on water points, often located at some distance, and do not have any kind of indoor plumbing or adequate waste disposal system.

It is not surprising that most towns and cities have never had a plan or been studied by planning experts. In Nigeria, for example, only Lagos and Kano could be said to have had adequate studies, and in Cameroon, as a second example, only thirteen of some thirty-eight towns over 10,000 have some kind of town plan. Improvements are also some-

[63] F. O. Okediji and O. Aboyade, "Social and Economic Aspects of Environmental Sanitation in Nigeria: A Tentative Report," *The Journal of the Society of Health*, Nigeria, II, No. 1 (January, 1967), 10.

times slowed by a multiplicity of authorities and conflicts in jurisdiction between modern and tribal sections of the town. But again, it is primarily shortage of funds which delays the provision of planning and of public services in most African cities.

HOUSING. The provision of housing, which is the most characteristic aspect of the urban setting, is the largest and most expensive urban need. One has only to visit an African city to see the deplorable conditions in some quarters—unbelievable crowding, grossly inadequate construction, bidonvilles or shanty towns on empty plots or on vast acreages on the margins. And there are very few major cities where these conditions are not worsening despite some heroic efforts to move against the problem. A number of examples will convey a notion of the difficulties.

With respect to the types of dwelling, the 1964 Libyan census found that of the 331,990 households in the country about 21.3 percent were shanties, 20.0 percent tents, and 3.4 percent caves. In three major cities in Senegal, including Dakar, from 80 to 97 percent of households were built of traditional materials in the mid-1950s. Abidjan had 80 percent traditional structures and Bouaké had 68 percent grass huts in 1955. In 1962 four-fifths of the dwellings in Ft. Lamy were traditional and only 1.5 percent were classed as permanent dwellings. In 1955, 40 percent of houses in Luanda were not built with permanent materials. In more than a few cities there are a substantial number of people without normal shelter of any kind, though nowhere are there thousands of people sleeping in the streets as in Indian cities. As early as 1948 the Nairobi Municipal Affairs Officer reported that "it was disheartening to see legitimately employed Africans sleeping under the verandahs in River Road, in noisome and dangerous shacks in the swamp, in buses parked by the roadside and fourteen to the room in Pumwami, two to a bed and the rest on the floor."[64]

High urban densities in the quarters occupied by Africans also suggest the dimensions of the problem. In the medinas of Casablanca, within the walls of Fès, and in the Casbah of Algiers densities ranged from 465 to 1,240 per acre in the period 1955–1960. The "old city" of Kinshasa had 121 per acre in 1956, which is still very dense considering the fact that there were no multi-story structures. Density per room is characteristically high and apparently increasing. In one *gourbi* of Tunis

[64] Report of the Municipal Affairs Officer, Nairobi (1948), p. 11.

60 percent of families lived in one room in 1960 with an average occupancy of 4.9 per room. In Dakar in 1955, 45 percent of rooms were occupied by three or more persons. In 1960 in Freetown two-thirds of families lived in one or two rooms with an average of three per room. 52 percent of the households in Nairobi had three or more in each room in 1955. And in Tananarive in 1960, half of the families lived in one room.[65]

The prevalence of bidonvilles or shanty towns in practically all major cities is another measure of the great need for housing. Data available for Moroccan cities indicate that over a half-million urban residents lived in bidonvilles in 1960, including 180,000 in Casablanca. So short was the supply of dwellings that enterprising entrepreneurs were making 40 to 80 percent profits annually on the rental of one room-tin and crate houses. One-third of the populations of Algiers, Oran, and Annaba (Bône) were estimated to be living in bidonvilles in 1954. In Tunis,

[65] Sources for these data include: Nerfin, "Towards a Housing Policy"; Milton Santos, "Vues Actuelles sur le Problèmes des Bidonvilles," *L'Information Géographique*, XXX, No. 4 (September-October, 1966), 143-50; Fredj Stambauli, "Urbanisme et Développement en Tunisie," *Revue Tunisienne de Sciences Sociales*, IV, No. 9 (March, 1967), 77-109; U.N., *Housing in Africa: Problems and Policies* (E/CN.14/HOUPA/3, 28 November, 1962).

Housing problems in African cities.
Some examples of bidonvilles or shanty towns.

Part of the former Sophiatown, near Johannesburg, South Africa, whose occupants have been moved to newly developed townships.

Urbanization in Africa

Shanty town near Lubumbashi, Congo.

Traditional housing on the periphery of Blantyre, Malawi. Such unplanned and unserved agglomerations are common and increasing in extent on the edges of many African cities.

Urbanization in Africa   285

110,000 people were counted in the bidonvilles in 1960, and some 90,000 lived in shantytowns in Lusaka in 1969.

How to meet the housing needs of the present and burgeoning populations of African cities is an almost insoluble dilemma. In no case to my knowledge is the housing program keeping up with the increasing population, and in some capital cities the only modern housing that has been constructed in recent years has been for government officials and foreign diplomats. Most governments simply do not have the requisite funds, and relatively few citizens have adequate resources, particularly the newcomer, who more than likely arrives penniless. Indeed, he cannot even afford a fair rent on minimum lodgings, which can easily require more than the average per-capita income of the country. Since food and minimal clothing requirements frequently absorb two-thirds to nine-tenths of his earnings, if he succeeds in finding a job, there is just not enough to provide housing as well.

Various efforts have been made to reduce the dimensions of the problem. A number of national and city authorities have contracted for the construction of low-cost houses or flats. The Moroccan state, for example, constructed 31,700 lodgings in the fourteen-year period to 1958, but despite this effort the surplus population was greater at the end of the period than at the beginning. Since 1959 several programs have been adopted, including financial and technical aid to individual builders, but the problem has continued to worsen.

The Ivory Coast provides a second example, interesting because the booming economy of that country permitted a far more ambitious program than could be mounted in most countries. Construction of low-rent housing in Abidjan began about 1951 when SIHCI, a mixed company, was given a subvention and a land grant and proceeded to build 912 one-to-three-room dwellings to mid-1955, in the two African districts of Treichville and Adjamé. Rents of between $7 and $18 a month were within reach of enough people to provide a waiting list three times the number of units available. In 1960 the Republic created SUCCI, another mixed company, which built 2,200 lodging units in four years. The emphasis was changed to creation of whole quarters, each with social services and a commercial center, and to multistory apartments, which the plan authorities deemed to be necessary despite the general prejudice against them. These two companies had constructed a total of 5,900 units housing about 36,200 people by 1965, at which time they were combined into SICOGI, with membership of several public and private organizations and aid from France.

At present about 1,000–1,200 units are being constructed annually in Abidjan. While this compares very favorably with programs in other cities and is augmented by construction of private companies, it is far from meeting the needs of the city which are for about 6,000–8,000 units solely to house the yearly influx. Replacement of existing substandard units is also required, as is indicated by the estimates made in 1965 that 24 percent of housing units in Abidjan were mud huts and 19 percent were wooden structures.

The housing situation in Kampala, Uganda, provides a third example. The densest African settlement has always been in Mengo, the capital of Buganda, which had a greater degree of urban development than any indigenous center in eastern and southern Africa, but other unplanned high-density quarters have sprung up in the swamps and low lying areas between the many hills and in several periurban areas, where the masses of unskilled, uneducated, poorly paid migrants build crude houses or thatched huts, "getting practically nothing in the way of increased material amenities such as roads, power and light, water, drainage, sanitation, religious and educational services or police protection."[66]

Richer Ganda constructed large houses in and around Mengo or smaller suburban dwellings set in gardens and banana groves, while the police, railways, and other authorities constructed modern accommodations for their employees. Cognizant of the need for improved housing, the government began construction of housing estates about 1950 for several economic classes, and units were either rented or sold under various tenant-purchase schemes. Southall estimates that in 1967 these officially sponsored estates contained no more than 10,000 persons, most of whom would not be classed as lower-income residents. In the meantime, unplanned housing continues to be constructed and many townsmen appear to "greatly prefer the homely and congenial, if materially rudimentary, accommodation which they could secure in the laissez-faire situation of Mengo and the periurban areas.[67] In 1968 the Uganda

[66] Aidan W. Southall, "Kampala-Mengo," in Miner, ed., *The City in Modern Africa*, p. 314. Most of the discussion on Kampala is derived from this source.
[67] *Ibid.*, p. 316.

*New housing in African cities. Despite some large-scale
programs designed to eliminate slum conditions, housing
problems are increasing in most cities.*

New Amboshidi, Nigeria. Housing
constructed to replace a village to be
inundated by the reservoir of the
Kainji Dam on the Niger River.

Kiosque housing estate in the middle
of Treichville, largest of Abidjan's
high-density residential areas.

government signed a $16.8 million contract with Israel for the building
of 2,000 prefabricated houses in Kampala.

South Africa has had a massive program for construction of low-
cost housing in the African townships, some of which are very large
indeed (e.g., Soweto, with a population of 377,249 in 1967). While
monotonous in the extreme, there is no doubt that they provide distinctly
superior facilities to the notorious slums and shanty towns they replaced.
In Soweto in 1967 each household averaged 6.1 persons and there was
a room occupancy of 1.5 persons; average rents were $7.35 per month.

The overwhelming difficulty associated with construction of modern houses to accommodate the rising urban populations of most African
cities is the inability of most residents to pay rent on even the lowest-
cost houses, which may average about $2,000. Nor can governments pro-
vide sufficient subsidies to provide manageable rents, and such subsidies

Urbanization in Africa

Part of Soweto, the largest African township in South Africa, outside Johannesburg.

are subject to objection in any case because they use funds derived from the poor. Tenant purchase schemes and housing loans do not solve the problem, because they are still characteristically beyond the means of all but the highest income groups.

It is frequently suggested that efforts to reduce the costs of construction and construction materials would go a long way to make modern housing more economic. Such efforts would include: reducing the thickness of walls, the height of rooms, and the average size of rooms; improving the productivity of labor through simplifying the work and developing worker skills; standardizing materials; and substituting local for imported materials. The last is desirable in saving foreign exchange but may not, in practice, reduce the cost of such materials. Continued research in these directions, including the possibility of improving local materials through such processes as cementing of locally produced mud bricks and better thatching, is desirable, but no break-through is to be expected from this approach.

A number of countries have used various forms of assisted construction whereby the individual builds his own house, the government providing all or part of the necessary materials, sometimes below cost, and possibly with some technical assistance. Under the Grévisse system in Elisabethville (now Lubumbashi), for example, the Belgians provided a finished foundation, all of the material, and a loan to the builder who completed the rest of the construction.

A system of growing importance involves the provision of a plot in a quarter which has been provided with more or less minimal amenities such as paths, drains, water points, and group sanitary facilities. The builder then constructs a house which may be required to meet minimal standards, which may, however, be placed below those legally required for urban construction. Morocco called this system one of "lot évolutif" or "lot économique"; elsewhere it is called a "site-and-service" scheme, while it has also rather unfortunately been termed one of "planned slums." Such systems, although certainly not meeting the needs in an entirely satisfactory manner, have the distinct advantages of lower cost, greater speed, provision of at least the basic necessities for a more healthy urban environment, and of permitting an individual to improve his dwelling as his income permits. The last is in strong contrast to high-density. unplanned construction, including bidonvilles, which can only be bulldozed if satisfactory housing is to be provided. A disadvantage of such schemes is that the low-density occupancy pattern may eventually result in cities of enormous size, à la Los Angeles.

The Zambian government, for example, is relying more and more on site-and-service schemes under which the prospective builder is given a plot in a section where services have been laid on, a loan of $50 to buy basic construction materials like cement and corrugated roofing, and assistance from local-authority experts who require maintenance of minimum standards.[68] In Lusaka, some 57,000 persons of a population of 120,000 were reported to be on the waiting list for housing in 1967; the stated goal was to raze all squatter compounds and replace them with site-and-service townships.

[68] Kelvin Mlenga, "Rural Urban Migration—a Perennial Problem," *Horizon,* December, 1967, pp. 14-19.

Urbanization in Africa

In South Africa, as a second example, site-and-service schemes became state policy in the early 1950s, local authorities being required to lay out 40 × 70 foot plots and provide basic services, the African being allowed to build a shack at the back of the plot leaving the front available for a permanent house. In Soweto, 35,000 sites were surveyed and 35,000 shacks were built; in 1967 only 68 shacks remained. Money for these schemes and for municipally built houses is secured from a tax on employers who do not provide housing, from profits on Bantu beer and a fifth of profits on liquor sold in the townships, and, for Johannesburg, some subsidization from general rates.[69]

The U.N. in a recent publication, previously cited, makes the following comment pertinent to the site-and-service system:

a poor man's dignity is not damaged by his poor house but by his poverty (a modern house exacerbates rather than eliminates the problem). But enable the poor man to get a job by helping him to live (no matter how poorly) where he can find one or, if he already has one, provide him with a piece of building land and advice where needed, and he will then make the best use of his opportunities and, slowly but surely, will cease to be poor. As he ceases to be poor, he will cease to live in a poor house.[70]

This prescription is similar to that of Herbert Gans for American cities. He writes that "the primary aim of rebuilding . . . should not be to rehabilitate houses or clear slums, but to raise standards of living of ghetto residents."[71]

Several conferences of urban experts have called for prime attention being given to housing and urbanization. But it is not always easy to see what other expenditures should be cut to divert the necessary funds. Some cuts can doubtless be made in administrative overhead or from prestige projects and palaces, which are in any case no longer so much in vogue as they were in the heady days of early independence. Substantial cuts from bloated military budgets could also be effected in a number of countries. But should cuts be made from education, public health, or agricultural and industrial development? Barber answers this

[69] Ellen Hellmann, "Soweto—Johannesburg's African City," an address to the South African Institute of Race Relations, Natal Region Branch, April 13, 1967 (mimeographed).
[70] "Urbanization: Development Policies and Planning," p. 127.
[71] *The New York Times*, magazine section, January 7, 1968.

question when he writes that "massive rural improvement is a necessary condition for healthy urban growth in the future" and that "restructuring the present urban centers . . . is both less essential and less demanding than an effective programme of agrarian uplift."[72]

In the long run, the only solution is a greatly improved production and productivity which can justify higher wages and generate higher incomes, which would imply that any diversion of funds from the productive sectors should be made with extreme caution. One is forced to the unwelcome conclusion that the problem is one which simply cannot be solved in the short run. Indeed, there is the nasty dilemma that the more attention given to the city the more attractive it becomes and hence the greater the migration to it is likely to be, thus perpetuating if not heightening the problem.

Nor is it easy to concur with the rising trend toward demanding control of all kinds of population movements: influx to the cities, distribution by city, and distribution between rural and urban areas. These plans are not only objectionable from the standpoint of infringement on human rights but appear to be entirely impractical politically and economically. As a Zambian minister put it, "no one can wave a magic wand and stop rural-urban migration."

Urban Growth in Africa and Elsewhere

The characteristics of urban growth and the problems of urbanization in Africa contain many elements comparable to those prevailing in other underdeveloped areas.[73] They also have similarities with urbanization in the West at a comparable period of growth, though some would question the validity of recognizing the comparability of these periods. Cities in the West had such similar earmarks as inadequate housing, unsatisfactory provisions for public health, endemic civil disobedience, increased crime and violence, and unpredictable employment opportunities,[74] and

[72] Barber, "Urbanisation and Economic Growth: the Cases of Two White Settler Territories," in Miner, ed., *The City in Modern Africa*, pp. 122, 123.

[73] See, for example, Glenn H. Beyer, ed., *The Urban Explosion in Latin America: A Continent in Process of Modernization* (Ithaca, Cornell University Press, 1967); T. G. McGee, *The Southeast Asian City* (New York, Praeger, 1967).

[74] See Hawley, "World Urbanization: Trends and Prospects," in Freedman, ed., *Population: The Vital Revolution*.

these characteristics now appear as current problems in many American cities.

Nor could anyone claim that the United States is meeting its urban problems any more successfully than Africa is at the moment. Indeed we merit severe criticism more than the understanding which is appropriate to the African scene, because we have the capacities in capital and one hopes in knowhow to attack the needs with vigor, but have not yet done so.

There are, however, a number of contrasts between urban growth in Africa and that of the West at the hypothetical comparable period.[75] They include:

1. The rates of growth, particularly of the major cities, are much more rapid in Africa. Some have achieved their present size in one-fifth to one-tenth the time required in Western Europe.
2. There is less correlation between the cities' rates of growth and measures of economic growth in their countries.
3. The growth of urbanization is often not paralleled by a comparable revolution in the rural areas.
4. A less favorable ratio of population to resources in rural areas means that the push factor is more important than it was in Europe.
5. The linkage of some cities with their domestic hinterlands is less developed, while the ties of these cities to the outside world and their dependence on it remain striking.
6. There is relatively less specialization in the African cities.
7. There are generally higher rates of unemployment. Here the European cities had the advantage of being able to drain off large numbers of people, who might have become redundant, to the New World. The African cities have no such convenient safety valve.
8. Differences in outlook and values may slow the adjustment to the city and reduce the tempo of its economic life, as, for example, the reliance on the extended family for support and the absence of the Protestant ethic with its emphasis on hard work, achievement, and success.

[75] See Hauser and Schnore, eds., *The Study of Urbanization*; Hawley, "World Urbanization: Trends and Prospects"; McGee, *The Southeast Asian City*; Kingsley Davis, "The Origin and Growth of Urbanization in the World," *American Journal of Sociology*, LX, No. 5 (March, 1955), 429-37.

9. There is a dual structure in many African cities.
10. Migrants to the towns differ in several important respects: almost all are unskilled, their level of educational achievement is relatively lower though above average as far as the source areas are concerned, and almost all arrive without capital resources.
11. Heavier responsibility is placed on governments, local and national, to provide for the urban residents. In the West, private enterprise normally met the needs for new housing, while local governments had a tax base adequate to provide the public services. Not so in Africa, where the demands on government are far more onerous and almost none are capable of meeting them. It is ironic, however, that big cities in the United States have now moved into a situation roughly comparable in these respects, in no small measure due to the flight to the suburbs and the migration of unskilled workers to the expanding slums.

Conclusion

The problems besetting African cities are, then, intense, complex, and probably growing in seriousness. Nonetheless, cities have made important contributions toward economic development. As the links between urban and rural areas continue to improve, their impact on rural living standards should gradually diminish the degree of self subsistence prevailing in the countryside and thus improve the rural-urban economic balance. The cities are also the foci of intellectual life, from which new ideas and technologies will slowly but surely diffuse into the rural hinterlands.

There is little doubt that the process of urbanization will continue at a rapid if not an accelerating rate. The complex effects of this growth need far more attention than they have gotten, because cities generate movements which influence cultural and political events very much more than the proportion of their populations would suggest.

At the same time, the problems of urbanization must be seen in context with the other problems facing Africa. The dilemmas then become distressing, at least for the short run. But a certain degree of hardheadedness will be required in the assignment of priorities based upon the desire for the long-run wellbeing of the individual countries and of the citizens within them.

Bibliography

Abiodun, Josephine Olu. "Urban Hierarchy in a Developing Country," *Economic Geography*, XLIII, No. 4 (October, 1967), 347-67.

Balandier, Georges. *Ambiguous Africa: Cultures in Collision*. Translated by Helen Weaver. New York, Pantheon Books, 1966.

Beaujeu-Garnier, Jacqueline. "Les Grandes Villes Surpeuplées dans les Pays Sous-développés," *L'Information Géographique*, XXXII, No. 4 (September-October, 1968), 159-66.

——, and Georges Chabot. *Urban Geography*. New York, John Wiley, 1967.

Beckinsale, R. P., and J. M. Houston, eds. *Urbanisation and Its Problems*. Oxford, Basil Blackwell, 1968.

Bernard, M. "Problèmes Urbains dans l'Afrique d'Aujourd'hui," *Industries et Travaux d'Outremer*, X, No. 100 (March, 1962), 249-51.

Blanc, Maurice. "L'Urbanisme Outre-Mer," *Industries et Travaux d'Outremer*, X, No. 100 (March, 1962), 134-37.

Breese, Gerald. *Urbanization in Newly Developing Countries*. Englewood Cliffs, Prentice-Hall, 1966.

CCTA. *Inter-African Conference on Housing and Urbanization*. London, CCTA/CSA, Pub. 47, 1959.

CCTA/CSA. *CSA Meeting of Specialists on Urbanization and Its Social Aspects, Abidjan, 1961*. London, CCTA/CSA, Pub. 75, 1961.

Davis, Kingsley. "The Origin and Growth of Urbanization in the World," *The American Journal of Sociology*, LX, No. 5 (March, 1955), 429-37.

Denis, J. *Le Phénomène Urbain en Afrique Centrale*. Brussels, Académie Royale des Sciences Coloniales, 1958.

Frazier, E. Franklin. "Urbanization and Its Effects Upon the Task of Nation-Building in Africa South of the Sahara," *The Journal of Negro Education*, XXX, No. 3 (Summer, 1961), 214-22.

Freedman, Ronald, ed. *Population: The Vital Revolution*. New York, Doubleday, 1964.

Gutkind, Peter C. W. "The African Urban Milieu: A Force in Rapid Change," *Civilisations*, XII, No. 2 (1962), 167-95.

——. "Urban Conditions in Africa," *The Town Planning Review*, XXXII, No. 1 April, 1961), 20-32.

Hamdan, G. "Capitals of the New Africa," *Economic Geography*, XL, No. 3 (July, 1964), 239-53.

Harroy, Jean-Paul. "The Political, Economic and Social Role of Urban Agglomerations in Countries of the Third World," *Civilisations*, XVII, No. 3 (1967), 166-85.

Hauser, Philip M., and Leo F. Schnore, eds. *The Study of Urbanization*. New York, John Wiley, 1965.

Holzner, Lutz. "World Regions in Urban Geography," *Annals of the Association of American Geographers*, LVII, No. 4 (December, 1967), 704-12.

——, E. J. Dommisse, and J. E. Mueller. "Toward a Theory of Cultural-Genetic City Classification," *Annals of the Association of American Geographers*, LVII, No. 2 (June, 1967), 367-81.

International African Institute. "The Housing of Workers in Urban Living Conditions in Africa: I and II," *Bulletin, Inter-African Labour Institute*, VI, No. 2 (March, 1959), 62-71; No. 3 (May, 1959), 58-83; No. 4 (July, 1959), 92-114.

Jenkins, George. "Africa as It Urbanizes: An Overview of Current Research," *Urban Affairs Quarterly*, II, No. 3 (March, 1967), 66-80.

——. "Urban Violence in Africa," *Afri-*

can *Urban Notes*, II, No. 5 (December, 1967), 37-38.

Jones, J. D. Rheinallt. "The Effects of Urbanisation in South and Central Africa," *African Affairs*, LII, No. 206 (January, 1953), 37-44.

Lambert-Lamond, Georges. "Aspects Économiques et Sociaux de l'Urbanisation en Afrique au Sud du Sahara," *Revue Économique et Sociale*, XVIII, No. 1 (January, 1960), 82-96.

Lee, R. H. "Urbanization and Race Relations in Africa," *Journal of Human Relations*, VIII, Nos. 3-4 (Spring-Summer, 1960), 518-33.

Lewis, L. J. "The Social and Cultural Problems of Urbanization for the Individual and the Family," *Rural Life*, V, No. 2 (1960), 3-13, 21.

Lewis, W. Arthur. "Unemployment in Developing Countries," *The World Today*, XXIII, No. 1 (January, 1967), 13-22.

Little, Kenneth. *Some Contemporary Trends in African Urbanization*. Evanston, Northwestern University Press, 1966.

——. *West African Urbanisation: A Study of Voluntary Associations in Social Change*. Cambridge, Cambridge University Press, 1965.

McCall, Daniel F. "Dynamics of Urbanization in Africa," *The Annals of the American Academy of Political and Social Science*, No. 298 (March, 1955), pp. 151-60.

McNulty, Michael L. "Urban Structure and Development: The Urban System of Ghana," *The Journal of Developing Areas*, III (January, 1969), 159-76.

Mayer, Harold M., and Clyde F. Kohn, eds. *Readings in Urban Geography*. Chicago, University of Chicago Press, 1959.

Miner, Horace, ed. *The City in Modern Africa*. New York, Praeger, 1967.

Nerfin, Marc. "Towards a Housing Policy," *The Journal of Modern Afri-*

can *Studies*, III, No. 4 (December, 1965), 543-65.

Oram, Nigel. *Towns in Africa*. London, Oxford University Press, 1965.

Pons, Valdo. *Stanleyville: An African Urban Community under Belgian Administration*. London, Oxford University Press, 1969.

Raymaekers, P. *L'Organisation des Zones de Squatting*. Paris, Éditions Universitaires, 1964.

Santos, Milton. "Vues Actuelles sur le Problème des Bidonvilles," *L'Information Géographique*, XXX, No. 4 (September-October, 1966), 143-50.

Simms, Ruth P. *Urbanization in West Africa: A Review of Current Literature*. Evanston, Northwestern University Press, 1965.

Sjoberg, G. "The Preindustrial City," *The American Journal of Sociology*, LX, No. 5 (March, 1955), 438-45.

——. *The Preindustrial City: Past and Present*. Glencoe, Free Press, 1960.

Smailes, Arthur E. *The Geography of Towns*. 5th ed., rev. London, Hutchinson University Library, 1966.

Smit, P. "Recent Trends and Developments in Africa: Urbanization in Africa," *Tydskrif vir Aardrykskunde*, II, No. 10 (April, 1967), 69-77.

Thomas, Benjamin E. "The Colonial Imprint on African Cities," paper presented to the 57th Annual Meeting, Association of American Geographers, Miami Beach, April, 1962.

United Nations. *Housing in Africa*. New York, 1965.

——. "Urbanization: Development Policies and Planning." *International Social Development Review*, No. 1. New York 1968.

——, Economic Commission for Africa. *Housing in Africa: Problems and Policies*. E/CN.14/HOUPA/3. November 28, 1962.

——. *Workshop on Urbanization in Africa*. Addis Ababa, 1962.

——, International Labour Office. *Afri-*

can *Labour Survey.* Geneva, Impri-
meries Réunies, 1958.

———, UNESCO. *Social Implications of
Industrialization and Urbanization in
Africa South of the Sahara.* Paris,
1956.

Van den Berghe, Pierre, ed. *Africa: So-
cial Problems of Change and Conflict.*
San Francisco, Chandler Publishing
Co., 1965.

Verhaegen, P. *L'Urbanisation de l'Afrique Noire: Son Cadre, ses Causes et
ses Conséquences Économiques, Sociales et Culturelles.* Brussels, Centre
de Documentation Économique et Sociale Africaine, 1962.

Wood, Eric W. "The Implications of Migrant Labour for Urban Social Systems in Africa," *Cahiers d'Études
Africaines,* VIII, No. 29 (1968), 5-31.

CHAPTER 5

SOME AFRICAN CITIES

THE PURPOSE of this chapter is to present thumbnail sketches
of a number of African cities, which will reveal something of the variety
that exists as well as certain characteristics which recur with frequency.
In each case brief attention will be given to such aspects as the attributes
and limitations of location and site, the historical evolution of the city,
the major features of its structure, and an analysis of its functions.

Included among the sketches are: several essentially traditional
towns (a composite one for Morocco, Kumasi, Ibadan, Kano, Addis
Ababa, and the Mengo portion of Kampala), a number of cities which
were colonial creations (Khartoum, Dakar, Abidjan, Nairobi, Kampala),
several which date to precolonial periods but whose later development
was strongly influenced by colonial authorities and interests (Accra and
Lagos), and a composite Copperbelt mining town.

It will be apparent from material presented in Chapter 4 that
many cities are undergoing dynamic change and extremely rapid growth.
This is particularly so for the mushroom cities. Balandier writes that
their faces have many common features:

Since most of them are under construction, their topography is constantly changing. They all present striking discordances between the vitality of their good neighborhoods and backwardness, more pronounced by comparison, of their "African towns"; between the outmoded, faded, ugly colonial style of buildings of the old era and the bold modernism of recent constructions; between the spaciousness that characterized them two decades ago and the verticality . . . of today. This upward thrust can be seen as the expression of a kind of pride in construction and the result of pressures exerted on the periphery of residential quarters by ever-increasing numbers of [African] citizens. In its physical aspect as in its human aspect, the city is seeking its form in confuson.[1]

MOROCCAN CITIES[2]

Morocco contrasts with most tropical African countries in having a considerably higher urban population and in the antiquity of many of its cities. In 1966 it had some ten cities of over 100,000 and one of over a million, Casablanca. As elsewhere, however, the growth of cities has continued apace despite a somewhat stagnant economy and large-scale unemployment in the cities. The interrelations of rural and urban problems are rarely as apparent as in Morocco; four-fifths of the rural population is devoted to traditional agriculture, 40 percent have no land or less than 1.25 acres, and 60 percent are estimated to be underemployed. Meanwhile the population is increasing by about 2.8 percent or over 400,000 per annum. While Morocco has a relatively high average per capita income ($169 in 1967), it is, like many sub-Saharan countries, characterized by the juxtaposition of a restrained modern economy

[1] Georges Balandier, *Ambiguous Africa: Cultures in Collision.* Translated by Helen Weaver (New York, Pantheon Books, 1966), pp. 175-6.
[2] See Hassan Awad, "Morocco's Expanding Towns," *The Geographical Journal,* CXXX, No. 1 (March, 1964), 49-64; J. P. Houssel, "L'Évolution Récente de l'Activité Industrielle de Fès," *Revue de Géographie du Maroc,* No. 9 (1966), pp. 59-83; Roger Le Tourneau, *Fès avant le Protectorat* (Casablanca, Société Marocaine de Librarie et d'Édition, 1949); Henry Maurer, "Les Problèmes Administratifs de l'Urbanisation au Maroc," in E.C.A., Cycle d'Études sur l'Urbanisation en Afrique, Addis Ababa, 25 April-5 May, 1962 (SEM/URB/AF/15, 19 February, 1962, multilithed); Moroccan Government, "L'Urbanisation en Maroc," in E.C.A., Cycle d'Études sur l'Urbanisation en Afrique, Addis Ababa, 25 April-5 May, 1962 (SEM/URB/AF/21, 8 March, 1962, mimeographed); M. Naciri, "Salè: Étude de Géographie Urbaine," *Revue de Géographie du Maroc,* Nos. 3-4 (1963), pp. 11-82; D. Noin, *Casablanca* (Rabat, Comité National de Géographie du Maroc, *Atlas du Maroc,* Notices Explicatives, Section IX, Planches 36a-b, 1965).

and a still preponderantly archaic economy, producing a profound disequilibrium. This is reflected in the urban sphere in the contrasts
between Casablanca and the ancient cities such as Fès and Marrakech,
between the new and old towns in such cities as Fès or Rabat-Salé, and
in the dense crowding in the medinas and bidonvilles as compared with
the low densities in the modern villas and apartment houses of the newer
cities.

The indigenous cities of Morocco have certain characteristics
which are common throughout the country. The city is usually situated
in a place which was selected with an eye to defense; it is typically fortified with crenellated walls and towers pierced with several monumental
gates. Within is the medina or Muslim city and an adjoining mellah, for
the indigenous Jewish population. Both consist of a disorderly net of

Characteristics of indigenous Moroccan cities

The site is frequently selected with an eye to defense (Xauen).

It is typically fortified, with several monumental gates piercing the walls (Rabat).

Numerous mosques are scattered within the medina; their minarets rise above the typical one- to two-story houses (Moulay Idriss).

narrow, tortuous alleys often covered with thatch screens to provide welcome shade. Houses abutting the alleys have several stories and few openings on them, though they may have interior courts. Their roofs are flat topped and are an integral part of the dwelling space, being used for sleeping in the summer. The medina is often divided by interior walls into wards housing specific tribal groups.

Usually there is one large mosque and a considerable number of smaller ones. Merchants and artisans occupy souks in portions of the medina and mellah. Some have multi-roomed shops; most occupy tiny stalls crammed with merchandise or serving as workshops. Merchants and craftsmen are often grouped according to specialization and the latter are formed into guilds, the most important of which are tanners, shoemakers, and potters, but which include joiners, smiths, jewelers, rug weavers, and workers in brass and copper. Undesirable crafts are often placed toward the edge (tanning, dyeing, butchering); near the mosque are candle makers, incense dealers, and booksellers, while beggars are also likely to be more numerous. Most of the indigenous cities also have

Merchants in the souks often occupy tiny stalls which may also serve as workshops; a candle shop near the main mosque in Fès.

Craft industries are well represented and are usually grouped in specific quarters; dyed wool being dried at Marrakech

a large open square, often situated near one of the principal gates, which serves as a market place, and, in the late afternoon and evening, as a place where a variety of entertainers, from dancers and drummers to snake charmers, hold forth.

The indigenous cities of Morocco have been remarkably well preserved, in part because Lyautey decreed at the outset of the French Protectorate that indigenous life was not to be disturbed any more than necessary. New towns were, therefore, developed outside the walls; they

Some African Cities

Undesirable crafts are relegated to peripheral areas; tanning vats at Fès.

Large open squares serve as market places and as stages for itinerant entertainers (Place Djemaa el Fna in Marrakech).

are characteristically well laid out with modern stores on one or two main streets and with residential sections occupied during the French period mainly by non-Moroccans. Around the periphery of the old cities there are likely to be bidonvilles and agglomerations of grass huts, while there are sometimes villages of troglodytes.

The ancient cities of Morocco are beset by numerous problems: a general decline in artisan employment, gross overcrowding, unsanitary conditions, lack of air, and inadequate health and other facilities. Just

Some African Cities

303

The port area of Casablanca.

how the teeming masses within them survive is something of a problem. The one bright light is the tourist boom now underway in Morocco, which has stimulated new hotel construction and helped to sustain at least some of the craft industries.

Casablanca provides the most appropriate example of a modern city in Morocco. While the city dates from 1830 it had a population of only about 25,000 when the French troops debarked there in 1907 and, with the exception of the Old Medina is essentially a product of the French Protectorate. The most important factor in basing the growth of Casablanca was the decision taken in 1913 to construct a major artificial harbor at the site; it developed rapidly thereafter as the main commercial center of the country, gradually extending and proliferating its lines of contact throughout the entire country.

The concentration of Europeans attracted to the city undoubtedly contributed to its rise to preeminence. In 1922 some 35 percent of the population were Europeans; while their percentage dropped thereafter due to the growing influx of Moroccans, their numbers rose to about 136,000 in 1952 when the total population was 682,000. After inde-

Some African Cities

Part of the CBD of Casablanca.

pendence in 1956 there was a considerable exodus of Europeans and of Moroccan Jews; the 1960 census recorded about 115,000 foreigners, mainly Europeans, and 73,000 Moroccan Jews. An estimate for 1965 placed the foreign population at about 79,000 or 7 percent of the total of 1,125,000 and the Moroccan Jewish population at about 45,000 or 4 percent of the total.

Casablanca is well located to handle a major share of Moroccan trade, at the center of productive Morocco, which has no other port with comparable facilities, and relatively close to important commercial agricultural zones, to the huge phosphate deposits at Khourigba, and to the capital at Rabat. Excellent roads radiate to other major centers and are more important in the carriage of general traffic than the railways. The latter are particularly important in the haulage of phosphates, which comprise about 70 percent of the export tonnage but which in fact contribute relatively little to the economy of the city.

The site features of Casablanca, particularly the small, very poorly protected bay, doubtlessly helped to account for the rise of Casablanca as a shipper of grains and wool in the nineteenth century but

Some African Cities   305

played no part in the new port which is entirely artificial. The topography is permissively favorable, rising as it does gently inland from the ocean front.

Casablanca is very spread out for a city of its size, stretching about twelve miles along the coast and six miles inland and covering an area slightly larger than Paris; it grew about 200 times in area in the sixty years to 1965. The internal structure, however, shows marked contrasts between the high-density medinas and bidonvilles and the suburblike sections occupied by villas (Map 27). Much of its growth was unplanned under a strong *laissez-faire* policy which permitted much speculation in land.

The oldest part of the city is the Old Medina enclosed in walls and comparable in most respects to the traditional indigenous cities. It housed about 100,000 people in 1960 with densities of over 500 per acre; as the city grew a large part of the indigenous Jewish population made it in considerable part the mellah but the dwellings of those who have left in the last ten or fifteen years have quickly been taken up by Moroccan Muslims. The so-called modern medinas were constructed between 1910 and 1950 but most are little better than the Old Medina. Indeed the earlier ones, such as that built outside the wall, have smaller plots, internal courts, and individual rooms and hence are more poorly ventilated and unhealthy. Water is rarely available in the dwellings. To make matters worse, huts have been built at the end of passages, in the courts, or on terraces, forming a veritable encrustation of shanties on the traditional structures. The medinas built after 1925 are a little better but still far from satisfactory. The New Medina has many of the same undesirable features, plus an additional one in its monotonous layout, but water is piped to about half the units and most have electricity. Densities vary from about 480 to 600 per acre. About 45 percent of the Muslim population of the city lived in the overpopulated medinas in 1960; half of the dwelling units had no kitchen or toilet, two-thirds had no piped water supply.

After the last war several "cités" were constructed, each housing 30–40,000 at densities intended not to exceed 140 per acre. They were designed to maximize the number of common walls, diminish the number of public roads that would require maintenance, and yet keep an adequate amount of open space. While the "cités" had more satisfactory features in many respects than the medinas, including their wider

Some African Cities

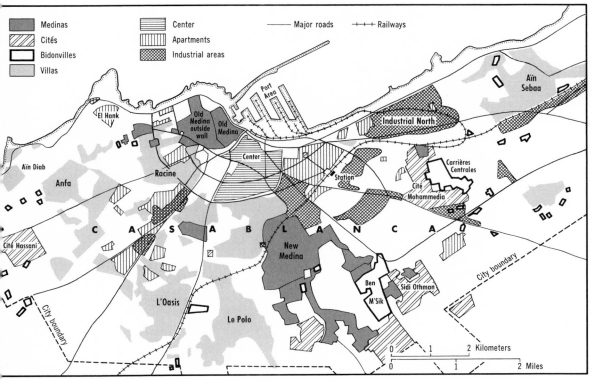

Legend:
- Medinas
- Cités
- Bidonvilles
- Villas
- Center
- Apartments
- Industrial areas
- Major roads
- Railways

Labels on map: El Hank, Aïn Diab, Anfa, Racine, Cité Hassani, L'Oasis, Le Polo, New Medina, Ben M'Sik, Sidi Othman, Cité Mohammedia, Carrières Centrales, Station, Center, Old Medina outside wall, Old Medina, Port Area, Industrial North, Aïn Sebaa, CASABLANCA, City boundary

Scale: 0 1 2 Kilometers / 0 1 2 Miles

MAP 27. Casablanca, Morocco. (After *Atlas du Maroc* and D. Noin.)

streets, centralized shopping areas, and greater provision of amenities, they proved too expensive to justify their extension after 1953. In the period 1953–1964 the government built an average of 3,000 "economic" lodgings per year in Casablanca but construction declined after the peak year in 1957. After 1959 a new housing plan was adopted which emphasized the layout of the "lot évolutif," the amelioration of bidonvilles by provision of sanitary installations, and construction of three- and four-story buildings with two-room apartments. In 1960, some 26 percent of Casablanca's Muslim population lived in "cités" and newer medinas not considered to be overcrowded.

Despite the relatively massive effort to provide modern, low-cost housing in Casablanca it has not proved possible to keep pace with the influx of migrants, and the number of inhabitants residing in bidonvilles

Some African Cities

307

Closeup of a shack in one
of the bidonvilles of
Casablanca.

increased from about 50,000 in 1940 to over 100,000 in 1950 and probably over 200,000 in 1969. A distinction can be made between those bidonvilles which have grown up spontaneously and those which have been provided with water points, public toilets, and street lighting, which house the bulk of the population in such quarters. Two bidonvilles in the city each had over 50,000 residents in 1960 and a half dozen others had over 3,000 each. Densities averaged about 400 per acre, which is remarkably high considering the nature of the huts; two-thirds of the plots, which measured about 16 by 20 feet, had more than one shack. About 21 percent of the city's Muslim population lived in bidonvilles in 1960. If one adds the indigenous population living in overcrowded medinas and those living in bidonvilles it is apparent that over 70 percent lives in houses which are mediocre at best. While this might be thought to be in sharp contrast to the type of dwellings characteristic of the countryside, in fact the mountain villages, grass huts seen on the meseta, and the *ksour,* or fortified villages of pre-Saharan Morocco are little if any better than the medinas and bidonvilles, the ksour being extremely picturesque but essentially honeycombs of hovels.

Some African Cities

An account of the indigenous housing in Casablanca fails to convey the general impression that the city gives, namely of a large agglomeration with many modern buildings, many rising in the CBD to 12 to 20 stories and most gleaming white in color, of a city more Mediterranean than Moroccan, more European than African.

Commerce is distinctly the basic function of Casablanca, although it accounted for only about 24 percent of those actively employed in 1960. It contains most of the banking and insurance houses in the country, the headquarters of large importers and exporters, and domestic wholesalers supplying all parts of the kingdom. These firms, together with the major stores, hotels, and theatres, are concentrated in the CBD near the Old Medina and the port area. Transportation employed about 10 percent of those actively working, with the port accounting for over two-fifths of that. Casablanca port handles about 70 percent of the total imports and exports of the country, and its level of traffic, about 10 million tons a year, puts it in the first rank among African ports. Plans call for doubling the harbor area and capacity of the port in the next five years, including installation of a new phosphate-loading quay capable of handling 100,000-ton ore carriers.

"Casa" is also by all odds the main manufacturing center of Morocco. About a third of those actively employed in 1960 were in industrial and artisan activities. Traditional crafts are of minor importance, most artisans being of the more modern type, such as plumbers, painters, electricians, and mechanics; they totaled perhaps a twelfth of all those employed. The city accounts for 57 percent of industrial employment and about three-fifths of Moroccan industrial output. Modern manufacturing, concentrated to a considerable extent in the industrial zone to the north, employs about 20 percent of the labor force. Its biggest development came with the large influx of capital and personnel in the first decade after World War II when industrial production increased 9 percent a year. Since 1952–1953, however, growth has been nearer 2 percent per annum. Only eight firms employ over 500 workers, but these account for a fifth of those engaged in manufacturing.

Casablanca is, in conclusion, the leading transport, finance, commercial, and manufacturing city of Morocco. And despite the more dirigiste policies of the present government the first ten years after independence still saw two-thirds of new investment focused on the city. Like many other primate cities in Africa, however, it suffers from severe unemployment, inability to house a large percentage of its population satisfactorily, and a continuing heavy influx of unskilled workers.

THE THREE TOWNS[3]

The location of the Three Towns—Khartoum, Omdurman, North Khartoum—at the confluence of two great rivers, the Blue and White Niles, is a most unusual one for Africa, though reasonably common for Europe and North America. While the location doubtlessly appealed to the Egyptians who founded the city by establishing a military post at Khartoum about 1829,[4] the rivers play only a minor and somewhat indirect role in the life of the conurbation today, providing water for irrigation and waste disposal but having little significance as transport arteries.

The Three Towns have, indeed, become more of a crossroads and rail center than a river port. The rail line runs northward from it via Atbara (which contains the main workshops of the Sudan Railways) to Port Sudan, and southward via Wad Medani to the junction of the major east-west line. Roads connect them with the major market town of El Obeid, Wad Medani and the Gezira-Managil Scheme, Kassala, and Atbara. Nonetheless, it has a crossroads location which is somewhat displaced to the north from what might be considered a more natural intersection of the north-south Nile axis and the east-west rain-grown crop belt running on a parallel with El Obeid, Kosti, and Gedaref. One wonders, indeed, if Kosti or Sennar might not have been preferable sites from the standpoint of nodality of transport routes; both are on the major east-west road and rail routes, Kosti is the northern terminus of the Nile River services to Juba, and Sennar has a more propitious location with respect to the several major gravity and pump irrigation

---

[3] J. V. Arthur. "Slum Clearance in Khartoum," *Journal of African Administration*, VI, No. 2 (April, 1954), 73-80; K. M. Barbour, *The Republic of the Sudan*, pp. 98-106; Saad ed Din Fawzi, "Social Aspects of Urban Housing in the Northern Sudan," *Sudan Notes and Records*, XXXV, No. 1 (June, 1954), 91-106; G. Hamdan, "The Growth and Functional Structure of Khartoum," *The Geographical Review*, L, No. 1 (January, 1960), 21-40; J. McLoughlin, "The Sudan's Three Towns: A Demographic and Economic Profile," *Economic Development and Cultural Change*, XII, No. 1 (October, 1963), 70-84; XII, No. 2 (April, 1964), 158-73; XII, No. 3 (October, 1964), 286-304; F. Rechfisch, "A Study of Some Southern Migrants in Omdurman," *Sudan Notes and Records*, XLIII (1962), 50-104; Sudan Census 1955-56—*Town Planners Supplement*, 2 v. (Khartoum).

[4] Remains of an ancient settlement indicate that the site had been occupied thousands of years earlier.

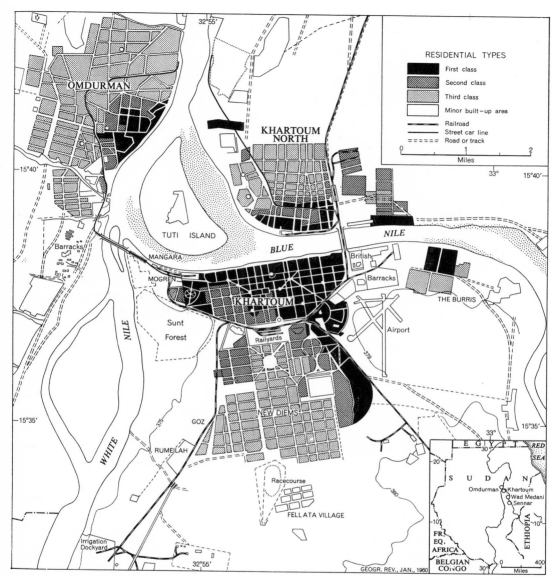

MAP 28. The Three Towns, Sudan. (From G. Hamdan, "The Growth and Functional Structure of Khartoum," *The Geographical Review*, L, No. 1, January, 1960, used by permission.)

schemes depending on the Blue Nile, as well as to the hydroelectric installation at the new Roseires Dam on that river.

The site conditions of the Three Towns are reasonably satisfactory with ample room for easy expansion. The river crossings do make for bottlenecks in internal traffic movements and some extremely flat sections create drainage problems. But the least attractive aspect of local physical conditions is the oppressively high temperatures prevailing in much of the year. These have given Khartoum a somewhat unsavory reputation among foreign travelers, whose only exposure to the city used to be stepping from a plane in the middle of the night into a blast of hot air to be handed a tall drink of lemonade in a suspiciously unclean glass while the plane stopped for refueling between Cairo and Entebbe. Today Khartoum is overflown on such flights, but it should be noted that the climatic conditions were in any case exaggerated, that adjustment to them is not too difficult, and that modern hotels and houses are provided with air conditioning.

Khartoum grew rapidly to 1846, when it was estimated to have 60,000 inhabitants, but cholera and other epidemics reduced the number to 20,000 about 1870, while the Mahdi suppressed it further by building Omdurman on the west bank of the Nile. After the British occupation in 1898 the conurbation developed into three towns: Khartoum, the administrative and modern commercial center (Map 28); North Khartoum, a European residential district and later the site of an industrial estate; and Omdurman, which has remained essentially a huge, indigenous town with thousands of baked-clay houses, one of the largest souks to be seen south of the Sahara, numerous mosques, and the tomb of the Mahdi. Remnants of his fortifications are visible along the Nile front of Omdurman.

The most favored sites in each town were the banks of the rivers. In Khartoum, the Blue Nile frontage was quickly taken up by the government for offices and residences of officials and, further to the east, by Gordon College (now the University). Behind the front came the shopping and commercial centers, the older high-class residential areas, and quarters occupied separately by such expatriate groups as Italians, Greeks, and Levantines. Newer residential areas have sprung up south of the rail line, both expensive villas housing the richer residents and members of the diplomatic corps and a huge gridiron colony of one-story mud hovels housing over 50,000 people. Khartoum is the only one of

312   *Some African Cities*

the Three Towns with more than a few multi-story buildings. Still further to the south is a shanty town occupied by perhaps 20,000 people, of whom the bulk are "westerners." The airfield is also situated to the south, having been moved some years ago from the less convenient site north of Omdurman on the west bank of the Nile.

The Three Towns have developed unevenly in the present century, with the more modern components growing more rapidly than Omdurman:

Population in thousands

	1903	1926	1956	1964
Khartoum	15	14	83	185
Omdurman	58	70	113	175
Khartoum North	4	14	39	80 (1965)
Total	77	98	235	440

They contain about half of the country's urban population in towns over 20,000, but they play a role in the life of the country out of proportion to this and other measures. In addition to being the political and intellectual capital of the country, the conurbation contains by all odds the largest commercial component and, with the exception of cotton ginneries, cement, and the railway workshops, it has practically all of the nation's modern industries as well as a substantial output of handicrafts. As stated by McLoughlin, "this is the highest per capita income region in Sudan. The population is sophisticated, cosmopolitan, and generally progressive. Nearly all educated Sudanese are employed here."[5]

DAKAR[6]

This prize city of former French colonial Africa has an exceptional situation at the extremity of the West African bulge, strategically

[5] McLoughlin, "The Sudan's Three Towns: A Demographic and Economic Profile of an African Urban Complex," p. 76.

[6] See M. Giraud, "Dakar, Un Siècle d'Urbanisme et de Travaux," *Industries et Travaux d'Outre-Mer*, V, No. 41 (April, 1957), 187-91; J. Lombard, "Connaissance du Sénégal; fasicule 5: Géographie Humaine," *Études Sénégalaises*, No. 9 (1963); Richard J. Peterec, *Dakar and West African Economic Development* (New York, Columbia University Press, 1967); Assane Seck, "Dakar," *Les Cahiers d'Outre-Mer*, XIV, No. 56 (October-December, 1961), 372-92; ———, "Les Industries Manufacturières de la Région Dakaroise," *Revue de Geographie de l'Afrique Occidentale*, Nos. 1-2 (1965), pp. 197-200; Derwent Whittlesey, "Dakar and the Other Cape Verde Settlements," *The Geographical Review*, XXXI, No. 4 (October, 1941), 609-38.

positioned with respect to both sea and air routes. About 2,560 miles from Bordeaux, 3,220 from Rio de Janeiro, and 4,150 miles from Cape Town, Dakar lies close to the sea routes used by all lines serving the Atlantic coast of Africa and the Eurafrican routes serving the Indian Ocean ports with closure of the Suez Canal in 1967, plus routes to parts of South America from Western Europe. Its large international airport

Part of the industrial section of North Khartoum, illustrating the problem of drainage resulting from the exceptionally flat terrain.

at Yoff is also conveniently placed with respect to traffic between Europe and South America, North America, and Africa, as well as major routes linking West Europe and West Africa. This location gives Dakar more than national significance and has accounted for the importance of several of its functions.

The hinterland of Dakar, however, is the principal limiting factor in both past and future development, the bulk of it being in low-productive desert and sahelian climatic zones, while even the relatively

314   *Some African Cities*

Part of the immense souks of Omdurman.

A crude oil-pressing mill in Khartoum.

small savanna zone has distinct limitations agriculturally and now provides only peanuts and their derivatives in significant tonnages. In the future, irrigation developments could change this picture, but increased production might cater less to foreign than to domestic markets, which now depend rather heavily on imported food, particularly rice, wheat, other grains, sugar, fruits, and vegetables. The only mineral produced in the hinterland contributing to the significance of port activities is phosphates.

Some African Cities   315

Aerial view of the main section of Dakar.

The hinterland of Dakar was adversely affected by dissolution of the French West African federation in 1959 and temporarily by the breakup of the Mali Federation which cut off trade with Mali (formerly Soudan) for three years. Trade with Mauritania, never very large, is being rerouted via Nouakchott and Port Etienne, while a possible Senegambian Union might divert some traffic from the Casamance region to Bathurst (a substantial tonnage of peanuts are now smuggled into Gambia for export at Bathurst). More important was the loss of the administrative function of a much larger area and increased difficulties of selling manufactured goods to other former French territories, each of which now wishes to develop its own industries.

Dakar is sited at the end of Cape Verde which gives natural protection to its port area from both west and north, and splendid exposure to refreshing winds. The port is, however, largely artificial, being enclosed by moles, but their construction and dredging of the harbor were relatively easy (Map 29). Indeed the port occupies the only site on the

Some African Cities

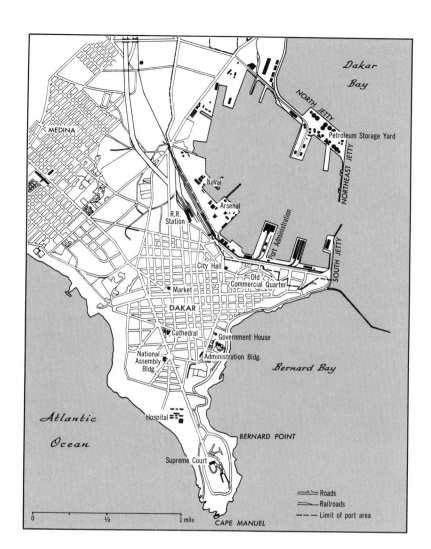

MAP 29.
Dakar,
Senegal.

Senegal coast capable of being developed into a large, modern port
without great expenditure.

The basalt plateau at the end of the peninsula protected the
harbor area from sands carried by the Canaries Current and provided
an attractive and airy site for the early residential and administrative
quarters, while lower, sandy areas provided for commercial and industrial installations and later extensions of the city. Climatically, Dakar

Some African Cities   317

enjoys conditions more comparable to those experienced by Mediterranean cities than to other coastal cities of West Africa, which helps to explain its attractiveness for European residents. One of the few disadvantages of the site relates to securing an adequate supply of water, a problem which has still not been fully solved.

Dakar is a relatively recent creation, having been founded only in 1857, after Ile de Gorée, where the French had been installed since 1675, proved entirely too small and inconvenient. Important contributions to its future growth included the provision of coal bunkering facilities and completion in 1885 of a rail line to Saint-Louis, whose difficult estuary was a major handicap, the decision made in 1898 to develop a naval base at Dakar, and selection of the city as capital of French West Africa in 1904. But the city grew slowly, from 1,556 in 1878 to 18,000 in 1904 and 25,000 in 1910. Continued competition from Saint Louis and from Rufisque, which had become the major peanut assembly point, reduced the significance of Dakar, but opening of the railway to Bamako in 1923 permitted bypassing Rufisque, which had far less attractive site features for port activities. By 1945, Dakar had become a city of 132,000 and it has since grown to 300,000 in 1955 and an estimated 600,000 in 1968. Plan authorities expect the population to reach 1.5 million by the early 1980's.

Important in the development of Dakar was its relatively large European and expatriate population, the largest of any West African city. This contributed to the exceptional size of its modern section, to a substantial consumption power in the city itself, and to development of industries catering not only to the domestic market but to other French territories in Africa. Indeed it has given the city a distinct Mediterranean flavor and characteristic modes of living mindful of Toulon or Nice. Whittlesey reported after visits in 1935–1936 that Dakar was "the only city, in the European (or Occidental) sense in the whole of West Africa." The expatriate population has remained relatively large following independence but one wonders just how long the honeymoon may last. If a large exodus did occur, the effect on the economy of Dakar would be depressing, at least in the short run.

At first, Dakar consisted of a port-commercial sector, a residential-administrative quarter on "le Plateau" dominating the port at an elevation of about 100 feet, and the African medina. The land between the Plateau and the medina was soon filled in by Syrian and Lebanese

Some African Cities

merchants. Residential densities in the medina were about eleven times as great as on the Plateau in 1945.

With the population growing at about 7.5 percent a year the authorities felt it necessary to staff a planning office in 1946. This office rejected the idea of creating satellite cities, and subsequent developments were the construction of many high-rise buildings on the Plateau, cleaning up of the medina and its enlargement along an arc formed by the autoroute to the new airport at Yoff, plus the development of several special quarters, including medical and university sections along the Atlantic coast where much space continued to be occupied by military installations. An industrial quarter has gradually extended northward from the port and will eventually link up with Rufisque, which is now essentially an industrial satellite of Dakar.

In more recent years several residential quarters have been constructed, mainly to the northwest, and housing estates are continuing to be built by private companies, by industrial firms for their own workers, and by SICAP, a mixed corporation which receives subventions from domestic and foreign sources. Nonetheless a large percentage of inhabitants continue to live in traditional structures and bidonvilles. A major study was approved in 1967 containing plans for all aspects of Dakar's expansion over the next few decades.

The major functions of Dakar are apparent from what has been written. Its port function is exaggerated, however, if one compares total tonnages handled with other African ports, primarily because petroleum bunkering plays a major role, but also because domestic exports are made up mainly of phosphates (66 percent of export tonnages) and peanut products (25 percent). In 1968 the port handled 5,277,000 metric tons, of which 50 percent was petroleum products (over 90 percent of which figured in both loadings and unloadings). Bunkering had been declining before closure of the Suez Canal, both because vessels no longer need as frequent refueling and because Dakar has strong competitors, particularly for shippers preferring non-franc currencies. Fishing installations have been enlarged in postwar years, particularly to handle the modern tuna-fishing component. The port's strategic location explains its selection as a naval base. In toto, Dakar stands as one of the finest and best-equipped port installations in tropical Africa, and one which has surplus capacity in relation to its actual traffic.

The administrative function is also somewhat overblown, and

after nearly a decade of independence the Republic had still not found it possible to reduce the bureaucracy in line with the reduced territory for which it is responsible. In 1969 there were still 35,000 civil servants, about 10,000 more than the number serving the burgeoning economy of the Ivory Coast.

Dakar is also the leading religious, cultural, and educational center of Senegal, and has been a favored locale for numerous African and international conferences in recent years. The University of Dakar is, however, still staffed mainly by French professors and is perhaps dangerously dependent on French subventions.

The number of commercial and financial concerns represented in Dakar is unusually large for tropical African cities and probably reflects the sizable expatriate population. Dakar also saw the development of the most diversified manufacturing complex in West Africa in the 1950s but it has expanded less rapidly than other centers since independence. Its export-oriented component is dependent upon France to an unhealthy degree, while many consumer-oriented plants are over-equipped in relation to the domestic market. There is also considerable unemployment, while many African residents subsist at extremely low standards. Stability of those who are employed, however, has been increasing steadily.

Dakar is, then, a little like post-World War I Vienna, the decapitated head of a much larger body, still living to some extent on past glories. Despite some of the more sophisticated studies and plans made for African countries, Senegal is having a difficult time engineering satisfactory growth and this is reflected in somewhat depressed conditions in Dakar amidst the glamour of a sparkling, modern, high-rise city.

ABIDJAN[7]

While Abidjan was in existence in the early part of the present century it is to all intents and purposes a product of post-World War II

[7] See S. Bernus, "Abidjan-Note sur l'Agglomération d'Abidjan et sa Population," *Bulletin IFAN*, XXIV, Série B, Nos. 1-2 (January-April, 1962), 54-86; A. Gelin, "Abidjan, Ville Champignon," *France Outremer*, XXXII, No. 307 (June, 1955), 19-22; L. Roussel, F. Turbot, and R. Vaurs, "La Mobilité de la Population Urbaine en Afrique Noir: Deux Essais de Mesure, Abidjan et Yaoundé," *Population*, XXIII, No. 2 (March-April, 1968), 333-52; Société pour l'Étude Technique d'Aménagements Planifiées, *Rapport sur le Plan d'Abidjan* (Abidjan, Ministère des Travaux Publiques, 1959).

Some African Cities

years, one of the newest and most rapidly growing urban communities in Africa. Its phenomenal rise is strikingly related to a reasonably simple technological improvement associated with its site which permitted Abidjan to take advantage of its location with respect to a potentially highly productive hinterland.

Abidjan is sited on the large, irregularly shaped, adequately deep, and well-protected Ebrié lagoon, which was separated from the Ocean by a relatively narrow sand bar (Map 30). Early efforts to create an entry from the sea which would have opened a harbor comparable to that of Lagos were unsuccessful, and Abidjan remained a town of modest proportions despite its position as capital of the Ivory Coast until engineers found a solution to the problem. The coast here is the epitomy of that of West Africa—straight, sandbar ridden, and subject to enormous movements of sand capable of closing or obstructing any entryway in a brief period. The solution involved cutting the 1.75-mile Vridi Canal through the coastal bar and creating a current to direct any

MAP 30. Abidjan, Ivory Coast.

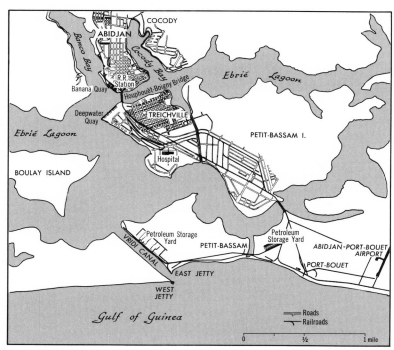

Some African Cities

sand deposited at the mouth to a deep offshore fosse known as the *trou sans fond* by extending the canal's west dike further than the eastern one and narrowing its seaward lip. Creation of this modern deepwater port, which was opened to ocean-going vessels in the early 1950s, gave the Ivory Coast its first such installation and the subsequent economic boom is a striking example of the stimulus that such a port can provide.

Of course development could not have occurred unless the hinterland had potentialities. That the Ivory Coast had such potentialities is illustrated by the facts that the country now ranks as Africa's leading coffee producer (and third in the world), its third producer of cocoa (fourth in the world), its leading shipper of bananas, and an important source of tropical logs and lumber. The bulk of these commodities are produced in the southeast within a reasonable distance of Abidjan, which is the focus of a very high percentage of commercial activity in the country. A new port being created at San Pedro in the west will not so much give competition to Abidjan as open up the heavily forested, little developed, but potentially productive southwest of the Ivory Coast.

A feature of the site and situation which contributes to the attractiveness of Abidjan is the extension of the lagoon system along 180 miles of the coast. Interconnection of the lagoons has made them tributary to the port and they are used for movement of logs and of manganese ore from Grand-Lahou. Selection of the city as terminus of the country's only railway and improvement of the relatively dense road net focusing on it have contributed even more importantly to its strength. Availability of a reasonably good hydroelectric site about 68 miles to the east at Ayamé permitted development of a low-cost source of energy for the bourgeoning metropolis, though it does not satisfy the full needs and a large thermoelectric station is to be built near the canal.

The immediate site of Abidjan has several disadvantages, including separation of the CBD and large residential quarters from the port-industrial sector on Petit Bassam Island, which has created traffic bottlenecks despite construction of new, multi-lane bridges. Penetration of the mainland by several bays again complicates traffic patterns, but provides numerous attractive sites for residential construction and permits unloading logs at several beaches for flotation alongside vessels in the harbor. Finally, some fairly steep-sided hills create problems, and swamps require draining as a health measure.

The growth of Abidjan may be revealed by a number of measurements. Its population was about 1,000 in 1910, 17,000 in 1936, and 50,000 in 1948, when Dakar had already reached 185,000. Opening of the Vridi Canal marked the turning point in the city's history. By 1955 it had grown to 121,000, ten years later there were 300,000 inhabitants, and in 1969 its population was about 500,000, a level above that predicted a few years earlier for 1975. More recent projections that the city will reach 1 million by 1980 and 3 million by the year 2000 may also prove to be underestimates, though the relatively small population of the country would tend to keep the growth below that of more densely populated countries. A second measurement is the growth of port traffic. Originally planned to handle 850,000 tons, the port had already reached 704,000 tons by 1953 and 1.4 million tons in 1959; in 1968 loadings and unloadings totaled 4.7 million tons. Exports of the country, the vast bulk of which are handled at Abidjan, increased in value from $150 million in 1958 to $425 million in 1968, which compares very favorably with

View from Le Plateau in Abidjan across the Houphouet Boigny Bridge to the main port area on Petit Bassam Island and the Ebrié Lagoon.

An older street on Le
Plateau, Abidjan.

At far right: Unloading
logs on one of the beaches
of the Ebrié Lagoon; the
logs are then floated
alongside vessels in the
harbor. Opening of the
lagoon to ocean-going
vessels greatly stimulated
the shipment of logs
from the Ivory Coast.

most African countries. Production of electricity in the country has seen
a remarkable increase from 1.8 million kwh in 1952 to 120 million in
1962 and 276 million kwh in 1966, when three-quarters were derived
from hydroelectric installations. Abidjan alone accounted for about
seven-eighths of electricity consumed and two-thirds of the outlets served
in the entire country.

Early construction at Abidjan was concentrated on "Le Plateau,"
situated on a peninsula between two bays of the lagoon. Here the CBD
was concentrated along the lower levels but it has since expanded to
parts of the Plateau. Government buildings were placed on the southwest
of the promontory and the rail line paralleled the western shore. European cottages and villas were built on the Plateau on attractive tree-
lined streets, but much of this area has been preempted for high-rise
apartment and commercial buildings which have been constructed with
such rapidity that the face of the city changes each year. Further inland
on the peninsula are several military and paramilitary camps and one of
the two largest workers' quarters, Adjamé.

Across a narrow waterway on the low-lying Petit Bassam Island

324 *Some African Cities*

are the major quays and other port installations, an extensive industrial area, and the largest of the high-density quarters, Treichville.

Newer residential areas extend inland from Adjamé, across the eastern Cocody Bay, and eastward on Petit Bassam Island. Plans call for additional quarters along two main axes, one paralleling the Ebrié lagoon on its north shores, the other extending inland to the northeast of the present center, the whole to cover an area considerably larger than Paris. A large national park has been preserved to the northwest. Despite impressive housing developments, site-and-service schemes, and much private building, the construction industry has not been able to keep pace with demands for offices, shops, hotels, and housing of all types.

Abidjan is first and foremost a port-commercial city. Its port facilities are modern and well-equipped with mechanical handling gear. There are special facilities for handling bananas, manganese, wood, and petroleum derivatives (a fuel dock on the east bank of the Vridi Canal and a post at sea for large tankers), plus a fishing port which was extended in 1964. Opening of the new port was particularly beneficial for goods requiring special handling such as logs, lumber, and bananas, for the fishing industry, and for handling of petroleum products (the railway was able to switch from wood-burning to diesel powered locomotives).

Serving the port is the 716-mile railway whose inland terminus is Ougadougou, capital of Upper Volta, which looks to Abidjan as its major ocean gateway. Mali also uses Abidjan for a portion of its shipments, but both of these countries are relatively poor and are only marginally responsible for the great growth of the city. Despite severe competition from road transport, particularly for the carriage of high-value coffee and cocoa and for bringing logs to the port the rail line trebled its passenger traffic from 1958 to 1966 and doubled its freight traffic in the same period.

It is not surprising that Abidjan's transport function has attracted a great deal of commercial activity—shippers, wholesale and retail outlets ranging from large department stores to luxury boutiques, banks and insurance companies, and head offices of companies operating in the Ivory Coast. In 1965 enterprises accounting for 83 percent of workers employed in the private sector in the country were centered in Abidjan.

Abidjan's political function is related mainly to its position since 1934 as capital of the country, but it is also the headquarters for several international groups including the Conseil de l'Entente. The substantial investments made since 1950 plus considerable aid from France have required important extensions to the bureaucracy, evident in several fine new buildings replacing or supplementing the former modest colonial structures. Abidjan is also the site of a new university, but it does not yet rate as an intellectual center of distinction. Here the emphasis is on production, profit, and pragmatism.

Finally, Abidjan has seen a dynamic growth in its manufactural sector which contains all of the usual early entries from beer to textiles, plastics to bulky metal products, and automobile and appliance assembly plants to a petroleum refinery. A 1967 study, prepared by two French agencies at the request of the Plan Ministry, proposed that by 2000 A.D. the city would require industrial estates twenty times the present size, with new plants to include two petroleum refineries, an integrated steel mill, naval construction facilities, petrochemical industries, a large tuna conservery, six breweries, and numerous other establishments.

ACCRA[8]

As in the case of Abidjan, site conditions played a very significant role in the history of Accra, but in a very different and somewhat indirect way. The major turning point in its evolution occurred when it was selected in preference to Cape Coast as the capital of the British Gold Coast holdings in 1876. The choice was apparently made primarily because of the more attractive climatic conditions, which are anomalous for the Guinea Coast, being much drier (precipitation averages only 27 inches per annum) and less humid. These features, plus the general absence of swamps, made Accra much healthier and more comfortable than other sites along the coast, particularly to the west. Indeed, horses could be kept at Accra, whereas they were likely to succumb to sickness after a few months at Cape Coast. The presence of a minor headland giving very partial protection to a stretch of beach may have been important in determining the exact position, which was the site of several forts protecting the trade in gold, ivory, and slaves.

Accra could certainly not claim to be well located with respect to the hinterland as it developed. The mineral-producing regions were concentrated in the southwest. The major forest zone was also toward the west, closer to Sekondi than to Accra. It was from that port that the first rail line was built and its neighbor, Takoradi, got the first modern port installation when its artificial harbor was completed in 1928.

But Accra had been selected as the capital, there was a tendency for commerce to concentrate close to the government offices, and the construction of a road system emanating from Accra to various parts of the south brought increased trade, particularly of high-value cocoa, which became the preoccupation of farmers and replaced the previously leading rubber, palm products, and kola nuts. The rail line from Accra was not begun until 1910 and did not reach Kumasi until 1923. But it

[8] See Ione Acquah, *Accra Survey: A Social Survey of the Capital of Ghana* (London, University of London Press, 1958); E. A. Boateng, "The Growth and Functions of Accra," *Bulletin of the Ghana Geographical Association*, IV, No. 2 (July, 1959), 4-15; Kwamina B. Dickson, "Evolution of Seaports in Ghana," *Annals of the Association of American Geographers*, LV, No. 1 (March, 1965), 98-111; "The Volta River Project: A Study in Industrial Development," in William A. Hance, *African Economic Development* (New York, 1958), pp. 87-114; D. Hilling, "Tema: the Geography of a New Port," *Geography*, LI, No. 231 (April, 1966), 111-25; E. C. Kirchherr, "Tema 1951-1962: The Evolution of a Planned City in West Africa," *Urban Studies*, V, No. 2 (June, 1968), 207-17.

strengthened the contributions of the road system as far as concentrating cocoa exports and a variety of imports on Accra was concerned. There is a marked contrast in the stimuli which led to development at Abidjan and at Accra; at the former, provision of superior port facilities was of prime importance in generating development, while at Accra development followed the establishment of government eventually making mandatory the construction of a modern port. In the case of Accra, construction of land transport routes preceded construction of the modern port, while at Abidjan opening of the modern port was succeeded by elaboration of the transport system in the hinterland and by massive growth of the city itself.

Nevertheless, history will probably record opening of the modern port of Tema as a major landmark in the development of Accra. This occurred only in 1962, but has already provided a major new stimulus to the economy of the enlarged city. Specific features of the Tema site were important in its exact positioning, but an overwhelming asset was closeness to Accra which, even as a surf port, was handling exceptionally large tonnages. Hilling notes the site features that made Tema attractive for a new port: deep water close to the shore, a rocky headland providing a "root" for the main breakwater, a steep, rocky seabed minimizing the effect of sand drift and reducing the necessity for dredging, extensive areas of level land suitable for industrial and urban development, and easy availability of suitable rock for construction.

Tema is frequently associated with the Volta River Project, and certainly this major hydroelectric scheme and construction of a large aluminum smelter at the port are important assets, but it was apparent that the port would have to be constructed irrespective of the project's adoption, and the decision to proceed with the port was in fact made some years prior to conclusion of contracts for the Volta Scheme.

Accra does have certain site features which are disadvantageous, including a perennial problem of water supply which will probably be solved by transmission over about 50 miles from the Volta River; drainage in the rainy season which usually sees flooding in the Korle Lagoon area; and periodic earthquakes, the last serious one, in 1939, having caused a few deaths and destruction of numerous old buildings over an extensive area of the city.

Settlements existed at Accra as early as 1500, when one was established by Ga people to facilitate trade with the Portuguese, and similar villages grew near the Dutch, Danish, and English forts built in the seventeenth century along a 3½ mile stretch of the coast. Activity flourished until cessation of the slave trade in the early nineteenth century, after which a period of lawlessness, instability, and depression ensued but gradually dissipated as trade in new products began to grow. During these several centuries Accra was only one of many trading posts; it became the most important installation on the coast only after transfer of the British Capital in 1876. Even then it was really only a village, its population in 1891 being only about 16,000. Growth in the present century was steady if not spectacular in pre-World War II years, the 1936 population having been 71,000. After the war there was a marked expansion in government activity, prices for cocoa brought previously unimagined wealth, and Accra began to attract a large number of migrants from both domestic and foreign areas. Its population grew from 136,000 in 1948 to 338,000 for the city and 492,000 for the agglomeration in 1960 and to an estimated 600,000 for the agglomeration in 1966.

The structure of Accra (Map 31) gives evidence of its role as a colonial capital in the government buildings in Victoriaborg, some distance from the old town cores, and in several residential areas built largely for expatriate officials. The oldest portions of the town, which are the poorest today, were built adjacent to James and Ussher Forts from which they took their names. They are inhabited mainly by Ga. These older sections are notorious for congestion; interdigitation of all kinds of functions is characteristic as is the juxtaposition of old buildings with gabled roofs and modern structures, which have usually been built on previously occupied land. Congestion in this section is increased by the narrow streets, the difficulty of establishing a reasonable traffic pattern, the large number of small shops and stalls which tend to preempt part of the street, and by the crowds which come and go, sharing the street with the vehicles. A notable feature of Accra is the very large part played by women in marketing and other activities. The more than 20,000 "market mammies" are a force not only in commercial but in political terms.

Efforts to decrease the concentration of activity in these areas and to relieve the pressure on them by providing space for neighborhood centers in new quarters of the city have not yet proved successful, nor have plans made two decades ago to eliminate the slums reached fruition. The older sections, however, cover only a fraction of the present area of Accra, which now extends about twenty miles along the coast and eight miles inland. Newer sections range from Tema, an entirely planned new town which is separated from the main agglomeration, through attractively laid-out residential sections, to unsightly and unsanitary shanty towns. Sectionalism by ethnic group is most clearly marked in the old Ga towns, but severe shortages of housing have tended to reduce the exclusiveness of other quarters on tribal, national, or racial lines.

The functions of Accra have been partially revealed in the discussion of its site and situation. In addition to the political, transport,

Part of downtown Accra, showing the juxtaposition of old dwellings, colonial buildings, and modern commercial structures.

and commercial functions, Accra has become the leading manufacturing center of Ghana. Small-scale operations are found in the old towns; industrial estates are located along the Ring Road and at Tema, which has been designated as the heavy industrial zone for the city. It already has the aluminum smelter, an oil refinery, and a small steel mill using scrap, plus a variety of lighter, consumer-product industries. Accra, too, is the headquarters for numerous financial, religious, publishing, and other enterprises, while Ghana's leading university is situated not far northward at Legon.

Accra is not a particularly attractive city; its unplanned growth, the survival of congested, dilapidated quarters, and the higgledy-piggledy

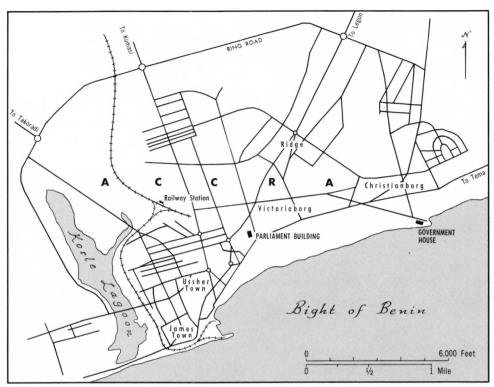

MAP 31. Accra, Ghana. (After Boateng.)

character of several sections give it an unfinished and sometimes un-
kempt appearance. But there is no question of its dynamic and spirited
life; its personality is far more attractive than its physical appearance.

KUMASI[9]

Founded by Osei Tutu about 1698, Kumasi owes much to its
position as capital of the Ashanti confederacy and seat of its ruler, the

[9] See Brian J. L. Berry, "Urban Growth and the Economic Development of
Ashanti," in Forrest R. Pitts, ed., *Urban Systems and Economic Development*
(Eugene, School of Business Administration, University of Oregon, 1962), pp. 53-
64; Gold Coast, Government Statistician, *Kumasi Survey of Population and Household Budgets, 1955* (Accra, Government Printer, 1956); K. A. J. Nyarko, "The Development of Kumasi," *Bulletin of the Ghana Geographical Association*, IV, No. 1
(January, 1959), 3-8.

Asantehene, and to the natural selection of the town as the administrative headquarters for Ashanti and the Northern Territories when the British took over in 1901. The key unit of the Ashanti was the village, but no other settlement was as large as Kumasi, which probably had 15–20,000 inhabitants about 1820 (estimates made by foreign travelers of 100–200,000 residents at this period are now considered as gross exaggerations). Following sackings by the British in 1874 and 1896 and the flight of numerous occupants, the town had only 3,000 persons in 1901, but grew to 24,000 in 1921 and 71,000 in 1948.

Kumasi enjoys an excellent location as a commercial center, and construction of railways and roads have enforced its nodal character. The rail from Sekondi reached the town in 1903, that from Accra in 1923. Roads in all directions were completed and improved in the 1920s and 1930s, stimulating the extension of cocoa farms and the collecting function of Kumasi. Berry likens it to an hour glass, funneling the economic produce of its area toward the two main southern ports; this is true particularly for the cocoa-producing areas of Brong-Ahafo and Ashanti, while the Great North Road to Tamale focuses practically all of the traffic and personnel movements to and from the northern region on Kumasi.

The original village was placed on the side of a large, rocky hill isolated by marshes on three sides, but it has since expanded across the marshes of the East and West Subin and the city is now fairly well divided into separate functional areas, each on a ridge rising 50–100 feet above the intervening valleys (Map 32). In the northeast is Menhyia with the Asantehene's Palace and houses of other notables; several new quarters mainly for Ashanti occupy the same ridge including the Asawasi housing estate, but most houses are built in traditional mud brick with corrugated metal roofs replacing the earlier thatch. In the extreme northeast is the Zongo, or stranger quarter, inhabited by a melange of foreign groups including many northerners, Voltaians, and Nigerians.

The East Subin valley is occupied by the railway and one of the main roads, along which commercial and industrial enterprises have been established, and which is the site of the very large central market. Numerous structures along the commercial avenues are combination stores and residences, with the upper floor(s) housing the families of the merchant or artisan. The swamp to the west of the river has been drained and is now occupied by the lorry park, the station, and a public

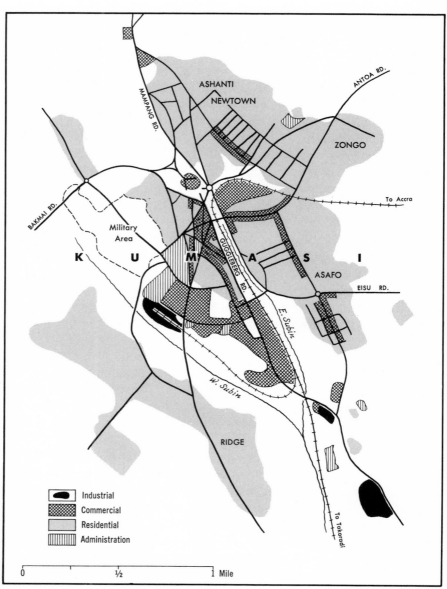

MAP 32. Kumasi, Ghana. (After Nyarko.)

Legend:
- Industrial
- Commercial
- Residential
- Administration

0   ½   1 Mile

Labels on map: ASHANTI NEWTOWN, ZONGO, MAMPANG RD., ANTOA RD., BAKMAI RD., Military Area, K U M A S I, GUGGLEBERG RD., ASAFO, EISU RD., To Accra, E. Subin, W. Subin, RIDGE, To Takoradi

Some African Cities

The central market and adjacent buildings, Kumasi.

park. On the hill between the two branches of the Subin is "commercial Kumasi," beyond which is the original European military and government quarter built around the old fort. It now includes offices of technical departments, barracks, and a hospital.

Across the West Subin to the southwest is the former exclusively European residential area and the newer political Kumasi, including the residential and office areas of political officials. As in Accra, the rapid growth of the city (from 71,000 in 1948 to 180,000 in 1960 and 249,000 in 1966, when the whole agglomeration had an estimated 301,000, or half the population of greater Accra) has broken down the earlier rigid separation of ethnic groups, though Ashante, Fante, and foreigners continue dominant in several quarters. Housing construction has not kept pace with the demand and numerous sections are composed of relatively crude structures built on traditional lines. Provision of public utilities has also frequently fallen behind needs, but this is not entirely related to the site features.

Some African Cities

335

The major problem of the Kumasi site has been the drainage of swamps and provision for runoff from the heavy rains. The swamps have, however, provided an opportunity to create green belts and parks within the city and thus to enhance its attractiveness. Whether it deserves the sobriquet "garden city" is open to question, but the prevalence of grassy areas and trees does provide an aura of cleanness and freshness not necessarily deserved in all quarters.

Kumasi is, in conclusion, primarily a vigorous commercial center, situated at the focus of numerous overland routeways. It is strengthened by its provincial administrative functions, by a fairly important industrial complex including a substantial handicraft segment, lumber mills, and some modern, market-oriented manufacturers. It is also the site of Ghana's second University, the former Kumasi College of Technology.

LAGOS[10]

The site of Lagos on a narrow island close to the coast was favorable for an early trading post because of its ready defensibility and its contact with a populous part of Africa with which trade was easily developed. Its site is much less attractive today because of the constrictions of its island character, while the pattern of expansion on the mainland has scarcely been what an intelligent planner would have recommended.

But Lagos was considered the best situated of any locale on the Bight of Benin in the nineteenth century, and it has grown to be the leading general cargo port on the west coast of Africa between Casablanca and Cape Town. Its hinterland covers much of Nigeria and

[10] See K. M. Buchanan and J. C. Pugh, *Land and People in Nigeria* (London, University of London Press, 1955), pp. 63-77; R. J. Harrison Church, *West Africa: A Study of the Environment and of Man's Use of It* (London, Longman's, Green and Co., 1968); O. Koenigsberger *et al.*, *Metropolitan Lagos* (Lagos, 1964); A. L. Mabogunje, "Urban Land-Use Problems in Nigeria," in Institute of British Geographers Special Publication No. 1, *Land Use and Resources: Studies in Applied Geography*, pp. 203-18; ——, *Urbanization in Nigeria* (London, University of London Press, 1968); ——, "The Evolution and Analysis of the Retail Structure of Lagos, Nigeria," *Economic Geography*, XL, No. 4 (October, 1964), 304-23; P. Marris, *Family and Social Change in an African City: a Study of Rehousing in Lagos* (London, Routledge and Kegan Paul, 1961); Babafemi Ogundana, "Lagos: Nigeria's Premier Port," *The Nigerian Geographical Journal*, IV, No. 2 (December, 1961), 26-40; Babatunde A. Williams and Annmaree H. Walsh, *Urban Government for Metropolitan Lagos* (New York, Praeger, 1968).

Some African Cities

Street scene in Lagos.

Niger as well, including regions producing the vast bulk of the country's cocoa, plus large tonnages of palm products, peanuts, cotton, hides and skins, and some wood and rubber. While the average income of Nigerians is low, that of the former Western Region is relatively high, and the Federation has by far the largest population of any African country.

While Lagos may be traced to the arrival of the Portuguese at the end of the fifteenth century, it remained not much more than a simple village into the nineteenth century despite its importance in the slave trade. It grew first on commerce carried on by Africans and then under the British, who occupied the town in 1851 to stop the trade in slaves and who established it as a colony in 1862. People flocked to it for the security offered, including freed slaves from Freetown and from Brazil.

Construction of the rail line, begun in 1895 and reaching Kano in 1912, and unification of Nigeria in 1914 brought more trade and administrative officers to the town, which had a population of 75,000 as

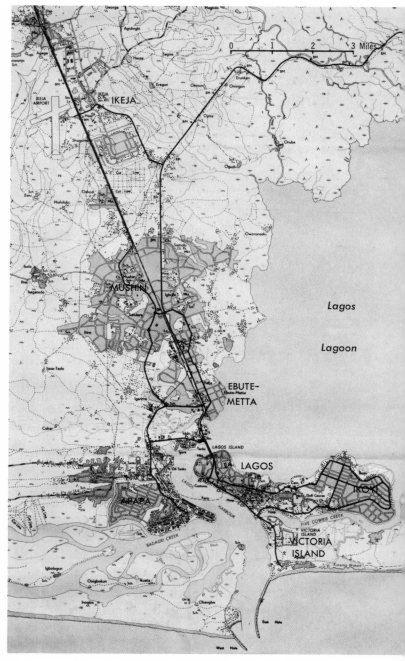

MAP 33.
Lagos, Nigeria.

338   *Some African Cities*

early as 1913. Other important steps which facilitated Lagos' exploitation of its location were the opening of its lagoon to ocean-going vessels in 1914, construction of new quays on the mainland at Apapa, and gradual formation of a rather dense road net in Yorubaland to the north. The difficulties of the site—its low-lying character, prevalence of swamps, inadequate drainage, the need to safeguard the protecting bar from sea erosion, and the difficulty of integrating its several parts because of the intricacy of the lagoon system and Lagos' being an island,—have gradually been ameliorated, but not completely and not without creating other problems.

The oldest and most densely populated portion of Lagos is on the island itself, where the original nodes were in the north in circles grouped around chiefs, a section which still contains grossly overcrowded slums, inadequately provided with public services (Map 33). In the center was the Brazilian Quarter, noted for buildings in the Portuguese style. The British preferred facing the harbor, and after their establishment of the colony the commercial houses and early customs quays were strung along the southwest waterfront, while the administrative buildings, missions, schools, a hospital, a race track, and the residences of the major officials were toward the southeast.

The northeast was occupied by Africans in the late nineteenth and early twentieth centuries and the east end of the island, which became Ikoyi Island by the digging of a canal, was made a European preserve with the usual houses set in gardens, golf course, and private clubs. But the European impact on Lagos stemmed far more from the technological introductions of an industrial society rather than from numbers of expatriate residents as was the case in Dakar; here the number of Europeans never exceeded 5,000 in the colonial period. Today, Lagos island still has the CBD, the administrative buildings, and most of the foreign embassies. Its legacy of narrow and twisting streets contributes to difficult traffic problems, heightened by illegal parking, use of the sidewalks for movable stalls, and consequent forcing of pedestrian traffic onto the streets. A major bottleneck occurs at the bridge connecting the mainland crossed by at least 200,000 workers a day, on foot, bicycle, and motor-bike, or by car and bus. The older colonial-period structures in the CBD are gradually being replaced by modern high-rise office buildings, but almost any street in the old section has a conglomeration of old and new, while unhealthy, congested slums penetrate it at

Old and new buildings in Lagos.

various points. Plans dating from the early postwar years to eliminate all of the slums on the island have not yet been carried out and are made difficult by property rights, preference for existing locations, and inability to finance replacement housing. In the meantime, good-grade housing is in extreme short supply, resulting in rentals at fantastic prices.

The first suburb of Lagos was created by construction of the railway at the turn of the century from Ebute Metta on the mainland. Here the railway workshops still constitute one of the main industrial establishments in the country. The residential portion of Ebute Metta, laid out on a gridiron pattern, houses various grades of railway employees. Opening of a new quay at Apapa in 1926 began development of another important section; the pier was extended in 1955 and again more recently. Apapa became not only the major port area for Lagos but the

Some African Cities

site of the first industrial estate and a new quarter for African housing.

Development also occurred inland from Ebute Metta along the rail line and major road all the way to the airport, creating one of the worst examples of ribbon development to be seen on the continent. Thus Lagos has developed as a great, elongated city stretching from the east end of Ikoyi west and then north for a distance of some 16 miles. True, it is constricted by the several lagoons and swamps of the region, but more effective planning might have reduced the dimensions of the problem. Political considerations have contributed to the pattern in that the boundary between the Federal Territory of Lagos and the former Western Region cuts across the northern extension, and the Western Region, wishing to take advantage of the pull of Lagos, promoted several industrial estates on its side of the border, at Ikeja and Mushin. At this end of the agglomeration a half dozen communities were beyond the control of the Lagos Executive Development Board and yet were entirely incapable of planning for proper growth until the Western Region Government coordinated such power under the Ikeja Town Planning Authority. Thus, while Lagos was the first Nigerian city to get a planning commission, the character of its growth would scarcely do credit to any rational scheme.

The main functions of Lagos are apparent. It is, first, a huge commercial-transport center, vibrant with activity. Second, it is an administrative center, not only as the political capital of the most populous country in Africa, but as headquarters of numerous commercial, financial, industrial, religious, and other organizations represented in the country. Numerous services cater to the more basic functions and to the inhabitants of what is probably the largest city in tropical Africa. As at Accra, women play a substantial role, particularly in retailing, and some carry credit lines with the large wholesaling firms running into five and six figures. While Yoruba make up about two-thirds of the total population, many other ethnic groups are well represented, including Hausa from the north and, both before and after the civil war, Ibo from the east.

A third function, and one showing rapid growth, is manufacturing, in which Lagos ranks as the single most important center in Nigeria. Despite high rentals, several industrial estates have been fully occupied and plans have been made for establishing others at several locales. The range of industries represented is quite wide by comparison with other West African cities. Its dominance within Nigeria is not so great as that of primate cities in other countries, however, because of the important role in development played by the former regional governments, each of which attempted to foster its own industrial buildup in a kind of "me-too" policy.

Despite the ugliness of its slums and the distorted character of its spatial arrangements, Lagos remains an exciting city, more African than many of its counterparts in other countries. It stands as a fine example of Balandier's phrase—a city "growing in confusion."

IBADAN[11]

The largest indigenous city in tropical Africa, Ibadan is the leading "Yoruba town," though it is not as classic an example of this unique group of agglomerations as many others. The heavy "urbanization" of Yorubaland in southwestern Nigeria has attracted the interest of numerous scholars, who variously call the agglomerations "towns," "rural cities," "city villages," and "agrotowns." According to the 1952 census, there were 120 Yoruba towns in western Nigeria with populations over 5,000 and six with over 100,000. Measured by size of town, not function, half of the region's population was urbanized as compared to 14 percent for the Eastern Region and 9 percent for the North. The 1963 census indi-

[11] See R. A. Akinola, "The Ibadan Region," *The Nigerian Geographical Journal,* VI, No. 2 (December, 1963), 102-15; ——, "The Industrial Structure of Ibadan," *The Nigerian Geographical Journal,* VII, No. 2 (December, 1964), 115-30; William Bascom, "Urbanization among the Yoruba," *American Journal of Sociology,* LX, No. 5 (March, 1955), 446-54; ——, "Some Aspects of Yoruba Urbanism," *American Anthropologist,* LXIV, No. 4 (August, 1962), 699-709; Archibald Callaway, "From Traditional Craft to Modern Industries," *ODU, University of Ife Journal of African Studies,* II, No. 1 (July, 1965), 28-51; P. C. Lloyd, "Craft Organization in Yoruba Towns," *Africa,* XXIII, No. 1 (January, 1953), 30-44; ——, A. L. Mabogunje, and B. Awe, *The City of Ibadan* (London, University of London Press, 1967); A. L. Mabogunje, "The Growth of Residential Districts in Ibadan," *The Geographical Review,* LII, No. 1 (January, 1962), 56-77; ——, "Urbanization in Nigeria—A Constraint on Economic Development," *Economic Development and Cultural Change,* XIII, No. 4, Pt. 1 (July, 1965), 413-38; ——, *Yoruba Towns* (Ibadan, Ibadan University Press, 1962); N. C. Mitchel, "Some Comments on the Growth and Character of Ibadan's Population," *Research Notes* (Department of Geography, University College, Ibadan), No. 4 (December, 1953), pp. 2-15; G. J. Afolabi Ojo, *Yoruba Culture: A Geographical Analysis* (London, University of London Press, 1966); W. Steigenga, "Ibadan, City in Transition," *Tijdschrift van het Koninklijk Nederlandsch Aardrijkskunding Genootschap,* LXXXII, No. 2 (April, 1965), 169-95.

Part of "Old Ibadan."

cated that at least ten Yoruba towns had populations exceeding 100,000.

Mabogunje believes that the tendency of the Yoruba to agglomerate aided rapid cultural diffusion, facilitated improved standards of living, and made it easier to develop commerce and achieve greater productivity, although he also states that the area is overurbanized in relation to the level of economic development and the capacity of local and regional governments to provide characteristic urban services.

Ibadan is one of the newer Yoruba towns, having been founded about 150 years ago. Around 1825–1830 the ancient Yoruba empire was defeated by the Fulani attacking from the north, leading to a flight southward and regrouping of the population in agglomerations, of which Ibadan was one. It first lost to Abeokuta but its selection as a military camp about 1829 provided a first important fillip; twenty years later it is thought to have reached 70,000. There followed a series of wars among Yoruba groups in which Ibadan opposed practically all the other towns; these wars ended only in 1886 under the British, who favored the Ibadan Yoruba on several occasions, in part because of their interest in developing trade with the coast, and thus contributed to Ibadan's surpassing Abeokuta in importance.

Early site advantages, which may have explained Ibadan's selection as a military post, were its placement on the edge of the grassland and the existence of large lateritic outcrops which afforded protection from Fulani cavalry and from hostile Egba in the neighborhood. Its location later proved convenient as a meeting place for traders from all over Yorubaland and as a collecting point for palm products carried to the coast for sale to the British. In the twentieth century its role was strengthened by arrival of the rail line in 1901, the focusing of several routes upon the city, helping it to become a major gathering and forwarding point for cocoa and palm produce destined for overseas markets, for kola nuts moving to the north, and for livestock shipped from

Street scene in Ibadan.

that area, and by its selection as headquarters of the Western Region.

Ibadan might be said to be two cities, "Old Ibadan," which has changed relatively little, and "New Ibadan," which was attracted to the west by the rail line and which has gradually spread in a great arc around the western part of the older agglomeration. The two tend to be kept separated by a range of hills running between them; they are adjacent but to a considerable degree exclusive. Mabogunje states that Ibadan

represents a convergence of two traditions of urbanism—a non-mechanistic, pre-industrial African tradition more akin to the mediaeval urbanism in Europe, and a technologically-oriented European tradition.[12]

Old Ibadan has most of the characteristics of an indigenous African city as described on pages 249–254 and is much like other Yoruba towns except that there is no palace (a reflection of its role as a military camp) and that it is not as old as many. Mabogunje writes that

the major features of the morphology of the older section . . . are its amorphous lay-out, its high density of housing and population which only resolves into some order in terms of "quarters," the indifferent though highly varied, quality of housing, its sprawl of rusty brown roofs with the towers of churches and

[12] Mabogunje in Lloyd, Mabogunje, and Awe, *The City of Ibadan*, p. 35.

A characteristic mud-brick, corrugated-metal-roofed house in Ibadan.

Part of the CBD of the "New Ibadan" with the old town in the background.

Housing estate for upper echelon officials on the outskirts of Ibadan.

Some African Cities

mosques providing the only breaks on the sky-line, and its "green-belt" eastern margin, made up largely of the football fields of the numerous schools, colleges and hospitals, which have grown up here only within the last decade.[13]

The old town has seen certain changes in the present century: a near demise of the old compounds and their replacement by or subdivision into smaller units and the filling-in of available spaces, both leading to greater densities; a substitution of metal roofing for thatch; and the cementing of some mud brick houses. About half of the housing is classed as slum, marked by overcrowding, abominable sanitary conditions (see p. 282), and much decay. Its structure makes provision of all kinds of urban services extremely difficult, which further contributes to the squalor and dilapidation.

The newer town was originally drawn to the west by the railway and its CBD is around the railway street. It contains a mix of old colonial structures and modern, high-rise office buildings; back streets are sometimes a jumble of buildings containing such diverse activities as garages, small shops, and sidewalk carts. Residential areas are varied and tend to be used by distinct groups: neighboring Yorubas; Hausa and Nupe in the northern Sabon Ngari; a suburb of decorated Brazilian style houses

[13] *Ibid.*, p. 48.

A small dyeing shop in Ibadan; craft industries of the more traditional type have tended to decline in recent decades.

with double rows of rooms opening on a common passage leading to the street and to a backyard with common kitchen, bathroom, and latrine facilities, and occupied in considerable part by white-collar workers; several former European Reservations with detached houses, now predominantly used by upper echelon Nigerian officials; and a public housing estate. The Universities of Ibadan and Ife, a few miles to the northwest, also established planned residential areas for their staffs. Ibadan's population reached an estimated 210,000 in 1900, 387,000 in 1936 when it was 2.8 times the size of Lagos, 459,000 in 1952, and 627,000 in 1963.

Ibadan's functions show greater variety than most other Yoruba towns, with a lower percentage engaged in agriculture than all such towns except Abeokuta and higher percentages in commerce and industry. Artisanal activities are characteristically important in Yoruba towns but they are undergoing radical change, particularly in Ibadan. Traditional crafts, whose workers are usually united by kinship ties, have tended to decline. Akinola, who distinguished between traditional crafts and small-scale modern industry on the basis of implements and techniques employed, indicated that four traditional crafts—weaving, dyeing, blacksmithing, and pottery-making—witnessed a decline in number of workshops from 715 in 1949–1950 to 94 in 1963–1964. But Calloway, combining the two categories, notes that blacksmiths, who enjoy high social standing, numbered 246 in 1963 or three times the earlier legendary number. New crafts are, in any case, far more important than the traditional ones. They include gold-, brass-, and tinsmithing, gunsmithing, carpentry, tailoring, barbering, shoemaking, and vehicle repairing. In 1964 there were some 5,135 craft and small-industrial establishments, 90 percent of which had been started since the Second World War; they employed 14,500 people. This count does not include artisans without permanent premises, such as carpenters, bricklayers, and construction workers, but it does include service activities such as barbering, gas station attending, photographing, and beauty parlor operation which are not truly industrial in character. Large-scale modern industry, defined as establishments employing over ten workers and using power-driven machinery, numbered 47 in 1964, of which only nine had over 100 employees.

Ibadan has also been the seat of government of the Western Region. The presence of Nigeria's most prestigious university adds a cosmopolitan element whose total economic contribution is probably quite substantial. Ibadan has, in conclusion, evolved much further than most Yoruba towns; it has the largest modern section, an important industrial sector, and a respected university. It does, however, have serious problems, including finding some way to rehabilitate the old town, substantial unemployment, and a deteriorating agricultural base in the immediate environs.

KANO[14]

One of the oldest cities of sub-Saharan Africa, thousand-year old Kano is probably the best example of a walled city of the Hausa emirates. Its exact site may have been fixed by blacksmiths using the iron-stone of Dala Hill, whose steep sides provided ready refuge in the event of attack. But it has been associated with trade throughout its history, possibly longer than any other city in West Africa. For some centuries it was a great caravan center at the terminus of trans-Saharan routes, the zenith of this period occurring in the fifteenth century. Trading led to the rise of artisan activity and the two functions reinforced each other in contributing to the wealth and fame of the city. Arrival of the railway from Lagos in 1912 expedited the development of trade via the Guinea Gulf, while later improvement of roads reinforced the nodal character of the city for a large part of the north.

The region around Kano has assets which have permitted support of dense populations, supply of produce to the rapidly expanding city and to southern Nigerian markets, and export of very large tonnages of peanuts, plus some millions of hides and skins. The agricultural hinterland is covered in the growing season with sorghum, millet, and other grains intersown with peanuts, the whole also carrying a large livestock

[14] See N. H. Doutre, "Kano—Ancient and Modern," *The Geographical Magazine,* XXXVI, No. 10 (February, 1964), 594-602; Gavan McDonell, "The Dynamics of Geographic Change: The Case of Kano," *Annals of the Association of American Geographers,* LIV, No. 3 (September, 1964), 355-71; E. P. Miller, *Kano: A Guide to the City and Its Environs* (Zaria, Gaskiya Corporation, n.d.); M. J. Mortimore, "Population Distribution, Settlement and Soils in Kano Province, Northern Nigeria 1931-62," in John C. Caldwell and Chukuka Okonjo, eds., *The Population of Tropical Africa* (New York, Columbia University Press, 1968), pp. 298-306; ———, and J. Wilson, *Land and People in the Kano Close Settled Zone* (Ahmadu Bello University, Department of Geography, Occasional Paper No. 1, 1965); B. A. W. Trevallion, *Metropolitan Kano: Report on the Twenty-year Development Plan, 1963-1983* (London, Newman Neame, 1966).

A portion of the old city of Kano.

population, partially supported on the stubbled fields. In the Kano close-settled zone, which contained nearly two million people living at densities in excess of 300 per square mile in 1962, some 80 percent of the area is cultivated, the relatively good yields and dense settlement aided by intensive practices and heavy application of animal manure and waste from the city. The region is advantaged not only by relatively good soils but by a shallow water table and availability of some surface water, land

Some African Cities

in marshes and along stream banks being used for intensive market gardening.

The environment does have disadvantages, including a somewhat arduous climate. Rainfall averages about 33 inches but the period October–May is practically rainless. Temperatures reach 115°F in April and May but may fall below 40°F in December and January, when the dust-laden harmattan produces discomfort. The high diurnal temperature range, as much as 40°, also requires adaptation. Providing for an adequate water supply has also been something of a problem, although Kano got one of the earliest piped-water systems of indigenous African cities with completion of the Challawa waterworks and electricity supply plant in 1931, bringing water from wells in the stream bed of the Challawa River to storage reservoirs on top of one of the two mesas in the city.

Like Ibadan and many other indigenous towns, Kano is now composed of two towns, old and new. The old town is enclosed within a wall which is said to have been thirteen miles around, eighty feet thick, and forty to fifty feet high at the time of the British occupation in 1903. Built of sun-dried bricks, much of the wall has since disintegrated, though more recently efforts have been made to preserve it and to reconstruct a number of the fourteen gates for their touristic interest. Early defense was further achieved by deep moats filled with thorn bushes and by enclosure within the wall of an area large enough to produce food during a sustained seige. These lands are now progressively being occupied by extension of the inner city.

Houses and most structures in the old town are built of sun-baked mud bricks; they are flat roofed, have few openings on the winding alleys, and many have two stories. Many borrow pits from which the mud was extracted are scattered throughout the town. Filled with water, they provide washing places, watering points for livestock, and, before piped water became available, for humans as well. They now are being filled in since they are hazards to health, require spraying against mosquitoes, and occupy space which can be used more effectively.

Old Kano contains one of the largest open markets in Africa. Many stalls are built of mud, others are bamboo structures, while numerous traders display their wares on the ground or carry them around the market on their heads. Only the meat shops have been required to build in more permanent materials which could be adequately cleaned. Other

The making of conical bricks used in construction of indigenous housing in Kano.
A Kano borrow-pit from which clay has been removed in the past for making bricks.

Some African Cities

The twin-minareted, main mosque of Kano; constructed of stone, it replaced an earlier structure built of less durable sun-dried bricks.

features of the walled city include the civic center, the palace of the emir bounded by a high mud wall enclosing a twenty acre space, and a mosque completed in 1948. The last, which replaced an earlier structure falling into disrepair, is a twin-minaretted structure in white stone and with a green-tiled roof, much in contrast with the ochre-colored houses and walls.

Numerous craftsmen carry on their work in dimly lit houses, in the market, or in special quarters of the city. Craft activities are very varied and have survived somewhat better than at Ibadan. The old city is occupied mainly by Hausa and Fulani; it has greater density and

Some African Cities   353

Merchants in front of small stalls characteristic of the Kano market, one of the largest in tropical Africa.

Wood and charcoal are brought to the city from considerable distances.

homogeneity, both in ethnic representation and in morphology, than the newer quarters outside the wall.

Like the additions to other indigenous towns those of Kano are divided into distinct quarters which often have correlations with ethnic groups and their specific functions. These include Nasarawa which was selected as the European Township and was excluded from the jurisdiction of the Emir under the system of Indirect Rule adopted by the British. This section contained the provincial administrative offices, the

354   *Some African Cities*

Building a peanut mound at Kano; in the foreground is the gravel base for another pyramid, designed to reduce the danger of insect attack.

Dye-vats at Kano; handicraft industries have retained very considerable importance in the old city.

railway station, a hospital, and the European Reservation housing expatriate officials, who also occupied Bompai to the northeast. The railway attracted commercial activity, a good deal of which is in the hands of Syrians and Lebanese plus some Indians and Arabs, who occupy a quarter between Nasarawa and the wall. North of that is Fage quarter, between the commercial section and the wall, originally occupied by Tuaregs from the Sahara who handled camel transport before this was largely displaced by trucks; it is now settled mainly by Hausa strangers.

Some African Cities

Northeast of that is the Sabon Gari in which numerous clerks, artisans, and traders from southern Nigeria, primarily Ibo, were housed, at least from its founding about 1925 until the riots of 1966 which led to the flight of most southerners. The Sabon Gari market was almost as large as the central market of the old city. East of that are several mission offices and hospitals. As in quite a few African cities, the road to the airport, which was one of the busiest in Africa until overflying somewhat reduced its significance, has tended to become the central boulevard of the new city. This is probably less related to existence of the airport than to construction of a better-than-average road artery. Finally, a new industrial quarter has been created to the east in which several score modern industries have been placed in postwar years, including a number of fairly large textile mills.

LUANDA[15]

The oldest of tropical African capitals, Luanda dates to 1575, though it was a town of no great significance until post World War II years. It is sited on one of the few good natural harbors on the Atlantic coast of Africa, which attracted early Portuguese navigators to it, though they used a bay just to the south until 1648 rather than the present harbor, which is a 1½ square mile bay protected by an elongated spit-type island connected to the mainland by a causeway (Map 34). Additional assets of the site are the predominantly calm seas which characterize this latitude, an attractive climate, moderated by the cool Benguela current, closeness to a source of low cost electricity on the Dande River, and presence of oil fields on the nearby coastal shelf.

Luanda proved to be well located with respect to some of the most productive regions of its hinterland, though these were poorly developed until the last few decades. Its domestic hinterland includes about a quarter of Angola, containing about two-fifths of its population.

[15] See F. C. C. Egerton, *Angola in Perspective* (London, Routledge and Kegan Paul, 1957), pp. 132-54; William A. Hance and Irene S. van Dongen, "The Port of Lobito and the Benguela Railway," *The Geographical Review*, XLVI, No. 4 (October, 1956), 460-87; Allison B. Herrick, ed., *Area Handbook of Angola* (Washington, American University, 1968); Irene S. van Dongen, "The Port of Luanda in the Economy of Angola," *Boletim de Sociedade de Geografia de Lisboa* (January-March, 1960), pp. 3-43; U.N., E.C.A. "Introduction aux Problèmes de l'Urbanisation en Afrique Tropicale, Cycle d'Études sur l'Urbanisation en Afrique" (SEM/URB/AF/1/Add.1, March, 1962).

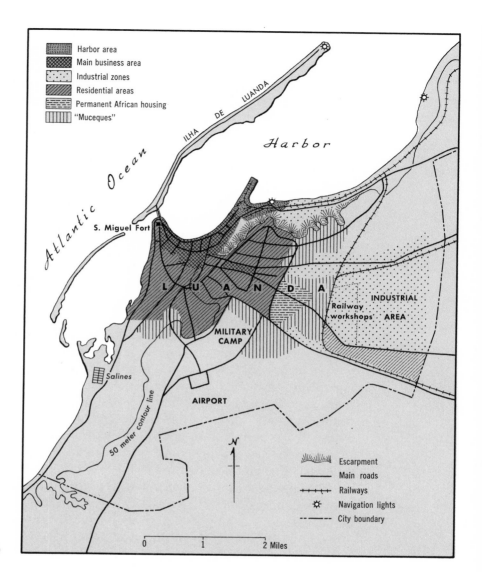

Harbor area
Main business area
Industrial zones
Residential areas
Permanent African housing
"Muceques"

Atlantic Ocean

ILHA DE LUANDA

Harbor

S. Miguel Fort

L U A N D A

MILITARY CAMP

Railway workshops

INDUSTRIAL AREA

Salines

50 meter contour line

AIRPORT

N

Escarpment
Main roads
Railways
Navigation lights
City boundary

0   1   2 Miles

MAP 34.
Luanda, Angola.
(After van Dongen.)

About 80 percent of the province's coffee, its leading export, a third of its sugar, all its cotton exports, and shares of a dozen other agricultural products pass from this zone across the quays of Luanda, plus small tonnages of manganese ore. The hinterland is tied to Luanda by a 280-mile rail line to Quela and a rudimentary though improving road net-

Some African Cities

work which is now handling the bulk of produce moving to and from the city. The port also handles cabotage traffic from outlying ports.

The immediate hinterland of Luanda is unattractive, except where it is possible to irrigate, because of the aridity induced by the cool Benguela Current. Unlike Lobito in central Angola, Luanda never served an extranational hinterland, though it was originally hoped that its rail line would be carried across the continent. It is conceivable that it may someday be connected with the Congo system, but that country would prefer to foster its own *route nationale* and is now examining the feasibility of connecting the Matadi-Kinshasa line with the Katanga line which would greatly reduce its dependence on Lobito.

Luanda was a small post and revictualing center for several centuries. Fort São Miguel, dating from the early seventeenth century, was the center of a community which saw endless convoys of slaves cross the beaches en route mainly to Brazil. At the cessation of this trade in 1836, Luanda, which had a population of about 6,000, suffered near collapse. Efforts dating from about 1850 to colonize the hinterland were unsuccessful and even after construction of the railway to Malange (1886–1909) very little trade developed; indeed portions of the hinterland were not pacified until 1917. Beginning in the 1920s, however, commercial agriculture gradually developed, first with sugar coming from plantations, then with cotton produced by Africans under control of European concessionaires, next with sisal and coffee, which became the ranking export in the early 1950s, and finally with minerals in the latter part of that decade. Angola now ranks second only to the Ivory Coast among African coffee producers.

Luanda's growth was relatively slow until development of the northern interior began, and particularly until the postwar coffee boom; its population was only 23,000 in 1923 and 61,000 in 1940, but reached 190,000 in 1955 and an estimated 250,000 in 1966. The presence of a large number of Portuguese in the city is an important contrast with most tropical African cities; this group has substantial consumption power, which contributes to the economy of the city and the import tonnages absorbed within the capital.

The structure of Luanda has been strongly conditioned by site features. After expansion of the original community absorbed the space available in the Upper City and commercial and transport activities utilized the Lower City along the margins of the bay, it was necessary to

Some African Cities

View of Luanda.

move to the top of the escarpment lying inland from the harbor, and it is in this area that a feverish expansion has occurred in the last two decades. Construction of a broad finger pier in 1945, the first deepwater facilities available in the port, permitted developing the attractive Avenida Marginal around the southern end of the bay, further enhancing the Mediterranean and Portuguese appearance of the city.

Most Portuguese residents live in villas and apartment houses reminiscent of the metropole; some modern housing estates have been built for Africans, but most live in *muceques* far from the city center where they have built their own houses, which vary from reasonably satisfactory structures to shanties built of packing crates and hammered-out pieces of metal. A large industrial zone has been laid out to the east and a modern airport lies south of the city but within the municipal boundary. The expansion of Luanda after the war was so rapid and so unexpected that earlier plans proved completely inadequate and most public services fell behind the growing demand.

Luanda's functions are comparable to those of other African port-capitals. In addition to its commercial and administrative functions it is Angola's most important industrial city, with a range of consumer-oriented industries comparing favorably with similar centers in other countries. The largest of these are a cement mill, a brewery, a textile mill, and a petroleum refinery. Long dormant, in part because of the poverty of the metropole, Luanda has been the recipient of a surprisingly large investment in postwar years. As the capital, main consuming center, first-ranking manufacturing city, and one of two major ports in Portugal's largest and richest overseas province, Luanda should see a continuing rapid growth; the greatest need is to work more effectively toward African participation in the advance of the city and the province.

ADDIS ABABA[16]

The highest important city in Africa, Addis Ababa probably owes its siting more to chance than any other factor. Although an authentically African city, it is much younger than most indigenous cities, having been founded only in 1886. The Shoa Emperors had the custom of changing capitals from time to time and Addis was roughly the hundredth of their capitals and the eighth of Menelik's. Mariam believes that Ethiopia's failure to develop an urban tradition was at the root of its stagnation and he attributes this failure to factors which were physical (the rugged terrain inhibiting commerce and sustaining subsistence farming as the predominant activity), cultural (the attitude of disdain for merchants, artisans, and anyone who worked with his hands, plus the strongly feudalistic society), and political (particularly the insecurity and turmoil caused by constant fighting among the various tribes and ras).

[16] See Edouard Berlan, *Addis-Ababa, La Plus Haute Ville d'Afrique* (Grenoble, Imprimerie Allier, 1963); Jacques Denis, "Addis Ababa: Genèse d'une Capitale Imperìale," *Revue Belge de Géographie*, LXXXVIII, No. 3 (April, 1965), 283-314; R. J. Horvath, *Around Addis Ababa: A Geographical Contribution to the Impact of a City on Its Surroundings* (University of California, Los Angeles, unpublished Ph.D. dissertation, 1966); ——, "Von Thünen's Isolated State and the Area Around Addis Ababa, Ethiopia," *Annals of the Association of American Geographers*, LIX, No. 2 (June, 1969), 308-323; Martin Johnson, "Addis Ababa from the Air," *The Ethiopian Observer*, VI, No. 1 (1962), 17-32; Municipality of Addis Ababa, *Report on Census of Population, 10-11 September 1961* (Addis Ababa, n.d.); Richard Pankhurst, "Notes on the Demographic History of Ethiopian Towns and Villages," *The Ethiopian Observer*, IX, No. 1 (1965), 60-83; ——, "The Foundations and Growth of Addis Ababa to 1935," *The Ethiopian Observer*, VI No. 1 (1962), 33-61.

It is possible that the presence of thermal springs at the site and that its deeply ravined character favoring defense were factors in selecting the exact site, though the distance an entourage could be expected to move in a day may have been just as important. In any case, the city has both advantages and disadvantages in its site and situation. The disadvantages include: the high elevation, unsuitable for persons with heart conditions; the high precipitation and humidity during the rainy season, conducive to pulmonary difficulties; the large number of flies, particularly at the beginning of the rains, causing ophthalmic difficulties, and the multitude of fleas which, if one is susceptible, are practically impossible to avoid; and the accidented terrain, which leads to problems of erosion and difficulties in road construction. Proper sanitation and other measures could eliminate some of these disadvantages.

Advantages include: the centrality of the city with respect to a large part of the country; its improving position as a crossroads center as major roads focusing upon it are upgraded and extended; its nearness to several climatic zones, permitting variety of production; a reasonably productive immediate agricultural hinterland; good sites for hydroelectric installations relatively near the capital; absence of the tsetse fly, of malaria and cholera; and visual attractiveness of the site and surrounding country.

Berlan suggests that a site near Nazareth, 60 miles to the southeast, would have been more attractive in several regards: a more varied agricultural potential with opportunity for intensive irrigation, closer proximity to hydropower sources, vast plains open for urban development at lower cost, a superior crossroads position, and more attractive climatic conditions with twice the sunshine and half the rain experienced by the city, but he concedes that no one contemplates moving the capital from its present site.

Menelik selected a hill above the thermal springs as the site for the first imperial quarter, called the Guebi. After specifying areas for the church and the market place lands were allotted to various nobles or ras, which resulted in the old town becoming an agglomeration of family properties known as *sefer* or camps. Johnson delineated some 150 *sefer* in Addis Ababa in the early 1960s, each somewhat separate if not isolated from its neighbors, each quite homogeneous ethnically and religiously. Few roads were laid out, and even today many houses are accessible only by path. These features, plus the well-wooded character

A nobleman's house in Addis Ababa. The city was originally divided into *sefers* or camps allotted to various nobles or ras.

A sales platform in one of the many small markets found in the numerous village-like quarters of Addis Ababa.

The scrap metal section of the main market of Addis Ababa.

362   *Some African Cities*

View of the Piazza in central
Addis Ababa.

The new city hall, Addis Ababa.

and the frequency of small local markets, make much of Addis more like
a group of villages loosely tied together than a metropolis of 630,000
(1966).

The Guebi and the market became poles of development in the
old city, and were joined at the beginning of the century by the first
paved road, forming an elongated arc. The eastern side tended to become
the administrative zone with large domains being allotted to foreign
embassies in the northeast; the west was more devoted to commerce.
Eucalypts were planted everywhere; they provide an important wood

Some African Cities

reserve and contribute to the characteristic well-wooded appearance of the city. In 1917 the rail line reached Addis from Djibouti and the station became a new pole of development. When Haile Selassie was crowned in 1930 the city had grown to about 100,000.

In the early decades after its founding the population fluctuated widely with each of Menelik's expeditions. When the imperial army was away, it was a city of women, children, priests, and the aged; each time he returned a large number of soldiers and slaves were added. The 1909 population of about 60,000 included about 25,000 slaves (15,000 blacks, 8,000 Sidamos, and 2,000 Gurage) and 35,000 free citizens (20,000 Galla, 13,000 Shoan Amharas, 1,000 Gojjamites, and 1,000 Tigreans). While slaving has been legally abolished and the feudal structure has become less rigid, the lower levels of society and descendants of the slaves still live at very low standards. They provide innumerable *zabanias*, or guards, *mamites*, or domestic maids and nurses, and "boys" performing a variety of menial tasks.

The brief Italian occupation, from 1935 to 1941, saw the creating of two new "cities." The Western one, devoted to commerce and industry, was centered on a new market set out in 1936, west of the former market, and a road from it to the station became the axis of a new zone of stores and schools dominated by the "Piazza." The other new city in the south was in principle reserved as the Italian community. Their preference for the lower, warmer, and less rainy side of Addis set a pattern which has continued to the present. In recent years a construction boom has seen many new high rise apartments constructed in this area, whose population doubled between 1961 and 1966.

The most striking new feature of postwar years has been development along an arc created by two boulevard systems tying the old airdrome in the southwest to the new palace situated to the northeast, opposite Africa House, headquarters of the E.C.A. and donated to the U.N. by the Emperor. Most of the apartment buildings and high rise offices and commercial buildings rise along these boulevards. Whether the opening of the new airport in the southeast will result in comparable development remains to be seen.

Despite an impressive amount of new building, especially in the last decade, Addis retains many of the qualities of an overgrown village. Each morning thousands of asses bring fuel, hay, and produce from peripheral areas into the city. Of 124,000 households counted in 1961,

90 percent lived in dwellings built of *chica*, or mud or wattle walls; two-thirds of these had no foundation and two-fifths had thatched roofs. While 30 percent had piped water, the percentage in some sections was as low as 4; another 45 percent relied on public taps each supervised by two women, leaving a quarter of the households without a safe water supply. From 23 to 75 percent of households were served by electricity, the total for the city being 58 percent.

The residents of Addis Ababa come from all parts of the country, though there is a recognizable correlation between numbers and the distance and difficulty of access from source areas. There is a high percentage of persons born elsewhere, but migration appears to be largely one-way, as 63 percent of those questioned in a recent survey had not left the city after arrival. A large number of disintegrated families were present, but over 97 percent stated that they preferred life in Addis to another residence.

The two main functions of Addis Ababa are commerce and administration. In addition to being the capital of a populous country striving to join the modern world after centuries of isolation and stagnation, it is the headquarters of the E.C.A. and the Organization of African Unity. Ethiopia's participation in numerous African conferences has played a significant role in eliminating the former sense of superiority and complacence; visits to other African cities have revealed the backwardness of Addis Ababa and doubtless stimulated desires to make it a more worthy showplace. In the commercial sphere the city, has, in addition to the large markets and shopping areas serving the city and its immediate region, numerous importing and exporting houses, and it gathers most of the country's main export, coffee, before forwarding it for shipment via Djibouti and Assab.

An unusually high number of persons are employed as domestics, while there are hundreds of teff and tej houses serving the favored national drinks and functioning as houses of prostitution. There are numerous artisans pursuing their traditional crafts, which are sometimes concentrated in specific quarters, plus several hundred small-scale "industrial" establishments such as grain mills and oil-seed processors. In the last decade a number of larger, modern, expatriate-managed manufacturing plants have been introduced, with a particularly notable development in the textile industry. Finally, Addis Ababa has numerous cultural organizations, including foreign mission and Coptic Christian establishments, the Institute of Ethiopian Studies, and the University of Ethiopia. It is a unique center, a peculiar mixture of the old and modern, of squalor and wealth, of village and city.

NAIROBI[17]

Founded when the Uganda railway reached it from Mombasa in 1899, the site for Nairobi was selected by railway officials because there was ample flat land at the boundary of the Athi Plains and the Kikuyu Uplands, water from the Nairobi River, and it was well situated for workshops serving the line at a point roughly midway between Mombasa and the proposed terminus at Kisumu on Lake Victoria.

Situated at an elevation of about 5,400 feet, Nairobi has a moderate climate usually described as healthful, though the city has not always enjoyed its present reputation as one of the healthiest centers in tropical Africa. The Nairobi swamp, black cotton soils which became waterlogged in the rainy season, and inadequate provision for drainage and sanitation led to at least four major epidemics in the first fifteen years. Provision of adequate water for a rapidly expanding community has also been something of a problem. Continuing advantages of the site include the large flat spaces available for expansion and the attractiveness of the upland sections for residential construction.

The location of Nairobi has proved to be a strategic one. While lands to the east and south are low-productive steppelands, Nairobi is at a point through which most of the high value agricultural exports of Kenya and all of those of Uganda must funnel. It is close not only to the productive Scheduled Areas (the former White Highlands) but to Kikuyu areas which are accounting for an increasing amount of commercial produce both for export and for the urban market. Kikuyu farmers have shown greater interest in supplying the agglomeration with market-garden produce than is true for groups adjacent to many African cities and their lands can produce a great variety of such crops.

[17] See L. N. Bloomberg and C. Abrams, *Report of United Nations Mission to Kenya on Housing* (Nairobi, Government Printer, 1965); City Council of Nairobi, *Nairobi, City in the Sun* (Nairobi, 1963); Dorothy M. Halliman and W. T. W. Morgan, "The City of Nairobi," in Morgan, ed., *Nairobi: City and Region* (Nairobi, Oxford University Press, 1967), pp. 98-120; Simeon H. Ominde, *Land and Population Movements in Kenya* (Evanston, Northwestern University Press, 1968); Soja, *The Geography of Modernization in Kenya*; R. W. Walmsley, *Nairobi: The Geography of a New City* (Nairobi, East African Literature Bureau, 1957).

Some African Cities

An asset of growing importance is Nairobi's location with respect to many of the major tourist attractions of East Africa, and tourism has had the most rapid increase of any economic sector in the country in recent years. No other city in tropical Africa benefits as much from tourist revenues as Nairobi.

Like so many African cities, Nairobi remained relatively small up to World War II. Three years after its founding it had a population of about 4,300, including about 100 Europeans. It received an important fillip when the colonial capital was moved to it from Mombasa in 1907. But it still had only 20,000 in 1920 and 33,000 in 1930. Migration of Africans for permanent residence was discouraged during at least the first half of its existence. In World War II it became an important military center and reached 100,000 sometime during it. Increased colonial interest in postwar years, accompanied by large-scale private investment, created boom conditions, the agglomeration reaching 200,000 about the mid-fifties and 315,000 in 1962. Independence unloosed a flood of migrants to the capital which has continued apace despite lack of housing, unemployment, and government efforts to stem the flow. The 1969 census gave the population of Nairobi as 478,000.

The main aspects of Kenya's structure were set very early; they reflect its early functions as headquarters of the railway, colonial capital, and commercial node, the presence of substantial Indian, Pakistani, European, and other expatriate groups, and rigorous separation of ethnic groups by residential area. Early growth was marked by minimal planning, conflict between the railway and government authorities, a good deal of land speculation, and not a little confusion. But Nairobi gradually acquired one of the more effective municipal authorities in tropical Africa, with a sufficient tax base to permit supporting some 6,000 employees and operating on a budget of about $15.4 million a year in 1969.

The center of Nairobi contains a number of distinct sections (Map 35), some rather sharply set off from others by function, some containing a considerable variety of activities. The railway area comprises a 200 acre block to the southeast, containing extensive sidings, the workshops, the main station, the large administrative headquarters building, and housing for some staff members. This large tract blocks the southern end of the CBD, creating traffic problems. Some day it may prove desirable to move at least portions of this complex to open lands to the east, or possibly parts could be overbuilt.

The railway section of Nairobi with part of the CBD on the right.

Northwest of the railway area is the City Square area, occupied by national and municipal government buildings. It, too, tends to create blocks to traffic moving to and from the commercial zones lying to the north and east. The shopping areas contain certain distinctive features, including a large covered city market, numerous streets lined with rows of Indian shops, and more modern shops tending to concentrate along and near Kenyatta Avenue and Government Road. The shopping area east of Tom Mboya (formerly Victoria Street) caters to those with a lower standard of living and contains numerous *dukas* comparable to

Some African Cities

Aerial view of Nairobi's CBD.

those found in large and small communities throughout East Africa, each selling a notable variety of nonperishable merchandise. Nairobi has a large and increasing number of modern, multi-story buildings, in part reflecting the somewhat constricted character of its CBD, in part the large number of business firms and organizations headquartered in the city, and, more recently, the need for more hotel space A variety of churches, mosques, and temples in the CBD reflects the cosmopolitan character of the city's population. At the northern end of the central area are the University College, Nairobi, and other organizations which

Some African Cities   369

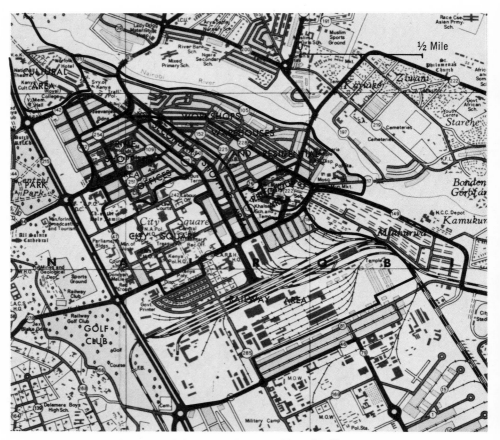

MAP 35. The Central Business District, Nairobi, Kenya.

justify defining the section as the cultural center of Nairobi. Slightly to the north is the Coryndon Museum.

About a twelfth of the total population of Nairobi resided in the central city in 1962; most important were Indians, including those who lived in the backs of or above their shops. The largest number of Asians, however, live in the Parklands-Eastleigh sections to the north of central Nairobi (Map 36), where gaudily colored, flat-topped, several-story houses are characteristic. Restrictions on expansion of these sections led to growth of another Asian quarter in South Nairobi. The major African residential area is in Eastlands, which housed 71 percent of the total of 155,000 Africans in 1962. There are about eight housing estates,

Some African Cities

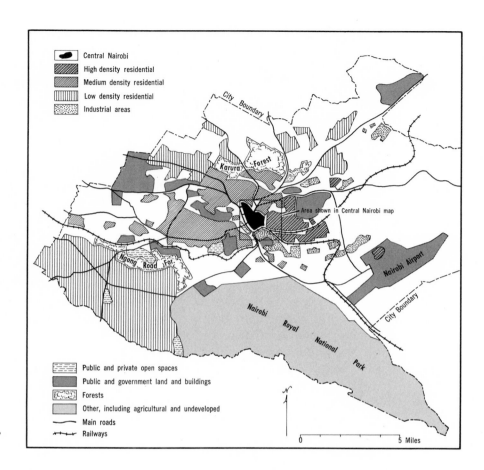

Legend:
- Central Nairobi
- High density residential
- Medium density residential
- Low density residential
- Industrial areas

- Public and private open spaces
- Public and government land and buildings
- Forests
- Other, including agricultural and undeveloped
- Main roads
- Railways

City Boundary

Karura Forest

Area shown in Central Nairobi map

Ngong Road For.

Nairobi Airport

City Boundary

Nairobi Royal National Park

0   5 Miles

MAP 36.
Greater Nairobi,
Kenya.

the first dating from 1929, in this section, which is by all odds the most densely populated part of Nairobi. Recent trends in African housing include replacement of the oldest section by a new group of apartments, greater emphasis on flats rather than single units, acquisition of a part of Eastleigh formerly occupied by Indians, and movement of some higher-salaried personnel into the old European sectors. Numerous Africans continue to live in former European and Asian sections on the properties of their employers.

One of the most striking features of Nairobi is the very large area formerly devoted exclusively to European housing, the vast bulk in attractive stone and tile-roofed structures set in plots of at least a quarter acre, well wooded and more often than not with fine gardens and lawns. The European section, which is more suburban than urban in appearance, is mainly on the higher lands west and north of the city core in what is called Upper Nairobi and in periurban estates beyond the former limits of the city. The existence of numerous ridges separated by deep, wooded valleys adds to the contrast with the flat sections to the east and explains a road pattern very different from the gridiron arrangement on the plains.

South and east of the central city is a large industrial sector, less than half of which has been occupied. Outside the old urban boundaries but within the much larger area delineated in 1963 are several forest reserves, the international airport opened in 1958 ten miles to the east of the center, and Nairobi National Park, a game park situated within sight of the CBD but possessing much of the interest and variety of much larger game parks in east and southern Africa.

Nairobi is well equipped with bus and taxi services and has an unusually high percentage of paved streets throughout. It is subject to increasing congestion, heightened by the filling in of empty lots and greater vertical construction of the inner city.

The major functions of Nairobi are essentially the same as in its first years—transport, commerce, and administration. In 1963 commerce accounted for 19 percent of reported employment and 52 percent of those employed in commerce in all of Kenya were concentrated in Nairobi. Transport and communications accounted for 13 percent of city employment, manufacturing for 13 percent and 29 percent of the total for Kenya, building and construction for 5 percent (49 percent of the national total), and services 45 percent, of which three-fifths were public.

Manufacturing is a more recent function than the first entries. Nairobi's industries, accounting for 41 percent of employees in manufacturing and 44 percent of value added by manufacturing in Kenya, consists both of plants processing domestic produce and of a wide range of plants catering to the domestic market. The latter were attracted by the largest single concentrated market in East Africa and by a climate of investment somewhat more attractive than other East African centers. Externalities have begun to develop to a significant degree, adding to the attractiveness of the city for manufacturing.

Nairobi is, then, one of the more modern cities of tropical Africa.

It probably has a greater variety of available services, cultural institutions, entertainment facilities, and manufacturing plants than any city in tropical Africa. It does have problems of congestion, inadequate housing, and unemployment, but its future should be somewhat brighter than many counterparts if for no other reason than the burgeoning tourist trade.

KAMPALA[18]

The capital of the ancient Kingdom of Buganda, which had one of the best-evolved hierarchical systems of East Africa, naturally drew the first European settlement in Uganda to it, just as it was the main objective for the Uganda Railway from Mombasa. In the late nineteenth century it probably had a population of about 3,000 in the Kabaka's enclosure and 10,000 in the surrounding African town, making it the largest such center among the several tribal kingdoms in East Africa.

Kampala is well located with respect to the fertile crescent of Uganda, the largest "island" of high potential in East Africa. Coffee, the major export of the country, cotton, sugar, and some tea are the major cash crops of southern Uganda, but it produces a great variety of subsistence crops, including the favored plantains, and supports a dense population at a relatively high level. One of the features of the Ugandan economy is the high percentage of commercial production coming from indigenous farms, in marked contrast with Kenya and Tanganyika with their continued heavy reliance on expatriate estates and plantations.

Kampala is connected with its hinterland by a system of radiating roads which compares very favorably with that of most African countries. The main roads are paved for considerable distances, while murram provides an excellent surface for secondary and tertiary roads in most parts of the country. Its main external connection from 1902 to 1931 was via Port Bell, six miles to the southeast, lake steamer to Kisumu in Kenya, and rail to Mombasa; in the latter year the rail finally reached Kampala, eliminating the necessity for transshipments.

The city has a somewhat unusual setting, on several dozen of

[18] See Peter C. W. Gutkind, *The Royal Capital of Buganda* (The Hague, Mouton and Company, 1963); Aidan W. Southall, "Kampala-Mengo," in Miner, ed., *The City in Modern Africa*, pp. 297-332; A. W. Southall and Peter C. W. Gutkind, *Townsmen in the Making* (Kampala, East African Institute of Social Research, East African Studies No. 9, 1957).

the low, flat-topped hills characteristic of the region (Map 37). Local relief is about 500 feet with the hill tops at elevations of about 4,300 to 4,400 feet. At Kampala, as elsewhere, the valleys were often partially occupied with papyrus swamps which helped to account for the endemicity of malaria and a high infant mortality rate. Their presence proved to have certain advantages, however. After draining, which was near completion by 1950, they provided additional land advantageous for the placement of schools, parks, and playing fields, the easy development of arterial roads, and, in the east, relatively flat land for rail sidings and an industrial quarter.

The structure of Kampala has two major features: its division into two sections, the indigenous town of Mengo and the municipality of Kampala, essentially the creation of the colonial administration; and the distinctive functions associated with many of the hills. While the Ganda had moved the *kibuga,* or residence of the Kabaka from hill to hill at least once in each reign, it was placed on Mengo Hill in 1884 and has not since been moved. Four other hills became religious centers, several having been allotted to Catholic and Protestant missions which arrived in 1877–1879. All of these ecclesiastical hills have marked similarities: a cathedral or mosque at the top, associated schools and hospitals, and residences of prominent members of the church, including officers in the Buganda government, several offices always being held by agreement by persons of particular faiths. Two other hills falling partially within Mengo have specialized functions, Makerere Hill with the University College and Mulago Hill with a large hospital. Their slopes are largely taken up with these institutions and residences of persons of various grades employed by them.

A duality has always existed between the old town and the municipality. The old town and its more recent extensions have remained under control of Ganda authorities and land, theoretically open only to Africans, remains under the traditional land-tenure system. The Ganda have zealously guarded their control, sustaining a separation from Kampala proper. The limited space in the latter made for high land values, which reinforced the separation. The duality created certain friction, the protectorate authorities frequently complaining of the lack of control and regulation in Mengo, where corruption was a logical result of the patron-client relations of the Ganda social system. Friction also developed between the Ganda and other Ugandan ethnic groups and was

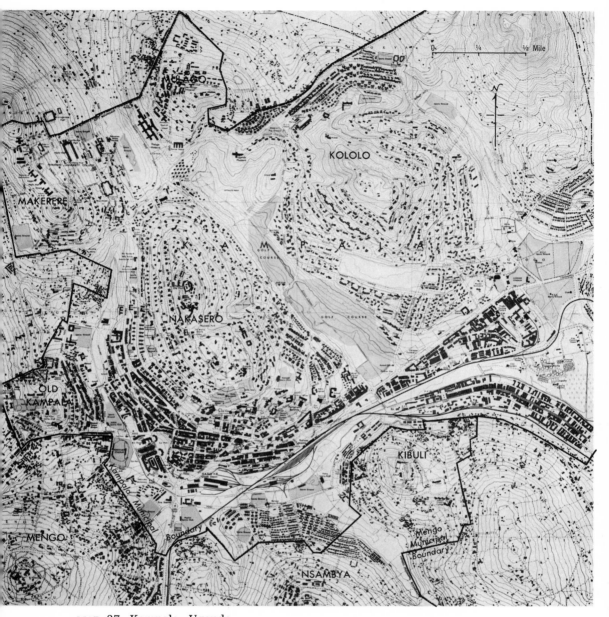

MAP 37. Kampala, Uganda.

exacerbated by juxtaposition of the Buganda capital and the national capital in the two parts of the agglomeration. The Kabaka was forced to flee the country in 1966 after the Obote government decided that the power of the Kabaka, who was also President of Uganda, must be reduced.

Kampala itself dates from 1890 when Lugard arrived as an agent of the Imperial British East Africa Company and erected a fort on Kampala Hill. Uganda was declared a British Protectorate in 1893 and a definitive agreement was signed in 1900. In 1905 the government offices were moved to Nakasero Hill a half mile to the east where a new fort had been constructed, and a flank of this hill soon became the CBD of the town, while the other flank was used for officers' residences. Kampala Hill, which came to be known as Old Kampala, became a predominantly Asian area, Asians having arrived with the British and taken over a large part of the nonindigenous commercial life.

As the town developed, several of the intervening valleys were brought into use. Katwe, lying between two main Ganda hills, south of Nakasero, and on the main road to the administrative capital at Entebbe, became the most important African commercial area in the Protectorate, with a large open market and some shops and offices. Lesser African

The central portion of Kampala.

Some African Cities

Old and new government and commercial buildings in Kampala.

shopping and residential areas developed in other valleys while, as has been noted, the railway and industrial areas were situated in valleys to the east. The axis between Kampala and Port Bell also filled in, often with temporary structures built by migrants from Kenya. Expansion of the colonial staff after the war led to construction of residences on Kololo Hill, northeast of Nakasero, for medium and higher grade officers and Indian auxiliaries. After 1950 several African housing estates were constructed toward the east of the municipality; these now house about 10,000 of the total population of about 175,000 for the agglomeration.

Kampala's functions are primarily commercial and administrative. Of secondary interest are its cultural and industrial functions. Finally, it is benefiting to an increasing extent, though on a much smaller

Aerial view of part of the plant area and part of the high-density Kantanshi Township of the Mufulira mine, Zambia Copperbelt.

scale than Nairobi, in the East African tourist boom. It is appropriate to conclude with a quotation from Southall, whose studies of Kampala have provided much of the data utilized in this summary:

Kampala-Mengo is interesting because it contains within itself most of the major factors, combined at different strengths, which are found in African cities of quite varied type, such as the older, more traditional West African cities and the newer, European-dominated cities of East and Central Africa. It combines both segregation and political dominance of a particular African tribe; it includes both European and African controlled land, traditional and modern roles, local African residents of long standing and high status, as well as thousands of temporary migrant labourers of many ethnic backgrounds.[19]

[19] Southall, "Kampala-Mengo," p. 326.

Some African Cities

COPPERBELT TOWNS IN ZAMBIA[20]

The Copperbelt of Zambia provides examples of urbanization developed on the basis of large-scale mining development. While many features would be characteristic of other mining towns, only Congo (Kinshasa) and South Africa have mining towns as large as those of Zambia. The copper industry permits that country to have one of the highest average per capita incomes on the continent. The industry accounts for 95 percent of Zambian exports, the value of copper exports having been $725 million in 1968, when Zambia ranked third after the United States and the U.S.S.R. in production of this metal. Production has grown from 214,000 tons in 1948 to 400,000 in 1958 and 727,000 tons in 1968, with increasing quantities being shipped as refined copper rather than copper matte. The industry provided 38 percent of the net

[20] G. Kay, "The Towns of Zambia," in Robert W. Steel and Richard Lawton, eds., *Liverpool Essays in Geography* (London, Longmans, Green and Co., 1967), pp. 347-61; ——, *A Social Geography of Zambia* (London, University of London Press, 1967).

Workers' housing at Roan Antelope Mine, Zambia.

Some African Cities

domestic product, 56 percent of total government revenues, and employed 16 percent of the paid labor force in the country in 1967, though the total number employed in the Copperbelt area was a considerably higher percent. The ratio of Africans to expatriates increased from 1.5 : 1 in 1960 to 9 : 1 in 1968 and in the same period turnover dropped from 28 percent per annum to 5 percent. Two large companies, Zamanglo and Roan Selection Trust, manage the operations, which are now jointly owned by the Zambia government and these groups.

The copper towns illustrate well the ability to overcome numerous obstacles if a high-value resource occurs in large deposits. Located far from the sea, the mining industry justified construction of rail lines, and later roads, most of which are now paved, and air facilities. The needs for electrolytic refining justified construction of the large Kariba installation on the Zambezi, even with the relatively long-distance transmission that was required, while expanding production has led to development of the Kafue station and a decision to go ahead with the north-bank plant at Kariba.

When sanctions were imposed on Rhodesia in 1966, massive international airlifts supplied Zambia with petroleum products, while millions of dollars have been expended in developing new sources of coal and alternate outlets for Zambia, which had previously depended on Rhodesia for all of its coal and for 98 percent of external shipments. Emphasis has been placed on transport via Dar es Salaam in Tanzania, including improvement of the Great North Road, installation of a pipe-line from Dar to Ndola, and construction, scheduled to begin in 1970, of a rail line to Dar. This will parallel in part the Tanzanian line, which has a different gauge.

In the area itself, which is on a monotonous, brachystegia-covered plateau, efforts were made to make living as attractive as possible for the large expatriate staff required to engineer and operate the massive and complex mining-industrial operations. Amenities of all kinds were provided: schools, hospitals, golf courses, tennis courts, tracks, swimming pools, clubs, theatres, etc. Particular attention was paid to health and disease control, with successful antimalaria campaigns being especially important.

The typical mining town is actually a twin town, composed of the mining town and the public town. The mining towns, built and managed by the companies, have three main components (Map 38): the plant

Some African Cities

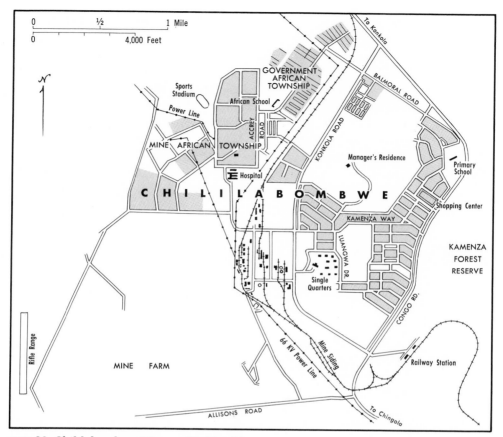

MAP 38. Chililabombwe (Bancroft), Zambia.

area, containing the mine headgear, crushing plant, concentrator, smelter, workshops, and, for some, an electrolytic refinery; the housing for higher-level workers, staff, and managers, formerly confined to expatriates (somewhat before independence these sections were desegregated and are now called low-density areas); and an African-workers or high-density area. A major change occurring over a period of years reflected the replacement of migrant workers by permanent employees and involved a greater provision of family housing as contrasted with hostels and bachelor quarters. The push to Zambianize the mines has led to mounting an impressive training program, for which space and buildings have been allotted in the plant area.

Some African Cities   381

The public town, which accommodates commercial, financial, and other private enterprises serving the mining community was also divided into expatriate and African areas. The African townships house nonmining employees; they were usually relatively small before independence but have since grown with considerable rapidity and now characteristically have a sizable squatter population.

In 1966 the six large Copperbelt towns had a total population estimated at 517,000. Kitwe ranked second to Lusaka, with an estimated 156,000 in 1968 as compared to 201,000 for the capital. Ndola, the second largest Copperbelt town (132,000 in 1968), is atypical in not being directly associated with a mine, though it is the site of copper and cobalt refineries. Its function was primarily administrative for some years, but it has become a large commercial community and has received most of the nonmining industries attracted to the area in postwar years. Population of the public towns has grown more rapidly than that of the mining towns in recent years, in part because of excessive in-migration, in part because improving productivity on the mines has kept the increase in labor force well below the increase in production.

The urban communities of the Copperbelt are not without their problems. The presence of a highly sophisticated, modern complex in a backward, largely illiterate, subsistence-oriented countryside rarely provides such striking contrasts as in Zambia. The long dependence on expatriates, partly enforced by white union demands, was bound to cause tension under an independent government subject to sometimes unrealistic demands for localization. While the transitional period, which must be a fairly lengthy one, has been handled with considerable skill by both the government and the companies, there has been a series of labor difficulties, strikes, and a number of racial incidents in recent years. High copper prices (unit values of Zambian copper increased from $25.62 per 100 pounds in 1961 to $45.82 in 1967) have led to demands for higher wages which have reduced the competitive advantage of the producers, which may cause serious adjustment problems when the world price declines; they also tend to reduce the government offtake and hence the ability to finance other developments within the country. In the long run, Zambia will face the same problems as South Africa in seeking to reduce the dependence on a single mineral; the reserves are, however, immense and probably capable of sustaining high-level output for many decades.

382

CHAPTER 6

POPULATION PRESSURE IN AFRICA

ONE OF THE MORE common generalizations regarding Africa is that, on the whole, it does not have population problems and does not suffer from population pressure. This myth has gained wide credence and is reiterated by numerous of our more prestigious authors, as may be seen in the following quotes: Melville Herskovits—"Subsaharan Africa . . . . is but lightly peopled, so that the question of the pressure of population against the land enters only in a few regions";[1] Walter Goldschmidt—"The African continent below the Sahara in general does not have any serious population problem";[2] Paul Bohannon—

[1] U.S. Senate Committee on Foreign Relations, *United States Foreign Policy: Africa* (A Study prepared by the Program of African Studies, Northwestern University, Washington, G.P.O., October 23, 1959), p. 18.

[2] Walter Goldschmidt, ed., *The United States and Africa* (New York, Praeger for the American Assembly, 2d ed., rev., 1963), p. 146.

"the people are probably less thick on the ground than anywhere in the world save the Amazon basin and the Australian outback";[3] Pierre Moussa—"Africa . . . , except in a few regions, is not overpopulated";[4] Andrew Kamarck—"There is little population pressure in Africa as a whole";[5] A. F. Ewing—"There is no serious over-population problem";[6] and several E.C.A. documents—"For the region as a whole the problem of human pressure on land resources may not be a serious one now or in the near future,"[7] "Except for a few regions . . . there does not seem to be much pressure of population on land,"[8] and "There does not seem to be much pressure of population on land in East Africa."[9]

It is true that various writers qualify their generalizations regarding lack of pressure by inclusion of such phrases as "except for." The exceptional regions most commonly noted are Rwanda, Burundi, and the adjacent Kigezi District of Uganda; Iboland in eastern Nigeria is sometimes added; occasionally someone remembers that Egypt is in Africa and includes it; less often, the Maghreb, Upper Volta, parts of northern Nigeria, the Sahara and its steppe fringes, many of the highlands of East Africa, southern Malawi, Lesotho, and the islands of Mauritius and Réunion are mentioned, but rarely more than two of them by the same observer. It begins to make a formidable list! And it still does not include areas which should be, including parts of Sierra Leone, Ghana, Togo, Dahomey, Cameroon, Zambia, Botswana, and the reserves of Rhodesia and South Africa.

There are a number of explanations for the rather widespread tendency to underestimate population pressure in Africa. First is the tendency to use continental or national density figures. How often one reads such statements as "Africa is not a densely-peopled continent. Its average density of 24 [it was about 29.5 in mid-1969] persons per

[3] Paul Bohannon, *Africa and Africans* (Garden City, The Natural History Press, 1964), p. 78.

[4] Pierre Moussa, in foreword to Andrew Kamarck, *Economic Development in Africa* (New York, Praeger, 1966), p. x.

[5] *Ibid.*, pp. 24-25.

[6] A. F. Ewing, "Some Recent Contributions to the Literature on Economic Development," *Journal of Modern African Studies*, IV, No. 3 (November, 1966), 343.

[7] E.C.A. *Demographic Factors Related to Social and Economic Development in Africa*, E/CN.14/ESD/12 (15 December, 1961), p. 4.

[8] E.C.A. *Economic Bulletin for Africa*, V (January, 1965), 39.

[9] E.C.A. *The Demographic Situation in Eastern Africa*, E/CN.14/LU/ECOP/2 (20 July, 1965), p. 3.

Population Pressure in Africa

square mile is lower than any other continent save Australia,"[10] or "Africa south of the Sahara, covered with savannas and forests, is privileged by comparison with Asia because of its low density of population."[11] Such statements are not helpful in understanding Africa, as was detailed at some length in Chapter 2. Of what use is a crude density figure for Algeria, for example, where 94 percent of the population resides on 11.7 percent of the area at a density 7.8 times the average for the country? Or what use in Kenya where, in 1962, 42.9 percent of the population lived at densities over 400 per square mile as compared to the crude density of 41.1? Even if the more useful "population density on arable land" is employed a feeling of security is achieved on the basis of continental comparisons. For example, the E.C.A. says that "For Africa this figure is 114 per square kilometer compared with the world average of 212. The former is lower than in other parts of the world, except Oceania, U.S.S.R., and North America."[12]

Secondly, statements regarding the lack of population pressure are frequently made with no reference to capacity of the land. The continental comparisons mentioned in the previous paragraph contain an inherent fallacy since they assume that carrying capacities are roughly equal from continent to continent. It is meaningless to compare densities on Africa's leached latosols with those on the rich alluvials of monsoon Asia and conclude that Africa does not suffer from population pressure.

A somewhat comparable fallacy which is surprisingly prevalent in the literature, at least implicitly, is that only a high density can mean population pressure. Som, for example, states that "there does not seem to be much pressure of population on land in Africa as measured by the density of population."[13] The point is that such a measure should not be employed. Rarely is it suggested that densities of 10 or 1 or even less per square mile may lead to as great or greater pressure on the land as those of 500 to 1,000 or more per square mile in favored areas. Look,

[10] E.C.A. *Seminar on Population Problems in Africa*, E/CN.14/ASPP/LF.2 (29 October—10 November, 1962), p. 36.

[11] Yves-Marie Choupaut, "L'Afrique Noire Est-elle Menacée par la Famine?" *Europe-France Outremer*, No. 440 (September, 1966), p. 8.

[12] "Recent Demographic Levels and Trends in Africa," in *Economic Bulletin for Africa*, V (January, 1965), 30.

[13] R. K. Som, "Some Demographic Indicators for Africa," in John C. Caldwell and Chukuka Okonjo, eds., *The Population of Tropical Africa*, New York, Columbia University Press, 1968), p. 190.

for example, at the vast, sparsely populated northeast of Kenya or at the Sahara, where most nomads and oasis dwellers are living at standards about as low as the average precipitation. While equating sparse population with lack of pressure is quite common, one occasionally runs into the reverse fallacy that all areas with high densities suffer from population pressure. A recent E.C.A. document on eastern Africa, for example, listed "the copper-mining belts of Zambia, the ports and industrial locations" of that region among the few areas where there was "much pressure of population on land."[14]

Third, relatively little attention has been paid in many parts of Africa to the subject of population pressure and this has, I believe, led to its underestimation. Contributing to the assumption of lack of pressure has been the recency of the population explosion in most areas and the belief that in some areas there had been a demographic crisis resulting from an actual population decline. Such views were common among Belgian officials and have been traced to an early gross exaggeration of the population in Congo. As late as 1961 a study in Cameroon concluded that numerous regions were experiencing a grave *crise de dénatalité* and called for concerted government action to increase the birth rate.[15] The estimated population increase in Cameroon in the period 1963–1967 was 2.2 percent per annum.

Fourth, all too many observers, and officials have been willing to accept a low standard for achievement, such as a peasantry which remains largely concerned with subsistence production while contributing only minimally to the money economy of their countries. Such a low-level goal is likely automatically to result in a different interpretation of the significance of population pressure.

Fifth, assessments of population pressure have rarely included any consideration of excess urban population, which is a growing phenomenon in Africa.

It is my contention that population pressure is substantially more pervasive and considerably more serious in Africa than has generally been accepted, and the purpose of this chapter is to examine the evidence contradicting and supporting this hypothesis.

[14] E.C.A. *The Demographic Situation in Eastern Africa*, p. 3.
[15] France, Direction de la Documentation, "La République du Cameroun," *Notes et Études Documentaires*, No. 2741 (19 January, 1961), p. 11.

Population Pressure in Africa

Problems of Definition and Measurement

There are numerous difficulties in defining the concepts of overpopulation, underpopulation, optimum population, population pressure, and population problems. Overpopulation and underpopulation imply that there is some more desirable population size for the given area, possibly the optimum population. But optimum population may be calculated by a variety of sometimes conflicting criteria including standard of living (which may itself be set at widely divergent levels), productivity, and national strength (the notion that the international power of a nation is directly related to its population size continues to attract supporters despite the general absence of supporting data). Furthermore, optimum population is an extremely elusive proposition in a dynamic world where economic, social, and technologic changes require a constant recalculation of the ideal. The concept also appears to imply that the population can or should be controlled somehow to reach the defined optimum at which point presumably no further growth should be allowed. The three concepts of overpopulation, underpopulation, and optimum population may also fail to recognize that the significance of population size is entirely different in an extractive economy than in one that is developed and diversified. They are, then, of questionable value or validity and are not used in this discussion, except insofar as the terms may be used in quoted sources.[16] The term "population problem" is so broad as to be meaningful only when the specific problem is more clearly defined; it may mean too few or too many people, too rapid or too slow population growth, the maldistribution of wealth, or the imbalance between population and food supply.

The concept used in this chapter is that of population pressure; the concern is with places where, under *existent* land use and technology and at the *present* stage of development, there is pressure on the land or in the urban environment, and with the population inhabiting these places. Because there is existing pressure it does not mean that there is overpopulation or that there will always be pressure. It does suggest that attention needs to be given to such things as soil protection, intensifica-

---

[16] See Notestein, "Some Aspects of Population Change in the Developing Countries," in 3 *Talks on Population*, pp. 11-21; J. I. Clarke, "World Population and Food Resources: A Critique," in Institute of British Geographers Special Publication No. 1, *Land Use and Resources: Studies in Applied Geography* (November, 1968), pp. 53-70.

tion, increasing production, creation of new and diversified employment opportunities—to economic development. The problem may, therefore, be far more related to economic, social, and political transformation than to population numbers per se. As Langdale-Brown states:

the physical capacity of an area only has reality in relation to the human population and its social organization and aspirations. Soils which are a prized resource to one tribe, such as the waterlogged peats of the Barotse plain from which the Lozi produce heavy crops, may be valueless or even a hindrance to another which has not developed the same skills or social organization.[17]

Similarly, William Allan writes that "shortage or abundance can only be defined in relation to environments and systems of land use."[18]

Various efforts have been made to develop methods of measuring environmental potential, which has obvious bearing at least on future carrying capacities, but such efforts have been plagued by problems of scale, of physical complexity, and of dissociating human responses from physical potentialities.[19] The two most useful methods which have been used to assess environmental potential are: (1) the agro-meteorological or agroclimatic approach, based on water needs of plants and on energy/water balance studies. While extremely useful, this approach requires intensive instrumentalization which is usually available only in a limited number of stations; and (2) the ecological approach, based on the principle that natural vegetation can provide a guide to a region's agricultural and forest potentials. It has been used with considerable effectiveness by Trapnell and Clothier in their studies of Zambia,[20] by Langdale-Brown and others in Uganda,[21] and by Allan,[22] who developed a formula for calculating the critical carrying capacity using, for any given area,

[17] I. Langdale-Brown, "Some Aspects of Land Capability Assessments in Africa," *African Affairs*, LXV, No. 261 (October, 1966), 307.

[18] Allan, *The African Husbandman*, p. 8.

[19] See Philip W. Porter, "The Concept of Environmental Potential as Exemplified by Tropical African Research," paper presented to Symposium on the Geography of Population Pressure on Physical and Social Resources, Pennsylvania State University, September, 1967 (mimeographed).

[20] See Colin G. Trapnell, *The Soils, Vegetation and Agriculture of Northeastern Rhodesia* (Lusaka, Government Printer, 1953); Colin G. Trapnell and J. N. Clothier, *The Soils, Vegetation and Agricultural Systems of North-western Rhodesia* (Lusaka, Government Printer, 1937).

[21] Langdale-Brown, Osmaston, and Wilson, *The Vegetation of Uganda and Its Bearing on Land Use.*

[22] Allan, *The African Husbandman.*

Population Pressure in Africa

the percentage of cultivable land, the relation between duration of cultivation and of fallow, and the acreage planted per capita per year. Useful for subsistence-farming conditions, it loses value when cash cropping is added to the traditional system. Valuable as these and other efforts to classify and quantify are, their shortcomings and the limited places in which they have been applied restrict their utility in the present assessment.

The estimates made later in this chapter are, then, based on an inadequate fund of data and mainly on subjective analyses. Many more maps, surveys, observations, and experiments will be needed before greater accuracy can be achieved. Fortunately, increased attention is being given to land capability studies, to research on unemployment and underemployment, to assessments of farming systems, and to other studies which will be helpful in understanding the extent and seriousness of population pressure. It is important to note, too, that numerous specialists have examined specific pressure regions; their studies are essential to the tentative summations presented here.

The Case for a Larger and Growing Population

Before presenting evidence supporting the thesis that Africa has a serious problem of population pressure, it is appropriate to examine the contentions that population densities and population pressure are too low. Colin Clark, for example, postulates that rapid population growth may, by stimulating transformations in farming systems, have a favorable impact on economic development,[23] and Boserup suggests that advanced agricultural techniques have only been adopted under compulsion of a slowly rising population density.[24] Both Clark and Boserup hold that population pressure is likely to expand world food production. Similarly, Smith and Blacker write that "In both East and Central Africa . . . economic development demands an expanding population";[25] the Royal East Africa Commission stated that "It has been the very sparsity of population in

[23] Clark, *Population Growth and Land Use.*
[24] See Boserup, *The Conditions of Agricultural Growth: the Economics of Agrarian Change Under Population Pressure.*
[25] Smith and Blacker, *Population Characteristics of the Commonwealth Countries of Tropical Africa.*

the past which has been one of the contributing causes of . . . arrested economic development";[26] and an E.C.A. document notes that, for many African countries, "a substantial increase in population would evidently be advantageous for economic development."[27]

With respect to the Clark-Boserup thesis, it is evident that indigenous farmers have often adopted new and more intensive techniques as their numbers have increased in a given area, witness the use of manuring, tethering and handfeeding of livestock, carefully planned rotations, provision of fodder crops and other practices on densely populated Ukara Island in Lake Victoria, the construction of more or less elaborate terracing in numerous densely populated mountain and hill areas, or some of the fantastically intensive methods used to provide water in numerous Saharan oases. Given a situation in which the land capability permits intensification it may be accepted that adjustments will gradually be made in response to the needs of a slowly rising population.

But two important qualifications reduce the validity of the thesis as a generalization which is applicable to Africa as a whole. First, it is questionable whether adaptations are likely to be made with sufficient rapidity in face of a rather sudden and rapid increase in population density in an area which had long experienced a stagnant growth pattern. Farmers on Madagascar, for example, do not appear to have adjusted as yet to the abrupt increase in population growth which began in the late forties. Second, land capability is already being utilized rather fully in many parts of Africa and there are numerous examples, as will be seen, of overuse and deterioration. Indeed, for every example of successful adaptation there appear to be many examples of failure to adjust to pressure on the land. Porter, commenting on the Boserup thesis as applied to Africa, writes:

in view of the poor environmental base over much of tropical Africa, it seems unlikely that the process of intensification could evolve in response to slow sustained population pressure except in the more favored parts. In semi-arid areas the process of intensification would overstep inherent environmental limits, lead to deterioration, and a lowered carrying capacity.[28]

[26] *East Africa Royal Commission 1953-1955 Report*, p. 37.
[27] E.C.A. *Seminar on Population Problems in Africa*, pp. 12-13.
[28] Porter, "The Concept of Environmental Potential as Exemplified by Tropical African Research," pp. 26-7.

An argument adopted by some proponents of an increasing population is that the small size of many African countries does not provide an adequate market to permit development of an integrated industrial complex. It is certainly true that there are a large number of small countries in Africa (Table 24), but it would be difficult to prove for any

Table 24

NUMBER OF POLITICAL UNITS IN AFRICA BY SIZE OF POPULATION, MID-1967

Population (millions)	Number of political units	Cumulative number	Population (millions)	Number of political units	Cumulative number
½	12		5–10	8	46
½–1	6	18	10–20	6	52
1–2	5	23	20–30	1	53
2–3	4	27	30–40	1	54
3–4	6	33	60	1	55
4–5	5	38			

SOURCE: U.N., *Demographic Yearbook 1967* (New York, 1968), pp. 98-103.

of them that the best way to stimulate industrial growth was to increase the population. It would be far more desirable in many cases to construct a larger market by uniting several national markets in customs unions.

Occasionally one sees the proposition that larger populations are needed to release people from agriculture and provide workers for industry and services; this argument appears to be based on historical experience of developed countries where very large numbers of rural residents migrated to cities as industrialization and economic diversification took place. While it is undeniable that there is a shortage of skilled workers in many African countries, there is overwhelming evidence to suggest that this proposition is not pertinent to Africa. There is already a surplus population in most large cities; there is large-scale unemployment and underemployment in both rural and urban areas; and migration to urban centers appears to be growing with little relation to their absorptive capacities.

It is sometimes argued that the low population of certain countries has inhibited their economic growth and development by preventing the exploitation of known resources. The point has been made with respect to such countries as Liberia, Gabon, and the Central African Republic. While the supply of labor may have been a problem in Liberia and Gabon in the early years it is no longer so serious and could readily be eliminated by the importation of foreign labor. And the problems of the Central African Republic appear to be more related to increasing productivity than increasing the total population, which would result primarily in enlarging the number of subsistence farmers. In any case, there is no law which requires that all visible resources be exploited as rapidly as possible. Even in Gabon there appears to have been an excessive rather than an inadequate exploitation of forest resources.

There are without question certain problems associated with a low population and with a dispersed population. These include: the difficulty of providing and maintaining an adequate infrastructure, particularly transport; the problems of providing educational, medical, and other social facilities, which are characteristically more costly on a per capita basis in sparsely populated regions;[29] the small base for support of an adequate central governmental structure; and the small size and lack of integration of the internal market.

Certain steps can be taken partially to offset these problems. Population can be agglomerated or aligned along key transport routes, which is such a marked feature of the population distribution of Gabon and, to a lesser extent, of the Central African Republic. Empty areas should not be considered as a vacuum to be abhorred, but rather as a reserve for eventual occupation if that proves economically desirable. Maintaining an adequate national bureaucracy can sometimes be aided by cooperation with adjacent countries in common services, such as geological surveys, power authorities, and research institutes. Some governments could also wisely cut down certain expenditures such as the number of foreign missions, the erection of excessively costly buildings, or military establishments out of proportion to needs and capabilities.

In any case, the problems associated with low population are frequently far less onerous and more readily solvable than those of countries with large numbers and population pressure. When labor is plentiful and cheap, the temptation is to use it poorly. When labor is scarce, more attention can be paid to the productivity of the individual,

[29] See, for example, Gilles Sautter, *De l'Atlantique au Fleuve Congo: une Géographie de Sous-peuplement: République du Congo, République Gabonaise* (Paris, Mouton, 1966) 2 vols.

Population Pressure in Africa

labor-saving equipment can more readily be introduced, and there is a greater possibility of achieving a standard well above the peasant level.

Evidences of Population Pressure

When one begins to examine the position of specific regions within Africa, rather than focusing upon continental densities, it soon becomes apparent that the problem of population pressure is a considerably more serious one than most observers have thus far admitted. In this section a number of examples will be given, sometimes for countries which are seen to have a pervasive problem, sometimes for specific regions within countries, for which quotations will be used from persons who have studied the regions closely. In the succeeding sections two summaries are presented, one a listing of the indicators of population pressure, the other a tentative depiction of pressure areas in Africa as of mid-1967.

NORTHERN AFRICA

The Maghreb. The three countries of the Maghreb—Morocco, Algeria, and Tunisia—all face serious problems of population pressure, marked by serious unemployment, large-scale underemployment, over-use of extensive rural areas, low rural standards of living, excessive migration to the cities, etc. The World Bank study of Morocco concluded that "with the population increase now in prospect, it will be very difficult to maintain economic standards"[30] and recommended that Tunisia, because of the inability of the land to support the population, plan to develop itself as an exporter of labor-intensive manufactured products, a difficult prescription for the country at its present stage of development and human achievement levels.

The situation in Algeria, however, shows all too dramtically how the pressure of population can compound the economic, social, and political problems of a country. This pressure has long existed, but under the French it went largely unrecorded and unreported for many years. It is evidenced by staggering unemployment, estimated to have increased from 800,000 in 1961 to 2 million in 1964 and about 2.7 million in mid-1968, at which time 50 percent of the working-age men and 80 percent of the women were unemployed. Only about 30 percent of the

[30] International Bank for Reconstruction and Development, *The Economic Development of Morocco* (Baltimore, The Johns Hopkins Press, 1966), p. 4.

working force was employed more than 50 days a year in mid-1968, according to *El Moudhajid*, the newspaper of the National Liberation Front. Employment of male migrant workers in France provides succor for many Algerian families. In mid-1964 no less than one-third of all wages paid to Algerians was earned by those working in France, but this source of employment and earnings appears to be facing increasing restrictions in the years ahead. In the meantime the extreme youthfulness of the population, with 56 percent of the total at the 1966 census being under 19 years of age, means that about 200,000 people are entering the job market annually, a figure which will double in another generation.

In the rural sector, thousands are landless and collective farms set up by the government, often on former European estates, are grossly overstaffed and lacking in qualified managers. The country is not now capable of feeding itself. In late 1967, the Education Minister said that three-fourths of the school children were affected by undernourishment. The United Nations, the United States, and others have provided substantial quantities of foodstuffs in recent years which have partially alleviated the shortages.

Many industries were shut down with the large-scale exodus of Europeans preceding independence and, while some new plants have been opened or are under construction, there has been little growth in the manufacturing sector. Thousands of landless peasants have flocked to the cities (an estimated 800,000 between 1960 and 1964 alone[31]), where jobs are not available and many occupy quarters which they neither are able to pay rent for nor to maintain. And the government bureaucracy and military forces are vastly out of proportion to the real needs of the country.

While Algerian economic problems are related in part to the traumatic events associated with the long Franco-Algerian war and the succeeding exchange of power, the surplus of people in all major sectors makes adjustment vastly more difficult, in turn making it politically impracticable to adopt measures required to rationalize the country's economy and to lessen the plight of millions of its people. Thus we see a truly frightening dilemma, paralleled perhaps only in Egypt among African countries, nascent in a few others (Morocco, Tunisia, Lesotho,

[31] William H. Lewis, "Algeria Changes Course," *Africa Report*, X, No. 10, (November, 1965), 14-15.

Mauritius, and Réunion), but possibly a harbinger of what could develop in many countries. It is interesting if somewhat precarious to speculate regarding the relation between population pressure and internal political unrest on the one hand and external adventurism on the other. Certainly conditions in Algeria are not conducive to stability, since they present almost an impossible task to the government. The intense pressure on the land in Egypt is also likely to encourage the government's assuming postures calculated to take the minds of the people off their immediate problems. The Civil War in Nigeria must also, in part, be related to the pressure of population in Iboland which stimulated large-scale migration to other regions, whose people came to resent their presence and their significance in various employment categories. Communal revolts in Mauritius and in several African cities with severe employment problems may also be connected in some measure to population pressure. The Mau Mau revolt in Kenya was undoubtedly a reflection of land problems among the Kikuyu, as are continuing intertribal frictions in that country and elsewhere.

Irene Taeuber writes with respect to the relations between population and political instability as follows: "Instability, subversion, and revolution are associated with frustrated advance. They also serve to retard the advance that is desired. . . . If population growth is a factor in economic failure, it is also a factor in the resulting political instabilities."[32]

Libya. Returning to northern Africa, the case of Libya is an interesting one, for on the one hand the oil boom has created a number of direct opportunities and a very much greater indirect stimulus to the economy, evidenced particularly in the burgeoning growth of Tripoli, to which numbers of foreign Africans have been attracted by the available employment. Yet many if not most of the Libyan population continue to live at the same level as before, plagued by the shortage of water, usable farmland, and adequate pastures.

The U.A.R. (Egypt). The population pressure in Egypt is striking. The average density of the occupied areas is about 2,060 per square mile. Before the High Dam was started there was about 0.20 acres of arable land per capita; upon completion of its associated irrigation works and the switching from basin-irrigation to multiple cropping there will be

[32] Irene B. Taeuber, "Population Growth in Underdeveloped Areas," in Hauser, ed. *The Population Dilemma*, p. 44.

about the same acreage of land per capita, the growth in population having matched the extension and intensification permitted by the dam. Implementation of the Jonglei Scheme in the Sudan would add perhaps 7 percent to the water available to Egypt; it too, therefore, would provide only stopgap relief. Nor will it be easy to provide alternate employment opportunities, while Egyptians are not prone to migrate even if there were countries prepared to accept them in large numbers.

In the meantime, there is evidence of extreme poverty, low dietal standards, frightfully unsanitary living conditions in both rural and urban settlements, high disease incidence, and excessive unemployment levels. It is difficult to see how any solutions other than the application of desalinized sea water to the desert areas, not now economically feasible, or the conscious restriction of population growth can do anything but temporarily alleviate the population pressure of Egypt.

Sudan. The lands along the Nile in Sudan north of Khartoum share many of the characteristics of the floodplain in Upper Egypt as far as man–land relations are concerned. Most of the inhabitants of this belt, possibly about a million, live on the verge of poverty, if not starvation. Land fractionation has a stranglehold on agricultural advancement

Typical village in Egypt. The average density of the occupied parts of this country is about 2,060 per square mile and evidences of extreme poverty are obvious in rural and urban areas.

Population Pressure in Africa

Grazing lands south of Khartoum, Sudan. Much of such land in the dryer parts of the country suffers from overgrazing.

and rural indebtedness is almost universal; disease rates are unusually high.

Much of the grazing lands in the dryer parts of Sudan also probably suffer from pressure, but available data do not permit delineating the areas or the populations involved with precision. Cunnison, writing about the Baggara, notes that "the official view is that the land is already overstocked" and that the "Humr themselves can see growing difficulties."[33]

The Sahara, shared by all of the countries of northern Africa, also suffers from population pressure. Most oasis communities have abysmally low standards of living and are a source of continuing migration to the coastal cities. With respect to the lands used for nomadic grazing a UNESCO study indicates that

[33] Cunnison, *Baggara Arabs: Power and the Lineage in a Sudanese Nomad Tribe,* p. 41.

Population Pressure in Africa 397

the demographic increase of the nomads is a recent development, much more apparent to the north than to the south of the Sahara. It can only upset the economic equilibrium of the nomadic graziers, for pasturage cannot be further extended: a surplus population, incapable of living on grazing, has been evident for more than twenty years in the northern Sahara. . . . The same transformations are certainly evolving in the southern Sahara.[34]

WESTERN AFRICA

Those countries of West Africa which contain large segments of the Sahara—Mauritania, Mali, and Niger—have comparable if somewhat less severe problems to those experienced by the Saharan portions of northern Africa. Parts of Senegal suffer from soil exhaustion, which has forced a shift in the peanut production zone. Upper Volta, particularly the Mossi country, experiences considerable demographic pressure; much of the country is characterized by poor and fragile soils which have been badly eroded and it "can not properly support its present population."[35] The north of the country has too many animals, as do most of the tsetse-free grazing lands of West Africa. Probably the whole of Niger, which Church describes as "a huge arid or semiarid, landlocked country, with so far no significant products to offer or other assets to develop,"[36] can be considered as being under pressure.

Sierra Leone, which has the fifth highest crude density for continental African countries has, according to Jarrett,

acute problems arising from the niggardly returns to be won from tropical soils, especially when they are overused. In many parts . . . a fallow of three years is all that can now be allowed, instead of the seven or ten years necessary for soil regeneration. Needless to say, soils in these areas are deteriorating, and the problem admits of no easy solution.[37]

The only part of the Ivory Coast which can be distinguished as experiencing population pressure is the Korhogo Cercle whose arable surface is totally exploited and apparently deteriorating.[38]

[34] UNESCO, *Nomades et Nomadisme au Sahara*, p. 100.

[35] Frank Lorimer, "The Population of Africa," in Freedman, ed., *Population: The Vital Revolution*, p. 211.

[36] R. J. Harrison Church, "The Niger Republic," *Focus*, XVI, No. 1 (September, 1965), 6.

[37] H. Reginald Jarrett, "Sierra Leone," *Focus*, VIII, No. 4 December, 1957), 6.

[38] "La Région de Korhogo en Côte d'Ivoire," *Industries et Travaux d'Outremer*, No. 161 (April, 1967), pp. 299-300.

Ghana. Considerable parts of northern Ghana show distinctive signs of population pressure. Zones of high density often abut empty areas; the latter are not reserves waiting to be opened up but rather depopulated areas from which the inhabitants have moved, most importantly because of widespread soil exhaustion followed by soil erosion. Hilton writes about various parts of the region as follows:

In North Mamprusi, owing to pressure of population . . . permanent cultivation is the rule. Compound farms are cropped almost continuously with the aid of household waste and animal droppings and subsidiary bush farms have in course of time fragmented and been built upon by overflow from the original compounds. Land shortage and concomitant problems are most acute in Frafra. . . . Animal population, as well as human population, presses on the land.[39]

In areas of concentrated pressure on the land [in the Upper Region], primitive farming methods and annual grass burning have led to widespread soil exhaustion, followed by soil erosion.

There is serious overcrowding in the Nandom, Jirapa and Lawra areas where, in parts of all these districts, topsoil has virtually disappeared.

The plight of the Kassenas in the Sissili area . . . could hardly be worse.

Tumu . . . may exemplify the last stage, when the exhausted land is no longer able to support any standard of living for its dense population, and the bush comes in once more. Twenty-five years ago Lynn pointed out that evidence available suggested that, once land has become completely exhausted, it takes many decades to recover. . . . soil exhaustion and erosion . . . , inevitable in the areas of dense population where old traditional agricultural methods prevail, [mean] that each generation faces a worse problem than the previous one.[40]

Hunter, writing about Nangodi in northeast Ghana, cites many of the characteristics noted by Hilton. He states that Nangodi is typical of a northeast Ghanaian area of 3,400 square miles with a total population of 469,000.

It is clear from field observations that, under the present system of agriculture and settlement, the land is overcrowded and overworked. There is insufficient land for adequate bush fallow, and many areas have already been stripped of their topsoil. Sheet-wash erosion is common, and gullying is not infrequent.

[39] T. E. Hilton, "Population Growth and Distribution in the Upper Region of Ghana," in Caldwell and Okonjo, eds., *The Population of Tropical Africa*, p. 279.
[40] *Ibid.*, pp. 281, 289-90.

Nutritional levels are at best precarious, and before each harvest there is an estimated forty-sixty percent deficiency in calorie intake.[41]

Togo and Dahomey. French and Belgian observers appear much more reluctant than others to accept and report on population pressure. Nonetheless it is clear that considerable parts of former French West Africa experience pressure on the land. In addition to the areas already noted, the southern parts of Togo and Dahomey are obviously grossly afflicted, while parts of the north of both countries also suffer from excessive pressure. The Samba in northern Dahomey, for example, are said to experience chronic under-nutrition; they live at densities four to seven times those of other northern regions, and despite undoubted pressure are reluctant to leave their home area.[42]

Nigeria. Nigeria has large areas with population densities well above the critical levels given by Allan for various major ecological types, but it is not clear that all of them are in fact suffering from population pressure. The evidence appears clearest for Iboland, Tivland, and the area around Sokoto, less clear for the close-settled zone around Kano and the densely populated parts of Yorubaland. With respect to the Eastern Region, Floyd writes that

within the Region there occur some of the most spectacular examples of soil erosion and 'badland' topography to be seen in West Africa. . . . Less pronounced though equally insidious sheet and gully erosion is widespread. . . . Soil deterioration and degradation . . . is well-nigh universal, due largely to overfarming and primitive, destructive methods of cultivation.[43]

And focusing upon the Nsukka Division he states that

with the encroachment of settlements and compounds upon the former outer farm land, smaller acreages of communally held land were available for the extensive methods of slash-and-burn, bush fallow farming. The period of rest and recuperation of soils also had to be shortened, with deleterious effects upon soil fertility, and a lowering of crop outputs per acre. In places, the soils became so impoverished as to be quite unsuited for farming, and they had to be left uncultivated for many years, fit only to produce thatching grass. In

[41] John M. Hunter, "Population Pressure in a Part of the West African Savanna: A Study of Nangodi, Northeast Ghana," *Annals of the Association of American Geographers,* LVII, No. 1 (March, 1967), p. 105.

[42] République du Dahomey, *Données de Base sur la Situation Démographique au Dahomey en 1961* (Paris, Ministère de la Cooperation/I.N.S.E.E., 1962), p. 85.

[43] Barry N. Floyd, "Soil Erosion and Deterioration in Eastern Nigeria," *Journal of the Geographical Association of Nigeria,* III, No. 1 (June, 1965), 33.

consequence, outer farm-land agriculture became increasingly unproductive, if not hazardous. In place of the much-prized yams, the more hardy though less respected cassava plant, tolerant of infertile sandy soils, became widespread, and indeed is now found ubiquitously in the Nsukka Division, even in communities such as Aku which formerly had a food taboo against the consumption of *gari*.[44]

Mabogunje notes many of the same signs of pressure for the densely settled parts of Iboland plus extreme fragmentation of holdings, a breakdown of both familial and communal control of the land, and frequent land disputes sometimes leading to fatal incidents.[45]

Writing about Tivland in the Middle Belt of Nigeria, Vermeer hypothesizes that when subsistence shifting agriculturalists willingly deviate from accepted, proven cultivation practices it is evidence that the system is being subject to extreme stress. Referring to the southern, densely populated areas of Tivland he states

rationalizations by the peasants themselves point toward the intimate relationships between population pressure and the quality of the environment under shifting agriculture: too many people on the land, diminishing soil fertility and crop yield, and a shift to rapidly maturing crops yielding well on degrading soils.[46]

In the densely settled nodes of northern Nigeria there is evidence that fertility of the land is reasonably well maintained in the vicinity of cities such as Kano, Katsina, and Zaria by heavy application of human and animal fertilizer from the city.[47] But,

[44] Barry N. Floyd, "Rural Land Use in Nsukka Division," in Department of Geography, University of Nigeria, *Nsukka Division: A Geographic Appraisal*, 1965 (mimeographed), p. 54.

[45] Akin L. Mabogunje, "A Typology of Population Pressure on Resources in West Africa," a paper presented to the Symposium on the Geography of Population Pressure on Physical and Social Resources, the Pennsylvania State University, September, 1967 (mimeographed), pp. 20-21.

[46] Donald E. Vermeer, "Population Pressure and Crop Rotational Changes among the Tiv of Nigeria," paper presented to the Annual Meeting of the African Studies Association, November, 1968.

[47] See M. J. Mortimore and J. Wilson, *Land and People in the Kano Close-settled Zone* (Ahmadu Bello University, Department of Geography, Occasional Paper No. 1, 1965); M. J. Mortimore, "Population Distribution, Settlement and Soils in Kano Province, Northern Nigeria 1931-62," in Caldwell and Okonjo, eds., *The Population of Tropical Africa*, pp. 298-306; and A. T. Grove, "Population Densities and Agriculture in Northern Nigeria," in Barbour and Prothero, eds., *Essays on African Population*, pp. 115-36.

the outlook for the more remote rural districts with high population densities appears to be less promising. A high proportion of the land in such areas is under cultivation every year, but supplies of manure are not adequate to maintain fertility. The cattle population has fallen considerably . . . and the condition of the soils is probably deteriorating. Communal grazing land has diminished with the extension of cropland and over large areas impoverishment and erosion are very serious.[48]

Prothero writes regarding Sokoto Province in Northern Nigeria that

the population is increasing in a physical environment which offers only marginal possibilities for development. . . . The fertility of the soil is rapidly depleted if unsatisfactory farming methods are used. As a result, population/ land relationships, particularly in northern Sokoto, are beset with many problems of overpopulation and land hunger. Movements of a considerable proportion of the population . . . are a manifestation of these problems.[49]

MIDDLE AFRICA

In former French Equatorial Africa and Congo (Kinshasa) much more concern has been expressed over the low population densities and what was considered a demographic crisis than about pressure on the land. Ziéglé suggests, however, that the fact that where people did live densities were not greatly different from other parts of tropical Africa should make one skeptical of the theses that the French [or the Belgians] found the region in a state of demographic crisis.[50] These theses were founded on early and entirely hypothetical estimations of population totals. Helping to explain the sometimes gross overestimations is the fact that early travelers were likely to move along rivers with many villages or to trek from village to village, bypassing the extensive empty or very sparsely populated regions.

There are regions in Middle Africa, however, which are generally conceded to suffer from population pressure. One of these is the Bamiléké country in the highlands of southwestern Cameroon, which experiences an average density of about 275 per square mile but whose densest regions have up to 860 per square mile. Despite favored ecological conditions the region is overgrazed, overcropped, and subject to serious

48 Grove, "Population Densities and Agriculture in Northern Nigeria," pp. 135-36.
49 Prothero, *Migrants and Malaria*, p. 21.
50 Henri Ziéglé, *Afrique Equatoriale Française* (Paris, Éditions Berger-Levrault, 1952), p. 52.

Clearing in a rain forest area of Cameroon. The supporting capacity of
the latosolic soils of such regions is low because of the necessity for
long fallow periods when the fields are permitted to revert to brush.

erosion; heavy emigration of Bamiléké has provided only partial relief.
Other regions of population pressure include the Mandara Mountains in
northern Cameroon, much of the desert and steppe lands of Chad, and
some of the densely populated parts of the highlands of eastern Congo.
Many of the major cities of the area have also attracted far too many
people in relation to the opportunities they provide. This is particularly
true of Brazzaville and Kinshasa.

EASTERN AFRICA

This very large and diverse region has examples of every type of
population pressure. Some cases are well documented, others are very
inadequately covered.

The Horn. The two main countries in the Horn, it will be recalled,
have not yet had a census. It is pretty obvious, however, that most of
the nomads and seminomads in Somalia and in the low-lying parts of

Ethiopia live at an incredibly low standard, suffering from a multitude of restrictive physical conditions and further inhibited by cultural attitudes and practices. About 40 percent of the area of Somalia is practically useless; the remainder is subject to periodic severe droughts. Overgrazing contributes to soil erosion, which is prevalent and increasing throughout the country. Stock and domestic water supply is often inadequate, sometimes leading to desperate conflicts over wells and water rights. Disease and insect pests are additional limiting factors.

It is not possible to assess the extent of population pressure on the highlands of Ethiopia, though evidence exists that it affects some regions, including gully erosion in several parts, sheet erosion in sections of Eritrea, which continue to support very dense populations, and farming on excessively steep slopes in a number of areas. Geological erosion reaches fantastic proportions in many parts of the country.

East Africa. The very large steppe areas of Kenya have many of the characteristics of Somalia; their low carrying capacities mean that any increase in livestock or human populations are likely to threaten

Hamlet in Eritrea west of Asmara. Evidences of pressure on the land are common in this part of Ethiopia.

Population Pressure in Africa

whatever precarious equilibrium exists. Morgan writes regarding the Masai country near Nairobi, a region which is somewhat superior to many of Kenya's steppe lands, that

although the pastoral areas have densities as little as one hundredth of the agricultural districts, the position [with respect to population pressure] may be equally serious. If the Masai way of life requires six cattle per person and in the drier areas ten acres are needed to sustain one beast, then a density of more than ten or eleven persons per square mile represents over-population.[51]

Panning water from the subsand section of the Bubu River in Tanzania. Most arid and semiarid grazing areas of East Africa are overgrazed.

Misused farming land on the flanks of the Aberdare Mountains in Kenya. Extensive parts of the highlands of Kenya, Uganda, and Tanzania suffer from excessive pressure on the land.

Extensive parts of the Highlands of Kenya also suffer from excessive pressure on the land. The Kiambu and Nyeri districts of Kikuyuland, for example, have extremely high densities; 30–40 percent of the adult males in one location of Nyeri are landless and in 1964 it was calculated that about 45 percent of all holdings in Nyeri District were smaller

[51] W. T. W. Morgan, "Agricultural Land Use," in Morgan, ed., *Nairobi: City and Region* (Nairobi, Oxford University Press, 1967), p. 87.

than three acres, the size generally considered to be adequate only for subsistence farming. In the Taita Hills

population growth has created one of the most congested parts of Kenya. By the 1950's pressure of population on cultivable land had reached critical proportions and considerable areas of steep hillside were being cleared for cultivation with disastrous results. Fragmentation has been one of the major problems of land use in the area.[52]

The Machakos District is another highland area suffering excessive pressure. While its population densities are much lower than in parts of Kikuyuland

the natural endowment and economic opportunities are so much less . . . that the real pressure on the land is at least as great. Indeed, the need for famine relief has been a distinctive feature of the Machakos economy for many years.[53]

Another report on the same district states that it is

characterized by periodic droughts and food shortages, by serious soil erosion, by overstocking and overgrazing, by extensive methods of agriculture which appear incompatible with a comparatively severe pressure of population on available land.[54]

The Nyanza Province of Kenya, a portion of the Lake Victoria basin, is another section of the country which has, in substantial parts, serious problems related to pressure on the land. Allan notes that the old farming systems have long since broken down as a result of population growth and cultivation of cotton.[55] He calculates that critical densities are about 100 to 120 per square mile for the district; most of it has densities two or three times that level. Ominde also calls attention to the pressure on land resources in central and north Nyanza which are "faced with the problems of diminishing space and falling yields of the basic cereal crops."[56]

Uganda, while having a greater percentage of its total area in

[52] Simeon H. Ominde, *Land and Population Movements in Kenya*, p. 49.
[53] W. T. W. Morgan, "Agricultural Land Use," p. 87.
[54] John C. de Wilde *et al.*, *Experiences with Agricultural Development in Tropical Africa*, II, 84.
[55] Allan, *The African Husbandman*, p. 191.
[56] Ominde, *Land and Population Movements in Kenya*, p. 34.

high-potential lands and considerable stretches which could undoubtedly support greater densities than they now have, has a number of subregions which have severe population pressure. Land pressure was noted with concern in Kigezi as early as the mid-thirties.[57] Turyagyenda, studying the subdistrict of Buhara, which had an average of 0.75 acres per family as compared to one of 1.2 acres per family for the whole Kigezi district, and some landless families, noted that, despite having 78 percent of the total under cultivation and despite such intensive practices as contour cultivation and swamp reclamation, the region was marked by a decline in food supplies and dietal quality, deterioration of soil fertility and erosion, low yields, a scarcity of firewood, and plots too small to permit efficient management.[58]

Other parts of Uganda which have excessive pressure on the land include sections of West Nile and Teso Provinces and Karomoja, the last being ecologically comparable to the adjacent parts of Kenya.

The pattern in Tanzania is somewhat comparable to that of Kenya. Occupied steppe lands are frequently overused and subject to periodic droughts and famine; some 400,000 persons live in the areas "where crop failures have been most frequent . . . Dodoma, Mpwapwa and Manyoni."[59] Only presence of the tsetse fly has kept some areas from being included in the list. Sukumaland, in the Lake Victoria region, is heavily overstocked and its lands are degrading; Bukoba District, on the western side of the Lake, now has "a formidable problem" of pressure on the land;[60] and conditions on Ukara Island, noted for the sophistication of its land-use practices, are deteriorating.

Many of the highlands of Tanganyika, which sustain a disproportionate share of the population, as they do in Kenya, display characteristic signs of pressure. Referring to the highland part of Kondoa District, Moffett wrote that

these hills were once covered with forest and had a great depth of soil, but

[57] Rachel Yeld, "Land Hunger in Kigezi, South-West Uganda," *Nkanga*, No. 3, (1968), p. 24.

[58] J. D. Turyagyenda, "Overpopulation and Its Effects in the Gombolola of Buhara, Kigezi," *Uganda Journal*, XXVIII, No. 2 (September, 1964), 127-33.

[59] Clarke Brooke, "The Heritage of Famine in Central Tanzania," *The Journal of the Tanzania Society*, No. 67 (June, 1967), p. 15.

[60] D. N. McMaster, "Change of Regional Balance in the Bukoba District of Tanganyika," *The Geographical Review*, L, No. 1 (January, 1960), 87.

when the forests were felled and the earth laid bare the very depth of the friable soil made it the more easily eroded and now nothing remains in many places but the rocky sub-soil, split by innumerable channels. . . . In places these channels are as much as fifty feet deep—veritable canyons—making reclamation of the land exceedingly difficult if not virtually impossible.[61]

The mountain areas of the Eastern, Northern, Tanga, and Southern Highland Provinces are similarly frequently characterized by dense cultivation on excessively steep slopes, denudation of forests, sheet and gully erosion, and increasing difficulties of supporting the existing population.[62]

Rwanda and Burundi have long been listed with the "few" parts of Africa having population pressure. Their crude densities are the highest of continental African countries. An annual report written in 1950 stated that

Ruanda-Urundi has always been a country of famine and shortage. The country as a whole is poor. The irregularity of the rainfall is a serious disadvantage. A too dense and ever-increasing population has to live on very rugged land, which erosion, either agricultural or geological, is impoverishing year by year.[63]

Harroy believes that the most appropriate solution to the population problems of Rwanda and Burundi is an internationally financed and supervised large-scale displacement of peoples to adjacent countries, though it is not clear where they will be accepted and whether such a mass emigration would be more than a temporary palliative.[64]

Central Africa. Congestion on the land affects large portions of the African reserves in Rhodesia. Prescott estimated a decade ago that seriously overstocked and overpopulated areas of Matabeleland included 20 percent of the total area containing 39 percent of the population and 32 percent of stock in the reserves. Additional large areas were considered to be lightly overpopulated, including 34 percent of Matabeleland.[65] Barry Floyd has detailed the problems affecting the Rhodesian

---

[61] John P. Moffett, ed., *Handbook of Tanganyika*, 2d ed., (Dar es Salaam, Government Printer, 1958), p. 154.

[62] *Ibid., passim.*

[63] As quoted in U.N., Department of Social Affairs, *The Population of Ruanda-Urundi*, p. 1.

[64] J.-P. Harroy, "Surpopulations en Afrique Centrale," *Bulletin des Sciences d'Outre-Mer*, VIII, No. 4 (1962), 524-30.

[65] J. R. V. Prescott, "Overpopulation and Overstocking in the Native Areas of Matabeleland," *The Geographical Journal*, CXXVII, No. 2 (June, 1961), 212-25.

reserves,[66] including pressure on the land. Phillips calculated that 70.0 percent of the African areas in 1961 were suitable only for extensive or semiextensive land use while only 10.6 percent could be used for intensive farming;[67] there have been some additions since then to the lands apportioned to Africans but not enough to allow for the increased needs. Kay believes that the African areas could not support the African population of the country and notes that "by 1950 resettlement of Africans . . . was made very difficult by excessive population pressure in most of the Native Reserves."[68] And Hamilton, focusing upon Chiweshe Native Reserve, 60 miles north of Salisbury, notes that it was carrying twice as many cultivators as the optimum in 1957 and that "overpopulation has created a situation in which . . . the limited advances that have been made . . . have failed to compensate for the overall failure to maintain soil fertility."[69]

Zambia is less plagued by pressure on the land than most African countries but several "islands" where it occurs can be delineated. Kay says that the "crowded areas fall into two groups: those associated with the important fishing grounds of the lower Luapula, Lake Mweru and the Bangweulu lakes and swamps, and those associated with important local settlements."[70] In the Bangweulu basin "pressure on favourable building sites is intense and villages are crowded one onto another, and local agricultural resources are frequently inadequate for the needs of the people in spite of intensive systems of cultivation based on cassava."[71] The Barotse floodplain is also not now providing the food needs of the population dependent upon it. The nodes of high density associated with the Ngoni in the Eastern Province have seen a breakdown in the mound system of cultivation due to "acute land shortage."[72] Kay

[66] See Barry N. Floyd, *Changing Patterns of African Land Use in Southern Rhodesia* (3 vols. Livingstone, Rhodes-Livingstone Institute, 1962); Floyd, "Land Apportionment in Southern Rhodesia," *The Geographical Review*, LII, No. 4 (October, 1962), 566-82.

[67] J. Phillips *et al.*, *Report of the Advisory Committee on the Development of the Economic Resources of Southern Rhodesia with Particular Reference to the Role of African Agriculture* (Salisbury, Ministry of Native Affairs, 1962).

[68] Kay, *The Distribution of African Population in Southern Rhodesia: Some Preliminary Notes*, p. 5.

[69] P. Hamilton, "The Changing Pattern of African Land Use in Rhodesia," in Whittow and Wood, eds., *Essays in Geography for Austin Miller*, p. 268.

[70] Kay, *A Social Geography of Zambia*, p. 49.

[71] *Ibid.*, p. 67.

[72] Allan, *The African Husbandman*, p. 99.

notes in general that "because of the very primitive nature of many African land-use systems and because of the mediocre potential of large parts of Zambia, the carrying capacity or critical population density with reference to soil conservation is very low in many cases."[73] And he opines that the critical population density "is greatly exceeded in parts of Zambia . . . . the partially devastated areas in the Native Reserves [bearing] witness to the impact of excessive population pressure on physical resources."[74]

Much of Malawi may be considered to suffer from pressure on the land, including not only the very densely populated highlands in the south but moderately populated sections of the north, where unfertile and rocky soils result in a low carrying capacity. Malawi's crude population density is third highest among continental countries.

Madagascar. The low crude density of the Malagasy Republic has led many observers to conclude that its population is too low. Thompson and Adloff, for example, write that "Madagascar is like a world apart, whose outstanding characteristics are its vast size, its isolation, the variety of its soils and climates, and its underpopulation."[75] Yet it is apparent that there is a shortage of land in the central highlands suitable for paddy rice, the favored food staple, that several of the densely populated east-coast nodes are experiencing pressure, that the southwest has periodic crop failures and famine conditions, that government planners are facing serious difficulties in some sections due to excessive numbers on the land, and that denudation of vast stretches of slope areas in the highlands by overgrazing and other malpractices has led to serious deterioration and to some of the worst erosion to be seen anywhere in Africa.

The Smaller Islands of Eastern Africa. Practically all of the island appurtenances of Africa have high to very high population densities and all of the island groups of the eastern African region—the Comoros, the Seychelles, and the Mascarenes—have serious problems of population pressure. The Seychelles and Réunion, one of the two main islands of

[73] George Kay, "Population Pressure on Resources in Zambia," paper presented to the Symposium on the Geography of Population Pressure on Physical and Social Resources, Pensylvania State University (September, 1967), p. 8 (mimeographed).
[74] *Ibid.*, pp. 6-8.
[75] Virginia Thompson and Richard Adloff, *The Malagasy Republic: Madagascar Today* (Stanford, Stanford University Press, 1965), p. 256.

Population Pressure in Africa

Extensive gullying in the highlands near Lac Alaotra on Madagascar, which has one of the worst erosion problems to be seen anywhere in Africa.

the Mascarenes, were treated in some detail in Chapter 2; attention here will be focused on Mauritius, the other main island of this group.

"Mauritius has the unenviable claim to fame of being one of the most densely populated agricultural areas in the world."[76] Uninhabited until the Dutch landed in 1598, its first permanent settlement dates from 1721. In mid-1967 it had a population of 774,000 on its 720 square miles, giving a mean density of 1,075 per square mile. Sugar is produced on about 90 percent of the cultivated land which comprises 50 percent of the island's area, which means that there is less than 0.3 acres of

[76] Great Britain, Colonial Office, *Report on Mauritius* (Port Louis, Government Printer, 1966), p. 15.

arable land per capita. The productive area is reduced by low rainfall in the coastal belt, by the rugged remnants of a former great volcano, by boulder-strewn fields which have not yet been cleared, and by urban and other settlements.

The population of Mauritius has grown from an estimated 60,000 in 1797 to 183,000 in 1851, 370,000 in 1900, 420,000 in 1944 and the present 774,000. Despite some reduction in the birthrate in recent years the crude rate of increase is still high, about 2.6 percent per annum and the population could easily reach 2 million by the year 2000. Mauritians have emigrated in some numbers, there being an estimated 12,000 in Durban and 30,000 in the whole of South Africa plus several thousand in Madagascar, Rhodesia, and Australia. But the present rate of emigration is only 0.37 percent and acceptance of Mauritians by other countries is likely to decline rather than increase.

The demographic revolution on Mauritius started in the late 1940s and occurred with great rapidity. The island's history to the Second World War was replete with disasters, including outbreaks of cholera and smallpox, cyclones, floods, and fire, while high endemicity of disease kept the deathrate high. The virtual elimination of malaria in 1947 removed what had been the single most important cause of death, contributing mightily to the decline in the deathrate from 23.8 per M in 1948 to 8.5 in 1967. The crude rate of increase grew from about 0.5 percent in 1944 to 2.5 percent in 1952, exceeded 3.0 percent in the early sixties but has since declined to about 2.6 percent.

That the population of the island is polyglot has led to severe racial animosities and fears which are probably exacerbated by the pressure felt by all groups. 67.7 percent of the total population are Indo-Mauritians, with three-quarters of that group being Hindus and one-quarter Muslims, each group being further fragmented into various sects. The so-called General Population, which consists of persons of European descent and of mixed and African descent comprises 29.1 percent of the total, of which about 1.7 percent are French and English; Sino-Mauritians account for the remaining 3.2 percent of the total population. The employment pattern has tended to follow ethnic lines with the Hindus being predominantly farmers, the Muslims traders and industrial workers, the Chinese shopkeepers and traders, and the General Population clerical, commercial, industrial, and professional urban dwellers. Conflict among these groups led to communal riots and bloodshed in the months preceding independence on March 12, 1968, and there appears to be little sense of national feeling cutting across ethnic lines.[77]

Evidence that these conditions have led to excessive pressure include a decline in the real national income per capita in the 1950s,[78] unemployment of about a quarter of the adult males, the imposition of employment requirements on the big sugar companies which has contributed to half of them operating in the red despite the sale of five-sixths of the output at the Commonwealth price of £45 per ton compared to £25 per ton received for the remainder sold on the open market,[79] and the obvious inability of the several intensification plans to meet the needs of the island economy and its inhabitants. The government itself concluded in 1966 that there was "no important outlet in sight for [the] surplus population," and "that effective measures must be taken to regulate births."[80]

SOUTHERN AFRICA

South Africa. The presence of population pressure in the Republic is to a considerable degree artificial, reflecting regulations which restrict Africans to the reserves. Demographically, the reserves are already under excessive pressure, their agricultural resources being entirely inadequate to sustain the present population even under the scientifically most rational land use methods.[81] The reserves are characteristically overworked, overgrazed, and more or less damaged by erosion. Almost none are self-sufficient in food production, several of the larger reserves depending on white farms for at least 25–40 percent of their food supply. In those areas which have been "planned," the government has found it impossible to remove all of the surplus families from the land, which has led to the splitting of units and the perpetuation of uneconomic holdings.

[77] See Burton Benedict, *Mauritius: Problems of a Plural Society* (New York, Praeger, 1965); P. O. Olusanya, "Implications of Population Growth in Nigeria: Some Lessons from the Mauritian Experience," *The Nigerian Journal of Economic and Social Studies*, VIII, No. 2 (July, 1966), 311-29; Philip M. Allen, "Mauritius on the Eve," *Africa Report*, XI, No. 5 (May, 1966), 16-24.

[78] Meade *et al.*, *The Economic and Social Structure of Mauritius*, p. 4.

[79] *The Times* (London), January 23, 1968.

[80] *Report on Mauritius*, p. 15.

[81] This, and the following discussion, is taken from the author's chapter in Hance *et al.*, *Southern Africa and the United States*, pp. 151-56.

Overgrazing in the Mqutu District of Zululand.

Erosion, abandoned wheatfields, and crumbling hills near Hofmeyr, Cape Province.

Bad land-use practices in South Africa have reduced usable land by an estimated 25 percent in the present century.

The opportunities for other employment are also entirely inadequate at the present time. In 1966, for example, only 33,007 Africans were in paid employment in the Transkei, which has received the greatest attention. This number compares with an estimated 258,000 employed outside the Bantustan, including 118,000 long-term if not permanent migrants. There are possibilities for development of mining,

414   *Population Pressure in Africa*

forestry, and other activities in a number of the reserves, but they do not add up to the creation of enough jobs even to approach the need.

The only possible answer to supporting the present population or an expanded population in the reserves would appear to be a large-scale development of manufacturing. But here the government places severe restrictions on the prospect by prohibiting capital investment not approved by the Bantu Investment Corporation or the Xhosa Development Corporation, whose resources are inadequate. The prospects are further inhibited by the shortage of available skills, the lack of resources, and the small size of the reserve markets. At present, the Transkei has only a handful of factories worthy of the name and these employ fewer than 500 Africans.

It is conceivable that the Border Industry program, designed to reduce the number of Africans permanently or semipermanently residing in white areas, could provide a considerable number of jobs for residents in the reserves who would commute daily to factories on the

Gully formation near Maseru, Lesotho, which probably has the most severe erosion problems of any African country.

Population Pressure in Africa

white side of the border. But the accomplishments of the program to date have not been particularly impressive, and if border industries were to stop the flow of Africans to the cities more new industries would have to be established in border locations than in the remainder of the white area. It must be concluded that the reserves will continue to experience pressure, probably increasing in scale, unless significant changes are made in present government policies.

Lesotho, Botswana, and Swaziland. Lesotho is, in many respects, very similar to some of the South African reserves, but the pressure of population on resources is, if anything, even more severe. Smit sees a "continuing shrinkage of the means of existence"[82] and Ward says that it has "virtually no hope of becoming viable or independent of South Africa and foreign aid."[83] The physical limitations of the country were given in Chapter 2 and need not be repeated here.

Botswana also suffers from pressure on the land, primarily because periodic droughts result in severe crop and livestock losses. In the lengthy drought of the 1960s tens of thousands would doubtless have perished had food not been provided by international and private famine relief agencies.

While Swaziland is the richest of the former High Commission Territories, portions of the Swazi rural areas have excessively high densities and are subject to soil deterioration and sometimes serious erosion. There are opportunities within the country to relieve the pressure but unless appropriate measures are taken land degradation may advance rather than diminish.

POPULATION AND FOOD CONSUMPTION

In addition to the evidence of population pressure recorded for specific countries and regions, attention must be given to the problem of food supply in Africa. The F.A.O. estimates that consumption in all parts of Africa is less than 2500 calories a day, the level which it considers necessary to prevent undernourishment; it also calculates that food production per person declined by 4 percent from 1939 to 1965. Others feel that the problem of hunger has been exaggerated, and Colin Clark has neatly exposed the propensity of F.A.O. and others to

[82] P. Smit, *Lesotho: A Geographical Study*, p. 21.
[83] Michael Ward, "Economic Independence for Lesotho," *Journal of Modern African Studies*, V, No. 3 (October, 1967), 368.

do just this.[84] So whether there is widespread undernourishment is open to doubt, but the evidence that there is malnutrition in extensive areas is too overwhelming to deny. To cite just one example, May notes for all of Nigeria before the recent civil war unsatisfactory hemoglobin rates, an incidence of anemia varying from 20 to 80 percent, numerous diagnoses of Kwashiorkor, and frequent nutritional diseases among children, and opines that there exists "a dangerous gap between the capabilities of the soil and people and the program devised for their development."[85] It is difficult to relate either undernourishment or malnourishment to pressure on the land because they exist both in areas subject to pressure and in areas where it does not exist. That there are connections in many cases, however, is suggested by the incidence of hunger in zones of erratic rainfall, by the apparently higher incidence of nutritional diseases in densely populated areas, and by the switch to simpler crop patterns and less nutritious staples in pressure areas. It is safe to conclude that some undernourishment and extensive malnutrition at least complicate the problem of relieving pressure on the land.

The Indicators of Population Pressure

Certain patterns appear again and again in areas suffering from population pressure, permitting the development of a typology of such pressure. Indicators that there *may* be excessive pressure in a given area include the following:
1. Soil deterioration, degradation, or outright destruction.
2. Use of excessively steep slopes and other marginal lands.
3. Declining crop yields.
4. Changing crop emphases, especially to soil-tolerant crops such as manioc.
5. Reduction in the fallow period and lengthening of the cropping period without measures to retain soil fertility.
6. Breakdown of the indigenous farming system.
7. Food shortages, hunger, and malnutrition.
8. Land fragmentation, disputes over land, landlessness.
9. Rural indebtedness.

[84] See Clark, *Population Growth and Land Use.*
[85] Jacques May, *The Ecology of Malnutrition in Middle Africa*, p. 67.

10. Unemployment and underemployment in rural and/or urban areas.
11. Certain types of out-migration.

The word "may" was italicized in the stem introducing this listing to emphasize that not all of the indicators necessarily signify pressure of population. Obviously some of them may reflect other conditions. For example, cultural taboos and practices with respect to food can explain malnutrition. And out-migration can be attributed to a variety of factors besides pressure in the source area. But all of the rural areas where population pressure now exists in Africa are characterized by one or more, usually more, of these evidential factors. Indeed there is a close relation and sometimes sequential relation among many of the indicators. Some of these relations were noted in various cases given in the previous section but the following quote provides a useful immediate example:

It is important to appreciate how an overcrowded area appears. . . . In the first place what was originally an effective rotation of bush fallow can no longer continue because the land is required by someone for cropping before it has been sufficiently rested. Faced with the need to support extra persons by obtaining extra crops from a limited area, most tribes have sought an answer by reducing the length of time during which the land ought to be rested. Such a solution is inevitably made at the expense of the yield. Moreover, a stage is reached when it is no longer possible to reduce the fallow period. There then takes place the final compromise. Extra persons can be accommodated by a reduction in the size of the holding of the individual. Subdivision begins and may go on until the holdings reach a sub-economic size. With increasing population, higher, not lower, yields are needed, yet there is no hope of producing a greater surplus per acre because of the fragmentation and subdivision of "shambas" into tiny, scattered holdings with different crops and fallows mixed up together. . . . The continuous cultivation leads to erosion and the hundreds of little boundaries become storm gullies.[86]

A Tentative Depiction of Pressure Areas

Map 39 has been prepared in an effort to make a preliminary estimate of the portion of Africa *now* affected by pressure of population. It divides the zones believed to be suffering from population pressure into low, moderate, and high-density areas. The map, which represents a

[86] *East Africa Royal Commission 1953-1955 Report*, p. 287.

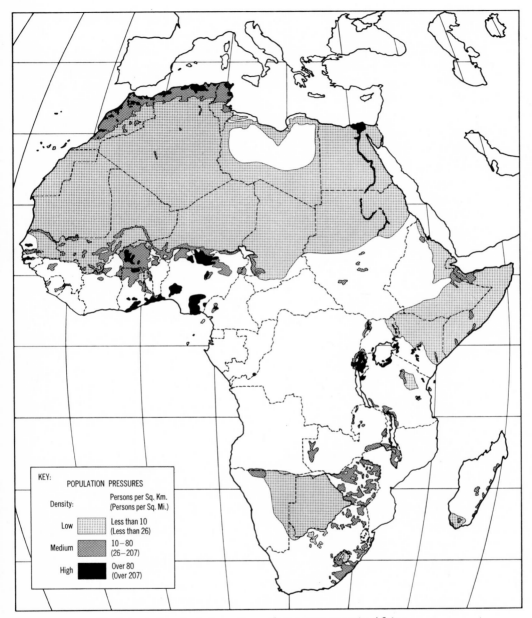

KEY:
POPULATION PRESSURES

Density:

	Persons per Sq. Km. (Persons per Sq. Mi.)
Low	Less than 10 (Less than 26)
Medium	10–80 (26–207)
High	Over 80 (Over 207)

MAP 39. A tentative depiction of population pressure in Africa as of mid-1967.

Population Pressure in Africa   *419*

revised and updated version of one originally presented in 1967,[87] is entitled "a tentative depiction" because there are areas both within and outside those delineated for which there are inadequate data to permit assurance of accuracy. The title of the map also includes the phrase "as of mid-1967" because it attempts to indicate pressure at a given time. If a similar map were prepared ten years from now some areas delineated on this map would probably be eliminated because technological and other advances will have removed the pressure. Other areas would doubtless be added, while the severity of pressure will have increased in many of the delineated zones. I believe that the depiction shown on the map and the statistical summations given below are conservative; indeed, if the general critical population densities estimated by Allan were employed, very much larger regions would be included. There perforce tends to be a bias toward areas which have been studied more closely and where the observer was conscious of the relationships between man and the land; this may have resulted in the exclusion of areas which are experiencing pressure but which have not been reported upon in literature examined by the author.

Some of the major findings include the following:

1. An estimated 47.1 percent of the area of Africa and 50.5 percent of its population was considered as experiencing population pressure in mid-1967 (Table 25). This scarcely fits the oft-repeated generalization that there is no serious pressure in Africa except for a few limited areas. The comparable figures for Africa South of the Sahara were 37.2 percent of the area and 40.2 percent of the population.

a. About 4.4 percent of the total population lived on about two-fifths of the continent considered to be affected by pressure at low densities, that is, under 26 per square mile.

b. Of the total population 16.1 percent lived in the 5.7 percent of the continent estimated to suffer pressure at moderate densities (between 26 and 206 per square mile). The average density in these regions was 79.9 per square mile.

c. Only 0.9 percent of the area of Africa was estimated to experience

[87] William A. Hance, "Population and Poverty in Africa," Presidential Address, Annual Meeting of the African Studies Association, New York, November, 1967, published as "The Race Between Population and Resources," *Africa Report*, XIII, No. 1 (January, 1968), 6-12 (map appears on page 10).

pressure at high densities (207 per square mile and above). But this area contained 56.2 percent of the population of northern Africa and 21.7 percent of that of subSaharan Africa. Thus 30.0 percent of Africa's total population was believed to have suffered from pressure at high densities in mid-1967, the average density being 912 per square mile. The rural component of this category had an average density over 500 per square mile.

2. Of the 52 political entities, 48 were found to have at least some portion of their areas and peoples living under conditions of population pressure. Twenty-three of these countries had more or less pervasive problems; others had only small sections under pressure.

3. A total of 164 million people were estimated to be inhabiting

Table 25

ESTIMATED PERCENTAGES OF AREAS AND POPULATIONS OF NORTHERN AFRICA, SUB-SAHARAN AFRICA, AND THE CONTINENT EXPERIENCING PRESSURE OF POPULATION AS OF MID-1967

	Northern Africa	Sub-Saharan Africa	Africa
Percent of area believed subject to pressure			
At low densities	64.6	30.7	40.5
At moderate densities	5.8	5.7	5.7
At high densities	1.2	0.8	0.9
Total	71.6	37.2	47.1
Percent of population believed subject to pressure			
At low densities	4.8	4.2	4.4
At moderate densities	21.6	14.3	16.1
At high densities	56.2	21.7	30.0
Total	82.6	40.2	50.5

areas experiencing population pressure. Of this total about 60 percent lived in high-density areas, 32 percent in areas of moderate density, and the remaining 8 percent in areas of low density.

These figures may appear not to be so serious when compared to parts of Asia, as is implied in frequent generalized comparisons of population distributions, densities, and pressures in Asia and Africa, but they are sufficiently impressive to suggest that there should be a much greater awareness of the problems associated with pressure than seems to be the case at the present time. Furthermore, it would be senseless to wait until demographic conditions in Africa parallel those of India or China today before concern is expressed or appropriate action is taken.

Opportunities for Relieving Population Pressure

Pressure on the land can be reduced in three major ways: by extension of agriculture to lands not now being used, by intensifying and rationalizing agriculture on present areas, and by absorbing an adequate number of people in non-agricultural activities. To assess the potentialities in each of these categories detailed studies of each country and each agricultural region would be required, studies involving all aspects of the economies, not just the agricultural attributes and limitations. The following discussion can, therefore, only present general considerations and some rather tenuous conclusions.

A number of elements make any assessment of the prospects of relieving population pressure difficult if not hazardous. First, there are many ecological environments which are very poorly understood, and acquisition of the data necessary to an analysis will require not only the collation of basic information on physical conditions but experimentation with individual crops, rotations, planting periods, fertility-maintenance, etc. Second, while it is desirable to measure as accurately as possible the scientific capacities of a given area, it is important to recognize that there is a distinction between the potential and the achievable, at least in the short run. Third, absorption of people in non-agricultural pursuits involves not only every aspect of a nation's internal economy but external forces which may be complex in the extreme.

EXTENSION OF AGRICULTURE TO NEW AREAS

Turning to the first of the alternatives noted, there are substantial areas in a variety of climate regions which could absorb a larger population than they now have. An E.C.A. study has estimated that only about 28 percent of the cultivable area of Africa is now under cultivation,[88] while Revelle calculates the potentially arable land at 1.81 billion acres and the cultivated land at only 0.39 billion acres, or 21.5 percent of the potential area.[89] Another E.C.A. paper concluded that "the great latitude for expansion in agriculture . . . makes potential population pressure in relation to natural resources relatively easy to cure by possible economic expansion, at least in the near future."[90]

Three parts of that conclusion must be queried: the latitude for expansion; the ease of the cure; and the "near future" qualification, which should perhaps have read "but not in the near future." The "great latitude for expansion in agriculture" is misleading for many areas. True, a few countries like Angola and Mozambique have only a small percent of their cultivable area in use and apparently have substantial areas with reasonably good soils and climatic conditions that could be opened up. True, there are more countries, such as Liberia, Gabon, the Central African Republic, the two Congos, and Madagascar, which have large empty areas, but these have predominantly very poor soils. Estimates of potentially cultivable but unused land are available for a number of countries: the Sudan, 98.8 million acres or 16 percent of the total; Senegal, 18 percent of the total; Togo, 23 percent; Angola, 48 percent; and Madagascar, 47 percent.

But the ability to solve the problem of pressure on the land and its attendant problem of maintaining an adequate food supply simply by extending the farmed area is not easy for many regions. This is so, first, because there often is no readily available land. This applies to all of the countries of northern Africa except the Sudan, for most of the sudan belt below the Sahara, for the Gambia, Sierra Leone, and large parts of West African countries from Ghana to Nigeria; it applies to most of the highlands of East Africa and much of the Lake Victoria Basin, to Malawi, Lesotho, and the Indian Ocean island groups as well as to the reserves of Rhodesia and South Africa. As de Wilde put it, "population pressure is increasing almost everywhere with high densities no longer restricted to particularly fertile areas or comparatively

[88] E.C.A., *Demographic Factors Related to Social and Economic Development in Africa.*

[89] Roger Revelle, "Population and Food Supplies: The Edge of the Knife," *Prospects of the World Food Supply—A Symposium,* National Academy of Science, September, 1966, p. 32.

[90] E.C.A. *Seminar on Population Problems in Africa,* p. 36.

remote areas in which people had sought refuge from tribal warfare."[91]

Second, even when land is available it is often very poor quality and low carrying capacity, is plagued by human and bovine trypanosomiasis, or requires great expenditure of capital and human recources to develop. Studies are more and more revealing the surprisingly low percentage of cultivable land in many countries. Zambia, for example, has a surprisingly low cultivable percentage, largely because of the seasonal water-logging of large areas; only about 2 percent of Niger is cultivable; 17 percent of Kenya is classed as agricultural land; little more than 5 percent of South Africa can be cultivated at present technologic and economic levels.[92] Extensions which would require heavy investment include irrigation projects, swamp reclamation, drainage schemes, and suppression of the tsetse fly or of the fly carrying onchocersiasis. The fact of the matter is that most of the good land of Africa is now under use and it is difficult to point to any particularly inviting areas just waiting to be exploited.

Third, extension of farming to new lands is difficult because it is not easy to engineer the landscape, much less the population, so that each ecologic zone has just the most rational population density. Most resettlement schemes in Africa have been notoriously expensive and unsuccessful. And there are also ethnic influences which militate against the dispersion of farming. These include the attachment of groups to their present areas, well illustrated by the inhabitants of the Bangweulu Basin in Zambia who, despite suffering from scarcity of good land,

refused to move; they like living amongst the crowds of their kinfolk, they like the open views, they like the water; they like their homes—and they would rather suffer considerable hardship than live elsewhere. Only when conditions are extremely adverse, as they now are on some of the islands within the swamps, will people voluntarily leave their homelands.[93]

There is also the existence of tribal "zones of influence," which prevent the mingling of people, the evening of population pressure, and sometimes the full use of good land.

Finally, the existence of unused land is only a temporary pallia-

[91] de Wilde *et al.*, *Experiences with Agricultural Development in Tropical Africa*, I, 21.
[92] Allan, *The African Husbandman*, p. 21.
[93] Kay, *A Social Geography of Zambia*, p. 67.

Population Pressure in Africa

tive. A century of population growth at the present high levels would bring African densities to twice those of Asia's population today.

INTENSIFICATION OF AGRICULTURE

There are almost always opportunities for increasing productivity of the land in Africa, but they should not be exaggerated, nor are they easy to achieve. First, there are severe limitations in many environments. An estimated 43 percent of the land area of Africa is waste, about three-quarters is not suitable for cultivation. Carrying capacities on the vast arid and semiarid lands of Africa, totaling about 60 percent of the whole, are very low, varying from zero to from 2 to 6 per square mile in pastoral areas to possibly 25–40 in the most favored parts. Miracle writes that

the zone of "seasonal hunger," with the current techniques of production, is probably among the least productive per capita in the world. It may not be possible to produce a surplus large enough to support the propensity of man to reproduce himself, given the unfavorable environment, the lack of specialization, and the primitive techniques of production. The level of production per capita may be so low that a surplus large enough to safeguard against years of pestilence is not possible.[94]

In the heavier rainfall areas with latosolic soils the problem becomes one of devising a system which can maintain or improve the fertility of the soil. While this may be scientifically possible it is seldom economically possible and the sequence—reduced fallows–declining fertility–reduced yields—detailed in earlier sections is all too common in Africa. Some experts even question the ability to use such lands intensively and persistently. Allan, for example, states that "it is improbable that any economic system capable of maintaining high population densities can be devised"[95] for the miserably poor soils so common in Africa, and Steel reminds us that "our knowledge of the potentialities and limitations of the tropics is still far from complete."[96]

Other difficulties are that increased productivity is likely to be

[94] Marvin P. Miracle, " 'Seasonal Hunger': A Vague Concept and an Unexplored Problem," *Bulletin de l'Institut Français d'Afrique Noire*, XXIII, sèr B, Nos. 1-2 (1961), 282.

[95] Allan, *The African Husbandman*, p. 391.

[96] R. W. Steel, "Some Problems of Population in British West Africa," in R. W. Steel and C. A. Fisher, eds., *Geographical Essays on British Tropical Lands* (London, George Philip and Son, 1956), p. 30.

achieved more slowly than is politically desirable and that subsistence agriculture cannot support fertilizers, insecticides, irrigation, and other inputs needed to increase the productivity of the land. Nor is the so-called "green revolution" likely to provide an easy answer to Africa's problems of pressure and food supply. Greatest gains from the new varieties of cereals are obtainable under irrigation and careful water control during the growing cycle; Africa does not have the requisite capital to extend irrigation rapidly and the total irrigable area would only be a tiny fraction of the continent. Madagascar, Sierra Leone, Sudan, and Egypt, with already existent irrigation systems or extensive paddy development, could, however, achieve rapid gains if the new high-yielding varieties are introduced under sensible management. More important in delaying realization of the green revolution in Africa is the need for farmers to adopt new methods.

> The new agronomic requirements are quite different [from traditional methods] as regards planting dates and planting depths; fertilizer rates and timing; insecticide, pesticide and fungicide applications; watering and many others. Unless appropriate extension measures are taken to educate farmers with respect to these new farming complexities the higher yields will not be obtained.[97]

Despite the difficulties of increasing output on existing lands, it is probably much more important in meeting population pressure than extension to new lands, and African governments would be well advised to concentrate their efforts in this direction and to allot a higher percentage of development funds to the agricultural sphere than they have been doing. There is, after all, evidence of improving agriculture in numerous areas and the high quality lands could be far more productive than they are. While the percentage of such lands compares unfavorably with most other continents the total acreage of potentially highly productive lands is still very large. These lands must, however, be zealously protected if their potential is not to be reduced or destroyed.

DEVELOPMENT OF NON-AGRICULTURAL SECTORS

Turning to the possibilities of reducing pressure on the land by developing other sectors of the economy, the evidence suggests that it

[97] Clifton R. Wharton, Jr., "The Green Revolution: Cornucopia or Pandora's Box?" *Foreign Affairs*, XLVII, No. 3 (April, 1969), 466.

Population Pressure in Africa

can rarely be counted upon to stabilize, much less reduce, population in the rural areas in the short run. As Christensen has pointed out:

agricultural population will continue to increase for another generation or two in the less developed countries. Agricultural population may double in some countries in the next 30 to 40 years, even though agriculture's share of total population decreases. Population growth rates . . . relatively high . . . together with the fact that agriculture now accounts for a large share of total employment in these countries, means that nonagricultural employment would need to increase 10 to 15 percent a year to absorb all of the net increase in number of workers. Yet employment opportunities in nonagricultural sectors cannot be expected to increase more than 4 or 5 percent a year.[98]

In Africa there are a number of factors which will make it particularly difficult for the nonagricultural sectors to ease pressure on the land in numerous countries. For the thirty-three countries in tropical Africa "only two sectors, agriculture and mining, appear capable at the present time of providing an adequate base for a satisfactory rate of economic growth," while "only seven countries would appear to have sufficient opportunities in new and expanded mining enterprises to make this sector the main hope for achieving an adequate growth rate."[99] Eight countries have "good" to "excellent" opportunities for forestry development, eleven have "good" to "excellent" chances for expansion of manufacturing while twelve countries in tropical Africa have very limited chances in this sector, and it will be many years before manufacturing becomes a major sector of any but a very few countries. It is estimated that twenty of the thirty-three tropical countries will find it "difficult" to "very difficult" to achieve an adequate rate of growth in the money economy in the short run, and that in twenty-nine of the countries it will be "difficult" or "very difficult" to achieve an adequate rate of short-run growth in the income of the hypothetical average man.[100]

To complicate matters, there is evidence that a higher relative investment is required today to provide a job in one of the nonagricultural sectors than was true when the developed countries were in a roughly comparable stage of development. Also, population growth is at a much higher rate than it was in western Europe, while levels of edu-

[98] Raymond P. Christensen, "Population Growth and Agricultural Development," *Agricultural Economics Research*, XVIII, No. 4 (October, 1966), 119.
[99] Hance, *African Economic Development*, p. 294.
[100] *Ibid.*, pp. 290-91 and Chapter 8.

cational achievement and average incomes in almost all African countries are much below those of western Europe at a similar period.

The difficulties of solving the problem of rural pressure by absorption in nonagricultural communities and pursuits may be revealed in a different way by looking briefly at certain aspects of urbanization in Africa. First, as was suggested by the quote from Christensen, urban growth, even at the present excessive rate, is not rapid enough to absorb the population increase of most African countries. A hypothetical example, based on characteristic rates of growth, will illustrate a kind of average situation for tropical Africa. Let us assume that our country has a population of five million, which is increasing by 2.4 percent or about 120,000 per annum. Of the total, 12 percent, or 600,000, live in towns and cities of over 5,000. Urbanization is increasing at double the rate of population growth for the country, or 4.8 percent per year; this would be 28,800 per annum, or less than one quarter the total population growth. After a third of a century the urban growth would still be just over half the total growth for the country; by then the population of the country would have increased to 10.9 million of which about 26 percent would be urbanized.

More important than this evidence, however, is the fact that the cities, or at least the primate and major cities are growing at a rate which appears to be excessive in relation to the employment opportunities in them or to the ability of the cities to provide for the new residents. Many countries seem more concerned to stem or reverse the flow to the cities than to look to them as ways of relieving pressure in rural areas. Indeed the cities themselves are frequently pressure points where conditions for many hundreds of thousands of inhabitants are probably worse than those experienced in rural areas under pressure. Clarke summarizes the position as follows:

Unfortunately, many cities in developing countries are consumption centres, with much greater growth of the service sector than of manufacturing industry. At the same time, they have experienced rapid mortality decline with no commensurate fall in fertility. Consequently, rapid urbanization is accompanied by the associated evils of unemployment, underemployment and acute housing problems. . . . Urbanization certainly cannot be said to offer a solution to the population/resource difficulties of developing countries.[101]

[101] Clarke, "World Population and Food Resources: A Critique," p. 68.

Population Pressure in Africa

Population Growth as a Specific Factor

One additional subject which requires discussion is the relation of population growth to economic development. Economic theory now holds that a high rate of population growth is itself an obstacle to advance, and this applies not only to those regions which are affected by pressure of population. This is so because the capital needs of the increasing population are likely to absorb a substantial fraction of investment capital thus reducing the amount available for "progressive" investment; it may be roughly estimated that it takes 3 to 4 percent of new investments to provide the additional population with essentially the same standards as the existing population.[102] Most African countries are simply not investing enough and would have great difficulty in investing enough to achieve desired rates of economic growth.

To illustrate the differences under varying rates of population growth, if one assumes a population increase of 3 percent and an increase in the national income of 5 percent, then there would be a 2 percent annual improvement in national per-capita incomes, requiring about 35 years to double per-capita incomes. If the population increase could be reduced to 1 percent per annum and the same 5 percent increase in national income were maintained then per capita incomes would more than quadruple in the same period.[103] Several studies have revealed that investment in reducing fertility can provide greater gains than any alternative investment, though it is obvious that investment in population control complements and does not replace expenditures on economic development.

Reduced birth rates affect the rate of growth of per-capita income in three obvious ways: the resulting lower population shares the national income, the lower number of children reduces the burden of dependency permitting the diversion of capital to promote faster growth of total income, and a long-term reduction in the labor force, an advantage in those countries with surplus labor or problems of unemployment and underemployment.[104] Zaidan proposed that additional benefits could be

[102] See, for example, Joseph J. Spengler, "Population and Economic Growth," in Freedman, ed., *Population: The Vital Revolution*, p. 67.

[103] See A. J. Coale, "Population and Economic Development," in Hauser, ed., *The Population Dilemma*; Simon Kuznets, "Population and Economic Growth," *Proceedings of the American Philosophical Society*, VIII, No. 3 (June, 1967), 170-93.

[104] George C. Zaidan, "Population Growth and Economic Development," *Finance and Development*, VI, No. 1 (March, 1969), 4.

better nutrition, health, and education which would affect positively the quality of the labor force.

These considerations, it should be noted, apply to any country, but the higher rates of population growth in underdeveloped countries mean that a considerably higher proportion of their GNPs has to be invested in order to keep per-capita incomes at a constant level, they have less capacity to invest, and the need to increase incomes is very considerably more urgent. The youthfulness of most African populations, reflecting the continuing high birth rates coupled with recent sharp declines in the death rates, mean that dependency rates are unusually high. About half of the population of the continent is in the dependent age group, compared to about 30 percent in the United States. Furthermore, the ratio of dependents to supporters may be expected to increase in the next decade or so. The high dependency rates call in particular for high investment in education and the larger numbers entering school at the lower level may make it far more difficult to finance higher and professional education, whose products are most needed for effective development.

It is no wonder that Spengler finds that an increase in population and population densities is a source of net economic advantage in but a few countries,[105] or that Kamark concludes that "it is no exaggeration to say that the policy on population adopted by an African nation may well prove to be the most decisive factor in deciding what kind of an economic future lies ahead for it."[106]

Conclusion

This chapter has queried the myths that Africa does not have population problems or does not experience rather widespread population pressure. Its conclusions are that the problems are considerably more serious than has generally been accepted and that pressure applies to about half of both the area and population of the continent. A brief examination of the opportunities for relieving pressure indicated that a variety of possibilities exists but that relief cannot be provided easily or cheaply. Finally, the theoretic advantages of reducing fertility rates were presented.

Solution or amelioration of the problems involves every aspect

[105] Spengler, "Population and Economic Growth," p. 67.
[106] Kamarck, *The Economics of African Development*, p. 27.

Population Pressure in Africa

of development planning. In the agricultural sphere efforts must range from emergency measures in areas where the traditional systems have broken down under overwhelming pressure, through rationalization of use on present lands, to careful assessment of the possibilities for extending farming to empty lands. Much more attention needs to be given to regional and subregional capabilities. Agriculture should probably be allotted a greater share of development budgets in most countries. At the same time potentialities of other sectors—fishing, forestry, mining, industry, tourism, and services—cannot be neglected because it is apparent that the most likely population-growth rates over the next decades will place increasing pressure on farming and grazing which should be relieved where possible.

With respect to population policies, it is obvious that it would be wise for almost all African countries to consider the rapid adoption of family planning as a first step in control. The conclusion from a 1966 conference on unemployment in Africa that "the introduction of medical practices which tend to reduce mortality rates is immoral, if it leads to a pauperization of the country,"[107] or Allan's statement that "probably the greatest 'sin' of the suzerain powers was the saving of life"[108] strike me as unacceptable. But it may be immoral not to give attention to birth control if it can prevent suffering and the future pauperization of the continent.

The plea, then, is for greater awareness of and attention to the population factor in economic, social, and political studies and plans. The problems besetting Africa are very great indeed, but the opportunities for development are also very great. Certain attitudinal changes provide evidence that the population factor is subject to change, possibly more speedily than one would have predicted a few years ago. As examples one can cite: the acceptance by Mauritians of birth control to a sufficient degree to affect birth rates; indications from a series of studies that the size of the ideal family is declining in some places; the correlations between improved education, higher incomes, and urban residence and lower fertility rates recorded in various surveys, and possibly the case of the Ugandan husband who was reported to have developed ulcers when he learned that all four of his wives were pregnant simultaneously.

[107] As reported in *African Urban Notes*, I, No. 3 (September, 1966), 21.
[108] Allan, *The African Husbandman*, p. 338.

Bibliography

Allan, William. *The African Husbandman*. Edinburgh, Oliver and Boyd, 1965.

Ardener, Edwin W. "Social and Demographic Problems of the Southern Cameroons Plantation Area," in Aidan Southall, ed. *Social Change in Modern Africa*. London, Oxford University Press, 1961, 83-97.

Barbour, Kenneth M., and R. Mansell Prothero, eds. *Essays on African Population*. London, Routledge and Kegan Paul, 1961.

Basutoland, Bechuanaland Protectorate, and Swaziland. *Report of an Economic Survey Mission*. London, Her Majesty's Stationery Office, 1960.

Bennett, Merrill K. "An Agroclimatic Mapping of Africa," *Food Research Institute Studies*, III, No. 3 (November, 1962), 195-216.

Biebuyck, Daniel, ed. *African Agrarian Systems*. London, Oxford University Press, 1963.

Boserup, Ester. *The Conditions of Agricultural Growth*. Chicago, Aldine, 1965.

Brooke, Clarke. "The Heritage of Famine in Central Tanzania," *The Journal of the Tanzania Society*, No. 67 (June, 1967), 15-22.

——. "Types of Food Shortages in Tanzania," *The Geographical Review*, LVII, No. 3 (July, 1967), 333-57.

Brown, Peter, and Anthony Young. *The Physical Environment of Central Malawi*. Zomba, Government Printer, 1965.

Buchanan, K. M., and J. C. Pugh. *Land and People in Nigeria*. London, University of London Press, 1955.

Christensen, Raymond P. "Population Growth and Agricultural Development," *Agricultural Economics Research*, XVIII, No. 4 (October, 1966), 119-28.

Chubb, L. T. *Ibo Land Tenure*. 2d ed. Ibadan, Ibadan University Press, 1961.

Clark, Colin. *Population Growth and Land Use*. New York, St. Martin's Press, 1967.

Clarke, John I. "World Population and Food Resources: A Critique," in Institute of British Geographers, Special Publication No. 1, *Land Use and Resources: Studies in Applied Geography*, 1968, pp. 53-70.

Coale, Ansley J., and Edgar M. Hoover. *Population Growth and Economic Development in Low-Income Countries*. Princeton, Princeton University Press, 1958.

——. "Population and Economic Development," in Philip M. Hauser, ed. *The Population Dilemma*. Englewood Cliffs, Prentice-Hall for the American Assembly, 1963, pp. 46-69.

de Wilde, John C., *et al*. *Experiences with Agricultural Development in Tropical Africa*. 2 vols. Baltimore, The Johns Hopkins Press, 1967.

Dumont, René. *False Start in Africa*. Translated by Phyllis N. Ott. New York, Praeger, 1966.

Eggers, Heinz, "Das Ovamboland: Sonderstellung und Probleme eines Dichtegebietes in Südwestafrika," *Geographische Rundschau*, XVIII, No. 2 (December, 1966), 459-68.

Etherington, D. M. "Projected Changes in Urban and Rural Population in Kenya and the Implications for Development Policy," *The East African Economic Review*, I (new series), No. 2 (June, 1965), 65-83.

Fearn, H. "Population as a Factor in Land Usage in Nyanza Province of Kenya Colony," *The East African Agricultural Journal*, XX, No. 3 (1958), 198-200.

Floyd, Barry. "Land Apportionment in

Southern Rhodesia," *The Geographical Review*, LII, No. 4 (October, 1962), 566-82.

——. "Rural Land Use in Nsukka Division," in *Nsukka Division: A Geographic Appraisal*. Department of Geography, University of Nigeria, 1965 (mimeographed).

——. "Soil Erosion and Deterioration in Eastern Nigeria: A Geographical Appraisal," *The Nigerian Geographical Journal*, VIII, No. 1 (June, 1965), 33-44.

——. "Terrace Agriculture in Eastern Nigeria: The Case of Maku," *The Nigerian Geographical Journal*, VII, No. 2 (December, 1964), 91-108.

Great Britain. *East Africa Royal Commission 1953-1955 Report*. London, H.M.S.O., Cmd. 9475, June, 1955.

Grove, A. T. *Land Use and Soil Conservation in Parts of Onitsha and Owerri Provinces*. Zaria, Gaskiya Corporation, 1951.

——. "Soil Erosion and Population Problems in Southeast Nigeria," *The Geographical Journal*, CXVII, No. 3 (September, 1951), 291-306.

Gulliver, P. H. "The Population of the Arusha Chiefdom: A High Density Area in East Africa," *The Rhodes-Livingstone Journal*, No. 28 (December, 1960), 1-21.

Hallett, Robin. *People and Progress in West Africa*. Oxford, Pergamon Press, 1966.

Hamilton, P. "The Changing Pattern of African Land Use in Rhodesia," in J. B. Whittow and P. D. Wood, eds., *Essays in Geography for Austin Miller*. Reading, University of Reading, 1965, pp. 247-71.

Hance, William A. *African Economic Development*. 2d ed., rev. New York, Praeger for the Council on Foreign Relations, 1967.

——. "The Race between Population and Resources," *Africa Report*, XIII, No. 1 (January, 1968), 6-12.

Harroy, J.-P. "Surpopulation en Afrique Centrale," *Bulletin des Seances de l'Academie Royale des Sciences d'Outre-Mer*, VIII, No. 4 (1962), 524-30.

Haylett, D. G. "Population Growth and Food Resources in South Africa," *South African Journal of Science*, LXIV, No. 10 (October, 1968), 369-74.

Hellen, Anthony. "Some Aspects of Land Use and Over Population in the Ngoni Reserves of Northern Rhodesia," *Erdkunde*, XVI, No. 3 (September, 1962), 190-205.

Hilton, T. E. "Population Growth and Distribution in the Upper Region of Ghana," Chapter 28 in J. C. Caldwell and C. Okonjo, eds. *The Population of Tropical Africa*. New York, Columbia University Press, 1968, pp. 278-90.

Hunter, John M. "Ascertaining Population Carrying Capacity under Traditional Systems of Agriculture in Developing Countries," *The Professional Geographer*, XVIII, No. 3 (May, 1966), 151-54.

——. "Population Pressure in a Part of the West African Savanna: A Study of Nangodi, Northeast Ghana," *Annals of the Association of American Georgraphers*, LVII, No. 1 (March, 1967), 101-14.

Joy, Leonard. "The Economics of Food Production," *African Affairs*, LXV, No. 261 (October, 1966), 317-28.

Kamarck, Andrew. *Economic Development in Africa*. New York, Praeger, 1966.

Kay, George. "The Distribution of African Population in Southern Rhodesia: Some Preliminary Notes," *Rhodes-Livingstone Communication*, No. 28. Lusaka, 1964.

——. "Population Pressure on Resources in Zambia," paper presented to the Symposium on the Geography of Population Pressure on Physical and Social Resources, Pennsylvania State University, September 17-23, 1967.

———. *A Social Geography of Zambia*. London, University of London Press, 1967.

Kuznets, Simon. "Population and Economic Growth," *Proceedings of the American Philosophical Society*, VIII, No. 3 (June, 1967), 170-93.

Lacoste, Y. "Problèmes de Développement Agricole dans la Région de Ouagadougou (Haute-Volta)," *Bulletin de l'Association de Géographes Français*, Nos. 346-47 (July-August, 1966), 4-18.

Langdale-Brown, I. "Some Aspects of Land Capability Assessment in Africa," *African Affairs*, LXV, No. 261 (October, 1966), 307-16.

———, *et al. The Vegetation of Uganda and Its Bearing on Land Use*. Entebbe, Government Printer, 1964.

Lorimer, Frank. "The Population of Africa," in Ronald Freedman, ed. *Population: The Vital Revolution*. New York, Doubleday, 1964.

McMaster, D. N. "Change of Regional Balance in the Bukoba District of Tanganyika," *The Geographical Review*, L, No. 1 (January, 1960), 73-88.

Mabogunje, Akin L. "A Typology of Population Pressure on Resources in West Africa," paper presented to the Symposium on the Geography of Population Pressure on Physical and Social Resources, Pennsylvania State University, September 17-23, 1967.

May, Jacques. *The Ecology of Malnutrition in Middle Africa*. New York, Hafner, 1965.

Meade, James E., *et al. The Economic and Social Structure of Mauritius*. London, Methuen, 1961.

Middleton, J. F. M., and D. J. Greenland. "Land and Population in West Nile District, Uganda," *The Geographical Journal*, CXX, No. 4 (December, 1954), 446-57.

Miracle, Marvin P. " 'Seasonal Hunger': A Vague Concept and an Unexplored Problem," *Bulletin de l'Institut Français d'Afrique Noire*, XXIII, Sér. B, Nos. 1-2 (January-April, 1961), 273-83.

Morgan, W. B. "Farming Practice, Settlement Pattern and Population Density in South-Eastern Nigeria," *The Geographical Journal*, CXXI, No. 3 (September, 1955), 320-33.

Mortimore, M. J., and J. Wilson. *Land and People in the Kano Close-Settled Zone*, Ahmadu Bello University, Department of Geography, Occasional Paper No. 1, March, 1965.

Notestein, Frank W. "Some Aspects of Population Change in the Developing Countries," in *3 Talks on Population*. New York, The American Assembly, 1965, pp. 11-21.

Ominde, Simeon H. *Land and Population Movements in Kenya*. Evanston, Northwestern University Press, 1968.

———. "Rural Population Patterns and Problems of the Kikuyu, Embu and Meru Districts of Kenya," *Proceedings of the East African Academy*, II (1964), 36-45.

Phillips, John. *Agriculture and Ecology in Africa: A Study of Actual and Potential Development South of the Sahara*. London, Faber, 1959.

Porter, Philip W. "The Concept of Environmental Potential as Exemplified by Tropical African Research," paper presented to the Symposium on the Geography of Population Pressure on Physical and Social Resources, Pennsylvania State University, September 17-23, 1967.

Prescott, J. R. V. "Overpopulation and Overstocking in the Native Areas of Matabeleland," *The Geographical Journal*, CXXVII, No. 2 (June, 1961), 212-25.

Prest, A. R. "Population as a Factor in African Development," United Africa Company, *Statistical and Economic*

Review, No. 30 (September, 1965), pp. 1-16.

Prothero, R. Mansell. *Migrants and Malaria.* London, Longmans, Green and Co., 1965.

"La Région de Korhogo en Côte d'Ivoire," *Industries et Travaux d'Outremer,* XV, No. 161 (April, 1967), 299-300.

Revelle, Roger. 'Population and Food Supplies: The Edge of the Knife," *Prospects of the World Food Supply: A Symposium,* National Academy of Sciences, 1966.

Scherer, André. "La Réunion." Paris, La Documentation Française, *Notes et Études Documentaires,* No. 3358, 1967.

Smit, P. *Lesotho: A Geographical Study.* Pretoria, Africa Institute, 1967.

Smith, T. E., and J. G. C. Blacker. *Population Characteristics of the Commonwealth Countries of Tropical Africa.* London, Athlone Press for the Institute of Commonwealth Studies, 1963.

Spengler, Joseph J. "Population and Economic Growth," in Ronald Freedman, ed. *Population: The Vital Revolution.* New York, Doubleday, 1964.

Stamp, L. Dudley. *Natural Resources, Food and Population in Inter-Tropical Africa: Report of a Symposium Held at Makerere College, September 1955.* London, Geographical Publications, 1956.

Steel, Robert W. "Population Increase and Food Production in Tropical Africa." *African Affairs,* Special Issue (Spring, 1965), pp. 55-68.

——. "Some Problems of Population in British West Africa," in R. W. Steel and C. A. Fisher, eds. *Geographical Essays on British Tropical Lands.* London, George Philip and Son, 1956, pp. 17-50.

Stevenson, Robert F. *Population and Political Systems in Tropical Africa.* New York, Columbia University Press, 1968.

Street, John M. "An Evaluation of the Concept of Carrying Capacity," *The Professional Geographer,* XXI, No. 2 (March, 1969), 104-7.

Taeuber, Irene B. "Population Growth in Underdeveloped Areas," in Philip M. Hauser, ed. *The Population Dilemma.* Englewood Cliffs, Prentice-Hall for the American Assembly, 1963, pp. 29-45.

Tardits, C. *Les Populations Bamiléké de l'Ouest Cameroun.* Paris, Collection l'Homme d'Outre-mer, 1960.

Titmuss, Richard M., and Brian Abel-Smith. *Social Policies and Population Growth in Mauritius.* London, Methuen, 1960.

Tondeur, G. "Surpopulation et Déplacement de Populations," *Bulletin Agricole du Congo Belge,* XL, Nos. 3-4 (1949), 2325-52.

Turyagyenda, J. D. "Overpopulation and its Effects in the Gombolola of Buhara, Kigezi (Uganda)," *The Uganda Journal,* XXVIII, No. 2 (September, 1964), 127-33.

Udo, R. K. "Land and Population in Otoro District," *The Nigerian Geographical Journal,* IV, No. 1 (July, 1961), 3-20.

——. "Patterns of Population Distribution and Settlement in Eastern Nigeria," *The Nigerian Geographical Journal,* VI, No. 2 (December, 1963), 73-88.

United Nations, Department of Social Affairs. *The Population of Ruanda-Urundi.* New York, 1953.

——, Economic Commission for Africa. *Demographic Factors Related to Social and Economic Development in Africa.* E/CN.14/ESD/12. December 15, 1961.

——. *The Demographic Situation in Eastern Africa.* E/CN.14/LU/ECOP/2. July 20, 1965.

——. *Seminar on Population Problems in Africa.* E/CN.14/ASPP/LF.2. October 29-November 10, 1962.

—— F.A.O. *African Agricultural Development*. New York, 1966.

——, UNESCO. *Nomades et Nomadisme au Sahara*. Paris, 1963.

——, F.A.O. *Africa Summary: Report on the Possibilities of African Rural Development in Relation to Economic and Social Growth*. Rome, 1961.

Vermeer, Donald E. "Population Pressure and Crop Rotational Changes among the Tiv of Nigeria," paper presented at the Annual Meeting of the African Studies Association, New York, November, 1967.

Wharton, Clifton R., Jr. "The Green Revolution: Cornucopia or Pandora's Box?" *Foreign Affairs*, XLVII, No. 3 (April, 1969), 464-76.

Yeld, Rachel. "Land Hunger in Kigezi, South-West Uganda," *Nkanga*, No. 3, 1968, 24-28.

Yudelman, Montague. *Africans on the Land*. Cambridge, Harvard University Press, 1964.

Zaidan, George C. "Population Growth and Economic Development," *Finance and Development*, VI, No. 1 (March, 1969), 2-8.

INDEX

Buhara, 407
Bujumbura, 102, 210, 232, 239, 241
Bukama, 151
Bukoba District, 171, 407
Bukavu, 265
Bulawayo, 107, 111, 216, 241, 264
Bullom Peninsula, 77
Burundi, 18, 32, 171, 203; population, 37; population distribution, 102; migration from, 152, 156, 174, 187; migration to, 184; urbanization, 220, 223 (*table*), 224, 232 (*table*); population pressure, 408

Cairo, 71, 212, 214, 216, 217, 231, 234, 235, 240, 312
Calabar, 213, 248
Caldwell, John C.: cited, 4, 20, 21, 142
Cameroon, 32, 83; 1963-1964 census, 12, 37, 228 (*table*); population distribution, 88-90; migration, internal, 130, 174; migrant labor in, 150-51; migrant labor from, 151; urbanization, 223 (*table*), 228 (*table*), 232 (*table*), 236; birth rate, 386; population pressure, 402-403
Cameroon, Mount, 89
Cape Coast, 213, 327
Cape Town, 121, 216, 224, 239, 241, 255, 265, 266, 366
Cape Verde, 316
Cape Verde Islands: 1961 census, 37
Carthage, 211
Casablanca, 68, 143, 210, 231, 234, 235, 240, 281, 284, 299, 304-10, 336; map, 307
Casamance, 185, 316
Censuses, 3-15; sample, 4, 13; cost, 5; enumerators, 5-6; problems of data collection, 5-9, 14-15; periods, 6-7; geographical problems, 7-8; data processing, 8; standardization of data, 8-9; difficulties in interviewing, 9-10; political orientation of, 10-11; difficulties in securing information, 11-15, 47-48; list of, by countries, 36-40; migration in, 140-42; *see also* Population data; for national censuses, *see under* countries
Central African Republic, 32, 89, 91, 132, 203, 392; 1965 census, 11, 37; 1968 census, 11, 37; 1958-1969 censuses, 12, 37; population distribution, 88; migration to and from, 150-51, 184, 186; urbanization, 220, 222, 223 (*table*), 232 (*table*); 1961-1963 census, 225 (*table*); villages and small towns, 225 (*table*)
Ceuta, 212, 267

Chad, 32, 89, 91, 203; 1963-1964 sample census, 9-10, 37; population distribution, 87-88; migration to and from, 151, 184; urbanization, 222, 223 (*table*), 228 (*table*), 232 (*table*); villages and small towns, 225 (*table*); 1962-1963 census, 225 (*table*), 228 (*table*); population pressure, 403
Chad, Lake, 89
Chagga people, 98, 176
Challawa River, 315
Chari River, 87, 89
Chelif Valley, 69
Chewa people, 193
Chililabombwe, 381 (*map*)
Chinese, 115-16, 136, 412
Chirundu, 107
Chiweshe Native Reserve, 409
Cholo Highlands, 111
Christensen, R. P.: quoted, 427; cited, 428
Cities and towns: range of population, 64, 86, 226, 228 (*table*), 229, 230; age, 211-14, 244-45; size, 218 (*map*), 224-36, 244; number over 100,000 population, 230 (*table*), 231 (*table*); large urban communities and primate cities, 230-36; primate cities, 231-32 (*table*), 233, 234 (*table*), 235, 237; growth of urban centers, 236-44, 292-94; population, *1900-1968*, 239, 240-41 (*table*), 244; growth rates, 242-43 (*graph and semilog*), 293; classification, 244-65; cultural characteristics, 245-59; expatriate-influenced, 246-47 (*map*), 248, 254-59; indigenous, 246-47 (*map*), 249-54; functional classifications, 259-65; capitals, 262, 264; physical sites, 265-68; problems of, 265-92; case studies, 298-382; indigenous, Morocco, 299-310; *see also* Urbanization; Villages and small towns
Clark, Colin: quoted, 389; cited, 416-17
Clarke, J. I.: cited, 76-77; quoted, 428
Clower, Robert W.: quoted, 135
Coale, Ansley J.: cited, 13; quoted, 18
Cocody Bay, 325
Colson, Elizabeth: quoted, 131, 173; cited, 194, 200
Commission for Technical Cooperation in Africa (CCTA), 9
Comoé River, 84
Comoro Islands, 32, 113; urbanization, 223 (*table*)
Conakry, 210, 213, 216, 232, 239, 240, 248, 267, 281

Index

Giuba River, 94
Giza, 55
Gleave, M. B., and H. P. White: quoted, 86-87
Gombélédougou, 82
Gondar, 213
Gorée, Ile de, 213, 267, 318
Grand Bassam, 267
Grand-Lahou, 322
Grove, A. T.: quoted, 402
Guinea, 32, 203; birth rates, 20; migrant labor from, 33, 146, 148; 1954-1955 census, 38; population distribution, 73, 76, 84; urbanization, 223 (table), 232 (table); see also Equatorial Guinea
Guinea, Gulf of, 79, 349
Gulliver, P. H.: cited, 47, 164-65, 187, 195
Gwelo, 216, 264

Hamites, 129
Harar, 212
Harris, Marvin: quoted, 179
Harroy, J.-P.: cited, 408
Hartz River, 268
Hausa people, 84, 89, 173, 178, 189, 341, 347, 349, 353, 355
Havelock, 123
Health services, 2, 27; shortage of doctors, 32 (table)
Hilton, T. E.: quoted, 399
Hippo Valley, 110
Hofmeyr, 414
Horn, 167; population patterns, 91, 93-94; population density, 92 (map); population pressure, 403-404; see also Afars and Issas; Ethiopia; Somalia
Hottentots, 129
Houghton, D. H.: quoted, 199
Housing, 281-92; see also Shanty-towns
Huambo, 152
Huíla, 152
Hunter, John M.: quoted, 399-400
Hutu people, 178, 184

Ibadan, 25, 214, 216, 236, 239, 240, 342-49
Iboland, 76, 78, 171, 384, 395, 400
Ibo people, 79, 174, 176, 178, 198, 356
Ife, 212, 348
Ijebu-Ode, 212
Ikeja, 341
Ikoyi Island, 339, 341
Ilesha, 212
Ilorin, 76, 212

Immigration, 14, 33, 79-80, 110; see also Migration
Indian Ocean, 94
Indian Ocean Island groups, 113-17, 176
Indians, 116, 133, 355, 412
Irély, 85
Iseyin, 212
Ivory Coast, 18, 32; immigration, 33, 80; 1957-1958 census, 38; population, distribution, 76, 85; labor recruitment in, 134; migration, internal, 148, 173; migrant labor in, 148-49; migration from, 183; urbanization, 221, 223 (table), 228 (table), 232 (table), 279; 1967 census, 228 (table); housing, 286-87; population pressure, 398
Iwo, 212

Jarrett, H. R.: quoted, 398
Jefara, 69
Jenkins, George: cited, 272-73
Jews, 143, 185, 300, 305
Jinja, 153, 215, 236, 264
Jirapa, 399
Johannesburg, 120, 136, 216, 217, 232, 234, 241, 270, 273, 284
Jonglei Scheme, 396
Jos, 178, 215, 248, 264
Jos Plateau, 150
Juba, 310

Kabrai region, 73
Kabré people, 149, 181
Kabwe, 216
Kabylia, 69, 144
Kaduna, 76, 215, 236, 248
Kafue River, 103, 380
Kairouan, 176, 212, 261
Kalahari Steppe, 64
Kampala, 153, 215, 216, 232, 234, 236, 241, 248, 264, 268, 287, 373-78; map, 375
Kano, 73, 82, 83, 84, 212, 216, 236, 240, 244, 254, 264, 282, 337, 349-56, 400, 401
Kaolack, 217
Karamoja District, 96, 407
Karamojong people, 168, 169
Kariba, 107, 217, 380
Kasai, 31, 151, 152
Kasama, 106
Kassala, 186, 310
Kassenas, 399
Katanga, 103, 105, 106, 135, 137, 151, 152, 184, 190, 264, 358
Katsina, 25, 84, 212, 401

Lulua people, 178
Lundazi, 106
Luo people, 175
Lusaka, 104, 106, 156, 157, 224, 232, 234, 236, 241, 256, 264, 268, 286, 290, 382
Luvale, 102

Mabogunje, A. L.: cited, 343; quoted, 345
Machakos District, 94, 184, 406
McLaughlin, Peter F. M.: quoted, 145
Madagascar, 32, 171, 390; 1965 census, 38; population density ranges, 45, 52 (table); 1955-1958 census, 52 (table); population distribution, 111, 113; population density, 1960, 112 (map); migration to, 129, 130; urbanization, 223 (table), 228 (table), 232 (table), 238 (table); 1960 census, 225 (table); villages and small towns, 225 (table); 1964 census, 228 (table); population pressure, 410
Maghreb, 133, 169, 173, 224, 384; population, 64, 67-69, 393-95; see also Algeria; Morocco; Tunisia
Mahé Island, 114-15
Maiduguri, 215
Majunga, 113, 216
Malange, 152, 358
Malaria, 28, 102, 124, 201
Malawi, 32; 1963 census, 12, 38; population density ranges, 52 (table); 1966 census, 52 (table), 225 (table), 228 (table); emigration, 110; population distribution, 111; migrant labor from, 111, 136, 139, 157, 192; migration from, 157, 158-59, 176; urbanization, 223 (table), 228 (table), 232 (table), 236, 238 (table); villages and small towns, 225 (table); population pressure, 410
Mali, 203; birth rate, 20; emigration, 33; 1956-1958 census, 38; population distribution, 73, 76, 84; migratory labor from, 146, 148, 150; urbanization, 221, 223 (table), 228 (table), 232 (table), 238 (table); 1965 census, 228 (table)
Malnutrition, 27, 28-32, 416-17, 418
Maluti Mountains, 122
Mamprusi, 399
Mandara Mountains, 89, 403
Manyoni, 407
Manzini, 124, 216, 236
Marampa, 77
Marasmus, 29
Marrakech, 239, 240, 244, 251, 253, 300, 302, 303

Marra Mountains, 145
Marsa Brega, 217
Masai Plain, 96, 97, 172, 405
Masai people, 132, 169
Mascarene Islands, 113, 203
Maseru, 210, 216, 232, 415
Mashonaland, 110
Massawa, 165
Matadi, 215
Mauritania, 32, 167; censuses, 38, 225 (table); population distribution, 76; migration from, 150; nomadism, 167; urbanization, 221, 223 (table), 231 (table), 238 (table); villages and small towns, 225 (table)
Mauritius, 32, 395; 1967 population estimates, 13, 412; family planning, 22-23, 413, 431; birth and death rates, 23 (table), 412; death rate and life expectancy, 26, 412; 1962 census, 38; population density, 64, 411; urbanization, 223 (table), 238 (table); population pressure, 411-13
Mbabane, 123, 216, 232, 236, 262, 265
M'Baiki, 278
Mboum people, 89
Mecca, 145, 176
Medical services, 27, 31; see also Health services
Medjerda Valley, 69
Meknès, 68, 212
Melhalla el Kubra, 264
Melilla, 212, 267
Mengo, 216, 287, 374
Menhyia, 333
Merina kingdom, 113, 129, 130, 213
Meru, Mount, 47, 97, 99
Metidja, 69
Middle Africa, 34 (map); population density ranges, 50 (table); estimated population density ranges, mid-1967, 61 (table); distribution and density of population, 1930-1967, 63 (table); population distribution, 87-91; migratory movements, 147 (map), 150-52; urban population, 220, 221 (map), 222-23 (tables), 228 (table), 238 (table), 241 (table); villages and small towns, 225 (table), 227; number of cities over 100,000, 230 (table), 231 (table); primate cities, 232 (table), 234 (table); population pressure, 402-403; see also Angola; Cameroon; Central African Republic; Chad; Congo (Brazzaville); Congo (Kinshasa); Equatorial Guinea; Gabon; Príncipe; São Tomé

Migrant labor, 130, 134-37, 173, 190-200, 203, 281; economic causes, 187; types of employment, 190-91; impact of, 192-202; families, 198-200; *see also* subhead *under* names of countries and regions

Migration, 1, 2, 3, 14, 32-33, 79-80, 128-208, 294; historical background, 129-40; measurement, 140-42; patterns of movement, regional, 142-61, 147 (*map*), 191, 204; theories and classifications, 161-66; environmental causes, 166-76; causes and motivations, 166-88; socio-cultural factors, 176-82; modernization as cause, 180-82; political causes, 182-86; economic causes, 186-202; impact, 191-202; future trends, 202-204; *see also* Migrant labor

Mining: migrant labor in, 137, 146, 150-51, 153, 156-57, 160-61, 190-91, 196, 197, 201

Minna, 248

Miracle, M. P.: quoted, 425

Mitchell, J. Clyde: cited, 180, 181, 187-88

Mlanje Highlands, 111

Moba people, 149

Moçambique, 213, 267

Moffett, John P.: quoted, 407-408

Mogadishu, 7, 212, 217, 232, 241

Mohammedia, 264

Mombasa, 100, 153, 212, 213, 216, 217, 236, 241, 248, 255, 267, 366, 367, 373

Monrovia, 76, 214, 232, 234, 239, 240, 267

Mopti, 212

Morgan, W. T. W.: quoted, 405

Morocco, 32, 210, 394; 1960 census, 5, 38, 48 (*table*), 225 (*table*), 228 (*table*); family planning, 21; population density ranges, 48 (*table*); population distribution, 64, 68, 299; population densities, *1960*, 67 (*table*); migratory movements, 142-44, 185; migration to, 144; urbanization, 223 (*table*), 228 (*table*), 232 (*table*), 238 (*table*); villages and small towns, 225 (*table*); unemployment, 276, 279; housing, 284, 286, 290; cities, 299-310; population pressure, 393

Morogoro, 155

Mortality and mortality rates, 1, 2, 3, 13, 26-32; *1960-1967*, 26

Mossi kingdom and people, 73, 84, 173, 179, 181, 192, 195, 212, 398

Moulay Idriss, 176, 211, 261, 301

Moundou, 88

Mozambique, 18, 32; emigration, 33, 110; 1960 census, 38-39; migrant labor from, 135, 136-37, 157, 197; migration from, 157, 158, 177, 185, 199; urbanization, 223 (*table*), 232 (*table*), 236

Mpulunga, 106

Mpwapwa, 407

Mqutu District, 414

Mushin, 341

Muslims, 19, 80, 144, 145, 176, 177, 184, 185, 212; in Algeria, 144-45; in Morocco, 300, 306, 307; cities, 247, 250, 251

Mwanza, 214

Mweru, Lake, 106, 409

Nacala, 236

Nairobi, 100, 153, 215, 217, 232, 234, 236, 241, 265, 268, 277, 280, 283, 284, 366-73; maps, 370, 371

Nakasero, 376, 377

Namib Desert, 64

Nandom, 399

Nangodi, 399

Nasawara, 354

Nasser, Lake, 145

Nazareth, 361

Ndebele people, 130

Ndola, 241, 380, 382

New Amboshidi, 288

Ngoko Valley, 90

Ngoni people, 106, 130, 132, 195, 409

Ngwenya, 123

Niamey, 82, 217, 232, 234, 239, 240, 244, 248, 278

Niari Valley, 278

Niger, 18, 32, 203; 1963 census, 39; migration from, 184; urbanization, 221, 223 (*table*), 228 (*table*), 232 (*table*), 238 (*table*); 1962-1963 census, 225 (*table*); villages and small towns, 225 (*table*); 1968 census, 225 (*table*), 228 (*table*); unemployment, 278; population pressure, 398, 417

Nigeria, 18, 32, 170; 1963 census, 5, 8, 10, 11, 12, 15, 39, 47, 55, 76, 342; 1952 census, 6, 11, 12, 229 (*table*), 261; 1962 census, 8, 10, 11, 39; 1960 census, 12; birth rates, 20; family planning, 22, 25; Nigeria-Biafra war, 29; population density ranges, 55, 63; population distribution, 73, 76, 79, 80-81, 85, 86; migration, internal, 131, 173-74, 176, 185; migrant labor from, 149, 150; migration to, 149, 150; urbanization, 222, 223 (*table*), 232 (*table*), 236; villages and small towns, 225 (*table*); unemployment, 277; population pressure, 400-402

39; migrants from, 185; urbanization, 223 (*table*)

Portuguese Territories: 1950 census, 12; migrations and expulsions from, 185

Powdermaker, Hortense: cited, 270

Prescott, J. R. V.: cited, 408

Pressure, population, 383-436; evidences of, by regions, 393-418; political unrest and, 394-95; and food consumption, 416-17; areas, 418-22, 419 (*map*), 421 (*table*); opportunities for relieving, 422-30

Pretoria, 120, 214, 241

Príncipe, 32; urbanization, 223 (*table*)

Prothero, R. M.: cited, 163-64, 166; quoted, 200-201, 402

Quela, 357

Que Que, 264

Rabat, 68, 211, 235, 240, 251, 300, 301, 305

Rand, *see* Witwatersrand

Read, Margaret: quoted, 193, 196

Regions of Africa: map, 34; *see also* Eastern Africa; Middle Africa; Northern Africa; Southern Africa; Western Africa

Reserves, native, 110; South Africa, 159, 271-72; Rhodesia, 193, 271, 409

Réunion, 32, 395; 1967 census, 12, 39; life expectancy, 26; 1961 census, 52 (*table*), 58; population density ranges, 52 (*table*), 58, 64; population density, 58, 116 (*map*), 117; population distribution, 115-17; urbanization, 223 (*table*)

Rharb plain, 67

Rhodesia, 32, 203; 1963 census, 12, 39; immigration, 33, 110; population distribution, 107-11, 108-109 (*map*); migrant labor in, 111, 135, 157, 158, 176, 190, 199; urbanization, 111, 223 (*table*), 228 (*table*), 232 (*table*), 236, 238 (*table*); migration to, 130, 156, 165, 176, 177; migration from, 137, 185; 1961 census, 157; 1961-1962 census, 228 (*table*); population pressure, 408-409

Rhodesia, Northern, *see* Zambia

Richards, Audrey I.: cited, 187

Rif Mountains, 68, 69

Roseires Dam, 312

Rouch, Jean: cited, 179-80

Rufisque, 213, 248, 264, 318, 319

Rwanda, 18, 32, 171, 203; population study, 39; population distribution, 102; migrant labor from, 135, 156, 174; migration from, 184, 187; urbanization, 220, 223 (*table*), 224, 232 (*table*); population pressure, 408

Sabi River, 107

Sabi Valley, 110

Sabon Gari, 356

Sabon Ngari, 347

Safi, 64, 143

Sahara Desert, 64, 68, 69, 75, 79; population pressure, 397-98

Saint-Denis, 213

Saint-Louis, 213, 266, 267, 268, 318

Saint-Paul, 213

Salisbury, 107, 111, 216, 217, 232, 234, 236, 239, 241, 257, 280

Sanaga River, 90

San Pedro, 322

São Salvador do Congo, 213

São Tomé, 32; urbanization, 223 (*table*)

Sapele, 248

Sasolburg, 120, 217, 264

Schaff, Alvin H.: cited, 275

Schapera, Isaac: cited, 180, 192

Scrimshaw, Nevin S.: quoted, 31-32

Ségou, 212

Segregation, 270-72

Sekondi, 213, 235, 272, 327, 333

Sena, 213

Senegal, 32, 83, 170, 203; population study, 39; population distribution, 73, 76; population, 1964 rural, 1967 urban, 73 (*map*), 76; labor recruitment in, 134, 137, 146; migration to, 146, 185, 186, 190; migration from, 150, 163; urbanization, 221, 223 (*table*), 228 (*table*), 231 (*table*), 238 (*table*); 1966 census, 225 (*table*); villages and small towns, 225 (*table*); 1960 census, 228 (*table*); housing, 283

Senegal River, 75-76, 80, 150, 171, 264

Sennar, 310

Sénoufo people, 84

Serengeti Plain, 97

Sétif, 212

Seychelles, 32, 203; life expectancy, 26; 1960 census, 39, 53 (*table*); population density ranges, 53 (*table*); population distribution, 114-15; urbanization, 223 (*table*)

Sfax, 69

Shanty-towns: bidonvilles, North Africa, 283-84, 286, 303, 307-308; Sudan, 313

Sidi bel Abbès, 214

Sierra Leone, 32, 133, 203; 1962 census, 6, 8; 1963 census, 12, 39, 49, 225 (*table*),

225 (table), 228 (table); migration to, 130, 133, 152, 155; migration, internal, 155-56, 163, 165; urbanization, 220, 223 (table), 228 (table), 232 (table), 236, 238 (table); villages and small towns, 225 (table); population pressure, 407-408; see also Tanzania

Tanga Province, 408

Tangier, 211

Tanzania, 18, 94, 97, 100, 101; 1967 census, 8, 12, 39-40, 50, 51 (table); population density ranges, 46-47 (table); migration from, 156, 163; migration to, 184, 185, 186; unemployment, 278; population pressure, 407; see also Tanganyika, Zanzibar

Taxation, 4, 15; head or hut, 134

Tema, 217, 231, 235, 255, 328, 330, 331

Teso Province, 407

Thiès, 248

Three Towns, 71, 145, 210, 221, 234, 276, 210-13; see also Khartoum; North Khartoum; Omdurman

Tibesti Mountains, 87

Tivland, 85, 172-73, 400, 401

Tlemcen, 212

Togo, 32, 203; 1958-1960 census, 40, 55, 225 (table); population density ranges, 55, 63; population distribution, 73, 76; migrant labor from, 149; migration to, 183, 184; urbanization, 223 (table), 228 (table), 232 (table), 238 (table), 261; 1968 census, 225 (table); villages and small towns, 225 (table); 1966 census, 228 (table); population pressure, 400

Tombouctou, 212, 248

Tonga people, 195

Tororo, 153

Toucouleur people, 163

Transkei, 121, 159, 414-15

Transvaal, 117, 136, 140, 268

Treichville, 286, 288, 325

Tribes, 14, 15, 78, 101, 131, 178; in migrations, 131, 177-80, 189, 198; of migrant workers, 163, 177, 201

Tripoli, 211, 231, 234, 235, 240, 395

Tsetse fly, 84, 88, 89, 94, 102, 107, 124, 168, 170, 361, 398, 407

Tsévié, 261

Tunis, 211, 214, 231, 234, 240, 281, 283-84

Tunisia, 32, 394; family planning, 21; population distribution, 69; migration to, 144; urbanization, 223 (table), 228 (table), 232 (table), 238 (table); 1966 census, 228 (table); unemployment, 279

Turkhana people, 168

Tutsi people, 178, 184

Ufipa Plateau, 99

Uganada, 18, 32, 94, 96, 100, 101, 170; 1958 census, 12; immigration, 33; 1959 census, 40, 153, 228 (table); 1962 census, 51 (table); population density ranges, 51 (table); forced labor in, 134; migration to, 135, 137, 153, 156, 174, 184, 186, 187; migrant labor in, 152, 175; urbanization, 200, 223 (table), 228 (table), 232 (table), 236; housing, 287-88; population pressure, 406-407

Uíge, 152

Ukara Island, 390, 407

Ulere Migrant Labour Service, 135

Umtali, 157, 216

Underemployment in cities, 279-81

Unemployment in cities, 276-79, 280-81, 393

United Arab Republic (Egypt), 32, 203, 394; family planning, 21, 24; death rate, 27; 1960 census, 40, 225 (table); population density ranges, 54; population distribution, 69-71, 70 (map); urbanization, 223 (table), 231 (table), 235, 238 (table); villages and small towns, 225 (table); population pressure, 395-96

United Nations Demographic Yearbook, 1967, 12-13, 18, 44 (table), 59, 60-61 (table), 63 (table), 391 (table)

United Nations Statistical Yearbook, 1967, 32 (table)

Upper Volta, 18, 32, 203; 1960-1961 census, 12, 40, 225 (table); emigration, 33; population distribution, 73, 84; population density, 1960-1961, 81 (map); migrant labor from, 148, 149, 173; urbanization, 221, 223 (table), 228 (table), 232 (table), 235, 238 (table); villages and small towns, 225 (table); 1966 census, 228 (table); population pressure, 398

Urbanization, 2-3, 203, 209-97; and birth rates, 20-21; levels of, 64, 86, 219-24; historical survey, 211-17; urban population for world regions, 1950, 219 (table); populations for Africa and world, 1920-1967, 220 (table); urban population in major African regions, 1960 and 1967, 220-24, 222 (table); urban population by African countries, 220-24, 1967, 223 (table); range of urban agglomerations, 226-29, 228 (table), 230 growth rates,

PICTURE CREDITS